Protecting Critical Infrastructure

Volume 2

Series Editors

Simon Hakim, Philadelphia, USA
Erwin A. Blackstone, Philadelphia, USA

For further volumes:
http://www.springer.com/series/8764

Robert M. Clark · Simon Hakim
Editors

Securing Water and Wastewater Systems

Global Experiences

Springer

Editors
Robert M. Clark
Environmental Engineering
and Public Heath Consultant
Cincinnati, OH
USA

Simon Hakim
Department of Economics
Temple University
Philadelphia, PA
USA

ISBN 978-3-319-01091-5 ISBN 978-3-319-01092-2 (eBook)
DOI 10.1007/978-3-319-01092-2
Springer Cham Heidelberg New York Dordrecht London

Library of Congress Control Number: 2013949430

Printed on acid-free paper

Springer is part of Springer Science+Business Media (www.springer.com)

We would like to dedicate this book to our families and many contributors to the manuscripts that appear in this volume. Bob is particularly appreciative of the support and encouragement from his wife Susan, his children Dan and Kathy, and grandchildren Thomas, Nicholas, Benjamin, and Eleanor. Simon would like to acknowledge the love and support he has received from his wife Galia, his children Liora, Rami, Doron, and grandchildren Dylan, Aslley, and Joshua

Contents

Chapter 1
Securing Water and Wastewater Systems: Global Perspectives

Robert M. Clark and Simon Hakim

Technical Terms and Abbreviations

ACHILLES	Spatial vulnerability identification tool
AMI	Advanced metering infrastructure
AMR	Automatic meter reading
AQUASEC-AUT	Austrian crisis management laboratory
CAIS	The cyber attack information system
CDC	Centers for disease control and prevention
CI	Critical infrastructure
CLOFs	Continuous rainfall and cloudbursts
COTS	Commercial off-the-shelf
DCS	Distributed control systems
DHS	Department of homeland security
ERP	Eastern route project
EU	European Union
FEIS	Failure experience improvement system
GAO	Government accountability office
GEDES	Risk of flood and landfill hazard tool
GeoSFM	The Geospatial Stream Flow Model
GLOFs	Glacial lake outburst floods
GRDC	Global runoff data center
ICS	Industrial control system
ICWater	Incident command tool for drinking water protection

R. M. Clark (✉)
Environmental Engineering and Public Heath Consultant, 9627 Lansford Drive, Cincinnati, OH 45242, USA
e-mail: rmclark@fuse.net

S. Hakim
Director Center for Competitive Government, Fox School of Business and Management, Professor of Economics, Temple University, Philadelphia, PA 19122, USA
e-mail: hakim@temple.edu

R. M. Clark and S. Hakim (eds.), *Securing Water and Wastewater Systems*, Protecting Critical Infrastructure 2, DOI: 10.1007/978-3-319-01092-2_1,

LDOFs	Landslide dam outburst floods
IT	Information technology
IWRM	Integrated water resources management
ML	Machine learning
MRC	Mekong river commission
MRP	Middle route project
NOAA	National ocean and atmospheric administration
ORTIS	Operational risk management tool and information system
PET	Potential evapotranspiration
PiReM	Pipe rehabilitation management
PLC	Programmable logic controllers
RTU	Remote terminal unit
SCADA	Supervisory control and data acquisition
SHARC	Coastal transport model
SRTM	Shuttle radar topography mission
TISN	Trusted information sharing network
US	United States
U.S. EPA	United States Environmental Protection Agency
Vewin	Association of drinking water companies in the Netherlands
WTP	Water treament plant
WQE	Water quality event
WQED	Water quality event detection
WSIAAG	Water services infrastructure assurance advisory group
ZuHaZu	Condition assessment of large-diameter transmission water mains

1.1 Introduction

There is general recognition that urban water systems are vulnerable to both manmade and natural, threats and disasters including droughts, earthquakes, and terrorist attacks. There is also growing concern over the possible effects of climate change on the availability of water supply. Natural disasters such as major storms, hurricanes, and flooding can have an adverse effect on water supply security. Several recent earthquakes centered in urban areas such as the earthquake that struck Kobe City, Japan in 1995 have demonstrated the disastrous effect that earthquakes can have on urban water systems.

Until recently, terrorism was not generally regarded as a serious threat to water supply, but, after the attacks of September 11, 2001 in the United States, government planners have been forced to consider the possibility that the nation's critical infrastructure may, in fact, be vulnerable to terrorist attacks. The rapid proliferation of telecommunication and computer systems, which connect infrastructure components to one another in a complex network, compounds this vulnerability (Clark et al. 2011).

There are many factors that may affect water security. According to Bakker (2012) water security may be defined "... as an acceptable level of water-related risks to humans and ecosystems, coupled with the availability of water of sufficient quantity and quality to support livelihoods, national security, human health, and ecosystem services". According to Bakker (2012) an estimated 80 % of the world's population faces a high-level water security or water-related biodiversity risk. It is possible to characterize water security threats according to the following categories (Bakker 2012):

- Direct threats to drinking water supply systems including natural or manmade threats such as terrorist attacks, earth quakes, storms, and flood.
- Lack of water resource availability, including impacts of economic growth and development. These could be impacts on growth and human livelihood from water-related hazards such as floods and droughts, water stress, and water scarcity, especially with respect to food and energy security.
- Threats from point- and non-point source pollution to water-related ecosystems, such as increased water consumption resulting in the increased use of ecosystem services and biodiversity loss, and the human impact on ecosystems.
- Impact of climate change resulting in increased hydrological variability, including increased amplitude and frequency of droughts and floods.

A central theme of these water security threats is the need to balance human and environmental water needs while at the same time safeguarding essential ecosystem services and biodiversity (Bakker 2012). Overarching, all aspects of water security are the concept of Integrated Water Resources Management (IWRM). IWRM emphasizes linkages between land-use change and hydrological systems, between ecosystems and human health, and between the political and scientific aspects of water management. According to Bakker (2012) this new water security agenda should "... include a conceptual focus on vulnerability, risk, and resilience; an emphasis on threats, shocks, and tipping points; and a related emphasis on adaptive management given limited predictability". This book attempts to address this broad spectrum of water security issues through a series of essays prepared by a group of internationally recognized experts. These essays are organized into the following five sections:

- Direct threats to water systems
- Water resource availability
- Threats due to ecosystem impacts
- Threats due to climate change
- Integrated total water management

1.2 Direct Threats to Water Systems

Among the six chapters that make up this section of the book are manuscripts that address the design basis threat to water supply security, cyber security, the use of machine learning for detecting "water quality events," the use and placement of contaminant warning systems, the potential for modeling the effect of natural disasters, and the need to have a quantitative framework for the regulatory evaluation of water utilities.

Water systems have historically proven to be vulnerable to contamination. For example, in 1993, Milwaukee Wisconsin experienced the largest waterborne disease outbreak in documented United States history. The etiological agent was determined to be the *Cryptosporidium* protozoan. Fox and Lytle (1996) and the Centers for Disease Control and Prevention (CDC) showed that this outbreak was caused by cryptosporidium oocysts that passed through the filtration system of one of the city's water-treatment plants. Over the span of approximately 2 weeks, 403,000 of an estimated 1.61 million residents in the Milwaukee area (of which 880,000 were served by the malfunctioning treatment plant) became ill. At least 104 deaths have been attributed to this outbreak, mostly among elderly and immunocompromised people, such as AIDS patients (Mackenzie et al. 1994). Water distribution systems have proven to be especially vulnerable to distribution system contamination as illustrated by waterborne outbreaks in Cabool and Gideon Missouri, in the US and an outbreak in Walkerton, Canada (Clark et al. 1996; Grayman et al. 2004). There are several steps that a water utility can and should take to protect itself against the sudden catastrophic effect of an earthquake. These steps include both technological changes as well as establishment of institutional mechanisms that will assist in mitigating the potential damage that occur from such an event.

Although there has never been a successful terrorist attack on an urban water system, there is major concern with regard to the potential of bioterrorism as a threat to public safety. For example, the U.S. Army Combined Arms Support Command evaluated 27 agents for the potential of "weaponization". Of 27 agents, 7 are listed as having the potential for being "weaponized" and 14 others are listed as either possible or probable weapons. A number of these organisms are listed as definite or probable threats in water (Clark and Deininger 2000). In addition, newly discovered or emerging pathogens may pose a threat to water supply systems. One such pathogen was isolated during a United States Environmental Protection Agency (U.S. EPA) study in Peru (Clark and Deininger 2000). Several chemical agents have also been identified that might constitute a credible threat to water supply systems. Although much is known about chemical and biological agents dispersed in air, almost nothing is known about these agents in potable water.

Water and wastewater utilities are also vulnerable to cyber threats. Growth in the use of the Internet throughout the world, since the 1990s, has dramatically changed the way that both the private and public sector organizations communicate and conduct business. It is becoming increasingly recognized that all countries need to prepare for the potential of debilitating Internet disruptions. Therefore, the US Department of Homeland Security (DHS) was assigned to develop an integrated public/private plan for Internet recovery, should it be impaired. The US Government Accountability Office (GAO) was asked to: (1) identify examples of major disruptions to the Internet (2) identify the primary laws and regulations governing recovery of the Internet in the event of a major disruption (3) evaluate DHS plans to facilitate recovery from Internet disruptions, and (4) assess challenges to such efforts (USGAO 2006). GAO found that a major disruption to the Internet could be caused by (Clark et al. 2011):

- A cyber incident (such as a software malfunction or a malicious virus).
- A physical incident (such as a natural disaster or an attack that affects key facilities).
- A combination of both cyber and physical incidents.

1.2.1 Design Basis Threats

Janke et al. (Chap. 2, this volume) discuss the potential for urban water systems to be vulnerable to both manmade and natural, threats and disasters including droughts, earthquakes, and terrorist attacks. They discuss the general principles and characteristics of water and wastewater system security and summarize current research as it relates to system security focusing on manmade threats to water systems in the United States (US). Most water supply systems in the United States consist of the common elements of a source, a treatment facility, and a distribution system. Distribution system infrastructure is generally the major asset of a water utility, even though most of its components are either buried or located inconspicuously. Drinking water distribution systems are designed to deliver water from a source (usually a treatment facility) in the required quantity, quality, and at satisfactory pressure to individual consumers in a utility's service area. In general, to continuously and reliably move water between a source and a customer, the system would require storage reservoirs/tanks, and a network of pipes, pumps, valves, and other appurtenances. This infrastructure is collectively referred to as the drinking water distribution system and is assumed the most vulnerable to terrorist attack. The use of computer technology in water and wastewater systems has become increasingly prevalent. A key aspect of computer use in drinking water utilities is an increasing utilization of Supervisory Control and Data Acquisition (SCADA) systems. SCADA systems are a type of industrial control system (ICS) which are computer controlled systems that monitor and control physical industrial processes. There are over 160,000 water systems in the US most of which serve less than 3000 people.

Janke et al. (Chap. 2, this volume) claim that water systems are vulnerable to a range of manmade threats including physical disruption, contamination, and cyber attack. Disruption of water service due to some type of physical destruction is often considered in the identification of water threats, but most studies rank such denial of service or disruption-based attacks below those of contamination in terms of magnitude of impact (cost, public health) and the length of time of the disruption. Quite simply, contamination threats represent the greatest risk to water systems and thereby the communities they serve. The ingredients of an intentional water contamination threat consists of: (1) a person(s) with intent and sufficient knowledge and resources (2) a contaminant or contamination source (3) the necessary equipment and knowledge to process, prepare, and release contaminant or contamination into a water system (4) identification of a suitable release location or locations in a particular water system at a particular date and time considering all the factors (e.g., accessibility, feasibility, and implementability) necessary for such an action. The motivation of the attacker is to cause harm, whether the harm is to cause public health impacts, infrastructure damage, or just fear. The authors believe that a utility can take physical, chemical, and institutional countermeasures to protect against terrorist threats.

There has been a great deal of research directed toward establishing water utility security. This research has been focused in the following areas: (1) laboratory testing of water quality sensors to detect contaminants; (2) development of methodologies and tools for assessing the consequences of water contamination events; (3) development of methodologies (algorithms) and tools for the optimal placement of sensors in a distribution system; (4) development of methodologies and tools for contamination event detection; (5) development of methodologies and tools for responding to contamination events, and (6) development of models for assessing public health exposures. Integration of distribution system models with Supervisory Control and Data Acquisition (SCADA) data systems. There is increasing interest on the part of US drinking water utilities in real-time models to enhance both system security and compliance with drinking water standards.

Clearly, the potential for severely damaging drinking water systems due to terrorist attacks is great. Planning to avert these attacks and how to deal with them if they do occur is critical for minimizing their deleterious effects. The integration of real-time modeling with the SCADA system framework has great potential for enhancing both water quality and water systems security

1.2.2 Cyber Security

Weiss (Chap. 3, this volume) discusses the vulnerability of Industrial Control Systems (ICSs) which operate industrial infrastructures worldwide including electric power, water, oil/gas, pipelines, chemicals, mining, pharmaceuticals, transportation, and manufacturing. These types of systems are commonly utilized throughout the global industrial infrastructures. Weiss (Chap. 3, this volume) asserts that security is like a three-legged stool which consists of physical, Information Technology (IT), and ICS security. Physical security is generally well understood and often addressed by experts coming from the military or law enforcement. IT security generally deals with traditional commercial off-the-shelf (COTS) hardware and software and connections to the Internet with experts coming from IT and the military. The third leg, ICS security is much less understood and is often not considered critical. ICS cyber security was formally identified in the mid-late 1990s and at the time was not considered critical. However, ICSs typically include Supervisory Control and Data Acquisition (SCADA), Distributed Control Systems (DCS), Programmable Logic Controllers (PLC), Remote Terminal Units (RTU), and field instrumentation. These types of systems are in daily use by most drinking water systems in the US. ICSs continue to be upgraded with advanced communication capabilities and networked to improve process efficiency, productivity, regulatory compliance, and safety. This networking can be within a facility or even between facilities continents apart. When an ICS does not operate properly, it can result in impacts ranging from minor to catastrophic. Consequently, there is a critical need to ensure that electronic impacts do not cause, or enable, mis-operation of ICSs. According to Weiss (Chap. 3, this volume) more than 200 control system cyber incidents have been documented globally in many industries including electric power, water/wastewater, chemicals, pipelines, manufacturing, transportation, etc. He cites two examples in which the failure of ICSs caused serious threats to wastewater infrastructure.

Based on the examples cited in this chapter, cyber security will play a critical role in ensuring water and wastewater utility security. Cyber security is not only an issue now; it is bound to be even more important issue in the future.

1.2.3 Water Quality Event Detection

Brill (Chap. 4, this volume) examines the implementation of machine learning for water quality event detection, and shows that several algorithms are likely to be used for this purpose. His work has special significance for the 52 water utilities which operate in Israel and distribute almost 2 billion cubic meters (70.6 billion cubic ft) of water. These utilities are also responsible for sewage treatment. Israel is generally considered to be a water short area and therefore 75 % of sewage is reused by agriculture after purification. The main objective of his manuscript is to focus on important managerial considerations when implementing machine learning (ML) for Water Quality Event Detection (WQED).

Machine learning is a mathematical technique that allows computers to evolve behaviors based on analysis of empirical data. Two different concepts govern machine learning technology for the purpose of WQED are based on either supervised learning or unsupervised learning. Supervised learning implies the involvement of an expert and unsupervised learning implies the use of empirical information. For both concepts, a mathematical procedure is used to build a model which is a relation between inputs and outputs. Inputs are quality or physical measurements of water. The application of the model results in a categorical determination which indicates if the situation is a water quality event (WQE) or not. The mathematical procedure which is used to create these relations is defined as an algorithm. The procedure of using the algorithm to construct the model is called "training the model".

The two necessary and sufficient conditions in developing and implementing machine learning algorithms for water quality event detection are that the events have occurred historically, and that these events have been classified by experts. If both of these conditions are met, then supervised learning is a better way in which to proceed. However, in the case of water systems, since automatic data recording is relatively new, and since many water utilities do not have a classified history of water events, another methodology should be considered. This alternative methodology is based on unsupervised learning. Based on this methodology, the system identifies rare events which should be further examined. Thus, the algorithm must learn the frequency of historical combinations in the data. Once this frequency is known, any rare combination which did not occur or have occurred with low frequency should generate an alarm and be classified by an expert for future reference.

The types of algorithms in discussed in this chapter in conjunction with real-time modeling can play an important role in protecting water utilities against security threats. This area has great potential for future research.

1.2.4 Contaminant Warning Systems

Shen et al. (Chap. 5, this volume) discuss the impact of nodal demand uncertainty on sensor placement in contaminant warning systems. Requirements for water supply system security in Canada are provincially mandated and are not federally regulated. The national strategy is based on collaborative efforts involving the federal, provincial, territorial, and critical infrastructure sectors. Approximately 89 % of Canadians obtain their water from public water supplies with most of the remaining population served by private wells. Less than 1 % obtain their water by truck. Most of the population served by public water supplies (typically 90 % of the 89 %) receive water from municipally owned distribution systems (approximately 1500 in number). The remaining consumers receive their water from many small distribution systems. The large, municipally owned systems are highly regulated by the individual provincial governments and require permits for construction, source water protection, and certification of operators. Small systems (numbering in the thousands, many with fewer than 15 connections) are significantly less regulated, and are more exposed to public health problems. The potential for contaminant intrusion into water distribution systems, as a result of either deliberate injection and/or an accidental event is attracting increasing attention. Accidental events such as cross-connection/backflow, pipe leakage/breakage, as well as the possibility of deliberate injections have the potential to contaminate water within the distribution system. Accidental contaminants may be mixed agents, chemical (e.g., diesel fuel), viral, bacteria, protozoa, parasites, and some unidentified agents. A contaminant warning system is generally considered to consist of: (i) online water quality sensors, which monitor surrogate water quality parameters (ii) an event detection algorithm (iii) a contaminant source identification algorithm, and (iv) an emergency response plan. An essential part of a contaminant warning system is a sensor placement design methodology. Specifically, hydraulic and water quality models are applied to simulate the movement of a contaminant intrusion event and require accurate nodal demand data. However, nodal demands are never known exactly and possess (possibly huge) uncertainty. As a consequence, the authors conclude that it is crucial to incorporate nodal demand uncertainty in sensor placement design.

As discussed by the authors, demand uncertainty is a critical issue in properly implementing water quality models. Much research remains to be conducted in this area but with the advent of Automated Meter Reading capability the potential is great.

1.2.5 Earthquakes and Natural Disasters

Samuels et al. (Chap. 6, this volume) discuss a model that was used to simulate the surface water transport of radioactivity released from the Fukushima Dai-ichi nuclear power plant, in Japan after the 2011 tsunami and earthquake. It provides an example of an approach that might have direct application to the failure of other critical infrastructure such as water supply systems during natural or manmade disasters. The tsunami that followed the 9.0-magnitude Tohoku earthquake of 2011 affected a huge stretch of Japan's northeast coast, sweeping far inland. It severely damaged the plant so the direct objective of this project was to model surface water transport of radioactivity released from the plant. The author's first step was to build a river network for Japan with associated flows and velocities. They accomplished this task using a combination of existing databases and models including:

- The geospatial stream flow model (GeoSFM).
- Digital terrain data from the Shuttle Radar Topography Mission (SRTM).
- Land use and soil data from the US Geological Survey and Food and Agricultural Organization.
- Rainfall and potential evapotranspiration (PET) data from the National Ocean and Atmospheric Administration.

Benchmarking of the hydrology (flow) model was done using data from the Global Runoff Data Center (GRDC). Results were compared with the observed data downloaded from the GRDC site. Average flow for the year was used for the comparison between the observed and simulated flows. The difference between the actual observations and the model appeared robust. The Incident Command Tool for Drinking Water Protection (ICWater) model was used for the radionuclide transport. Hydrologic inputs for ICWater were provided by GeoSFM. Radionuclide data was available for air, dust, soils, leafy vegetables, and at a few places for drinking water in a 50 km radius around the plant. A two-step approach was used for radionuclide transport modeling to determine the following parameters: (1) time of travel a long major river reaches in the 50 km radius zone and (2) dilution of radioactivity in the 50 mile radius zone. Under these conditions, travel times ranged between 5 and 22 h and dilution factors ranged between 3 and 14 depending on the river being modeled. Surface radionuclide runoff and transport was modeled for Cesium-137 in two watersheds north of the plant. This simulation was performed using an actual rain event on 4-24-2011. The rain file contains a global estimate of 24 h accumulated precipitation in GIS GRID format with a spatial resolution is 0.25°. The ICWater model was run for 24 h and the downstream trace from the simulation and breakthrough curves were calculated for two rivers that discharge to the ocean. Export files were created to provide a source term to the SHARC coastal transport model.

It is clear from this manuscript that earthquakes and natural disasters can dramatically degrade critical infrastructure. The types of modeling effort discussed in this chapter can be very useful in understanding the degree to which this degradation might occur. These models can be very useful in mitigating some of the effect of these types of disasters.

1.2.6 Developing a Quantitative Regulatory Framework

Beuken et al. (Chap. 7, this volume) describe efforts in the Netherlands to develop a quantitative regulatory frame work for the provision of drinking water. The Dutch drinking water sector, is characterized as publicly owned, but privately managed. It embraces the idea of self-regulation in drinking water in the Netherlands which is supplied by ten water companies one of which is municipality based, while the other nine are limited liability companies. The associated municipalities and provinces are shareholders in these nine limited liability companies. The changes that led to the reassessment of the existing water management structures are:

- Several disasters that received a great deal of public attention, such as the Tsjernobyl nuclear meltdown.
- The expansion of drinking water networks due to a merging of water companies and closing of smaller water plants.
- A more critical attitude on the part of customers toward water supply interruptions.

Drinking water companies in the Netherlands are responsible for providing sufficient potable drinking water at adequate pressure. Drinking water quality is based on legal standards involving a large number of parameters. The legal standards governing quantity and pressure are much less detailed than those for quality. To deal with these changes, the Association of Drinking Water Companies in the Netherlands, the Vewin, established a committee in 1989 with the objective of developing a method for evaluating drinking water reliability. In 1994, the committee presented a regulatory framework that requires every water company to develop a Supply Plan. The plan describes how drinking water is to be supplied under normal conditions and how water would be supplied in case of a disturbance. The plan also prescribes a quantitative approach for assessing the reliability of the supply system.

The authors describe 15 years of experience with the quantitative assessment of water supply system reliability in the Netherlands. They draw the following conclusions:

1. The framework for reliability and the corresponding quantitative assessment as applied by water companies provides useful insights and guidance for improving the reliability of water supply systems.
2. During the past 15 years in the Netherlands some companies experienced major problems, however, none have resulted in interruption of supply for more than 24 h.
3. The framework for reliability is a product of self-regulation by the drinking water sector.
4. The quantitative assessment of reliability should be regarded as an instrument to evaluate the robustness of the total supply system against failure.
5. The framework is not meant to be rigidly applied and in certain cases it should be possible to deviate from the guidelines but these deviations must be approved by an independent body.
6. A specific module for reliability assessment has been developed within the most commonly network calculation model.
7. The reliability assessment applies to incidents that effect larger parts of the system and have a duration of more than 24 h. Further improvements for increasing the level of service are to be expected with the development of asset management techniques.

1.3 Water Resource Availability

This section contains two chapters that address water scarcity and water resource availability in widely separated parts of the world. Even though the two chapters deal with widely separated geographic areas they represent issues and problems that are typical in many countries. The first chapter discusses a bi-national agreement involving Peru and Ecuador that is intended to improve water quality in the Puyango-Tumbes watershed despite a long history of border disputes. The agreement has the potential to solve a lingering water pollution issue in the watershed. The second paper deals with the serious issue of water scarcity in Asia and the implications this scarcity may have for Australian security over the next 10–50 years.

It was estimated that in the period 1990–2025 there were and will be water scarcity problems in 118 countries. Approximately one billion people face absolute water scarcity because they live in arid regions that lack sufficient water resources. Another 348 million people face severe water scarcity because they live in regions where there are water resources but require high cost water acquisition developments to tap this potential (Bruins 2000).

Global water use has more than tripled since 1950 mainly based on the availability of surface water resources such as rivers and lakes and groundwater. Water use is generally categorized as follows:

- Domestic or urban water use
- Agricultural water use
- Industrial water use

Irrigation is the largest consumer of freshwater using over 70 % of all developed water supplies in the world. Most often increasing water demands are met by constructing large engineering projects which involve building dams and the storage of surface water. Some very large surface water projects include the Sardar Sarovar Dam in India and the Three Gorges Projects in China. The number of large dams in the World had increased from 5000 in 1950 to about 38,000 by 1995. There is also an increase in drilling groundwater wells. For example, in India the area irrigated from groundwater is greater than all other surface water irrigation combined. Throughout the world groundwater resources are being over utilized. In northern China, groundwater tables are declining at approximately one meter (3.28 ft.) each year and in the Punjab area of India groundwater tables are declining 20 cm (7.87 inch) each year. Rivers and lakes will eventually dry up as groundwater disappears (Bruins 2000).

The United States is generally blessed with adequate water resources. The contiguous US receives an average of approximately 75 cm (30 inches) of rainfall per year although there is a great deal of variability across the country (Clark and Sethi 2002). However, throughout the West and Midwest large acquirers have declined substantially because extraction rates exceed recharge rates (NRC 1999). Other areas of the world are not as fortunate as the US with respect to water resource availability.

1.3.1 Puyango-Tumbes River Basin

The Puyango-Tumbes watershed drains into the Pacific Ocean, and is one of the most important watersheds along the Pacific coast in South America. The watershed encompasses a large diverse land area, parts of which are devoted to agriculture and is an important source of water for irrigating crops. It is especially important for crops grown in the downstream valleys located on the north and south sides of the river. However, portions of the watershed are severely polluted especially in the lower portions of the basin, particularly in areas close to the city of Tumbes (Peru). Bioaccumulative contaminants and biomagnified pollutants, such as heavy metals, have been identified and these pollutants pose a significant risk to the health of persons. Another fundamental human health risk is due to the fact that people living near the river drink water from this source. Residents in the city of Tumbes drink water from the Tumbes River, that has been treated with a conventional filtration system, however, a number of pollutants adhere to water colloidal elements, which are difficult to remove with conventional treatment. Despite these problems a bi-national agreement involving Peru and Ecuador has been signed to improve water quality in the Puyango-Tumbes watershed. This agreement was reached despite a history of border disputes between these two countries. For a period of more than 150 years battles were fought between the two countries over disputed territory, particularly in the Amazon region. The last war of the mid-1990s, was followed by a signing of a comprehensive peace accord on October 26, 1998. Since the signing of the agreement, trade between the two countries has increased substantially and other kinds of cooperative endeavors occurred, including an agreement to join in efforts to improve the water quality in the Puyango-Tumbes basin. Puño (Chap. 8, this volume) presents the basics of a plan to achieve a sustainable society based upon an optimal and rational use of natural resources, with a special attention to water. Success in achieving this objective would provide multigenerational security through the wise use of land, water, and natural resources in the basin.

1.3.2 Water Scarcity in Asia and Its Water and Border Security Implications for Australia

Tularam and Marchisella (Chap. 9, this volume) have examined water scarcity in Asia and the implications that may result for Australian security. The authors selected India, Pakistan, Nepal, Bangladesh, China, and Indonesia for their study based on geography, economy, population, and water stress. The authors concluded that if the current levels of population growth, urbanization, industrial and agricultural development, and related water use are maintained there will be a serious risk to water scarcity throughout the region in the next 10–50 years.

Australia is considered the driest inhabited continent on earth, with limited naturally renewable water sources. Coastal Australia is abundant in naturally renewable resources and it is forecast that the coastal regions will be able to satisfy the minimum required level of fresh water for a long time interval, assuming no population increases. However, with even moderate population growth the length of time these resources will remain adequate is considerably shortened.

Australia represents a stable and attractive alternative for many residents in the study area. Therefore, the authors review the push and pull factors that are evident in nearly all recorded incidents of large-scale human migration. Push factors are the factors that bring about the desire to leave a current place or region of dwelling on a permanent basis. Pull factors bring about the desire to relocate to a new, specific place, or region of dwelling on a permanent basis. General water scarcity is a significant push factor while the abundance of water resources in the coastal area is a significant pull factor. The authors conclude that water scarcity in Asia will likely result in threats to both short- and long-term border security for Australia.

1.4 Threats Due to Ecosystem Impacts

This section contains one paper that deals with the frequently unrecognized impact that water resource management decisions and demand for water supply can have on ecosystems (Liu and Yang 2102). The construction of large dams and water diversion projects can have serious ecosystems impacts. Large-scale water diversions and construction of dams and reservoirs change water flows, nutrient flows, and water chemistry, all of which contribute to alteration and fragmentation of riparian habitats in rivers and streams. Water diversion projects can also change downstream hydrological regimes causing sediment deposition which can create a convex riverbed profile which means that the slope of the river bed increases from upstream to downstream.

The problem of ecosystem impact is illustrated by some of the water supply issues facing China. For example, China faces a severe water crisis and its government has started an ambitious water conservation plan. There is a tremendous pressure for development including greater demand for water at the expense of ecosystem protection. It is possible that these plans may not achieve water sustainability and may cause unintended and undesired environmental and socio-economic consequences. In addition to water resource management challenges within China there are some serious issues between China and other nations that share transboundary waters with China (Stone 2012). There are headwaters for 12 major international rivers in China, and it is therefore important for China to take a more active role to reduce its water footprint and to cooperate on the use of international rivers. For example, although the Mekong River Commission (MRC) consisting of Cambodia, Lao Peoples Democratic Republic (PDR), Thailand, and Vietnam have been operating since 1995, China has been uninterested in joining because doing so would compromise its sovereign control and limit unilateral action. Thailand has been reluctant to admit China because it did not want MRC which already includes Lao PDR and Vietnam to be dominated by communist countries.

1.4.1 Water Diversion Projects in China

Chen and Wegner (Chap. 10, this volume) discuss the many challenging water resource problems in China. China which has the world's largest population is 109th in terms of the average per capita availability of water resources. There is, however, considerable variation in water availability across the country; in South China the annual per capita fresh water resource is 3000 m^3 (105.9×10^3 ft^3) or more, while in the north it is about 500 m^3 (10.24×10^3 ft^3). As much as one-fifth of the world's population lives under water scarcity. Because of the water shortage in northern China, groundwater in some locations has been pumped out at a very rapid rate, with the result that subsidence of ground surfaces has occurred. Since the 1970s, the amount of water pumped has exceeded the amount replaced from rainfall by 120 billion m^3 (4,237 billion ft^3); as a result, the water table in the North China Plain has fallen steadily. It is estimated that the national annual water shortage in China is 40 billion m^3 (1,412 billion ft^3) of which about 6 billion m^3 (211.9 billion ft^3) is ascribed to urban areas. Among a group of 668 large cities, more than 300 face water shortage problems, with 110 of these cities characterized as having severe water shortages. In an effort to address these water shortage problems, China has implemented a number of water diversion projects. These large infrastructure projects are designed to alleviate the conflict between supply and demand by diverting water from areas with plentiful water resources to areas with scarce amounts of fresh water. Since the 1950s and especially since the 1980s, China has been constructing, 20 large inter-basin water diversion projects with others in the planning stage. The most celebrated of these projects is the South-to-North Diversion Project (nanshui beidao gongcheng), but there have been a number of other lesser known diversion projects. However, large-scale diversions has caused negative impacts. Water diversions change water flows, nutrient flows, and water chemistry, all of which contribute to alteration and fragmentation of riparian habitats in rivers and streams. Fragmented areas are often too small and disconnected to support mammals and birds that need large territories for food sources and reproduction requirements. Habitats for aquatic species are also fragmented, in this case by dams and other types of structures associated with water diversion projects. Other kinds of environmental impacts that may occur are the creation of anoxic conditions in hypolimnetic water, alteration of water temperatures, eutrophication, increased concentration of mercury in fish, intermixing of alien invasive aquatic species, increased erosion of downstream sediments and associated contaminants, and changes in metal concentrations in outflow. Water diversion projects can also change downstream hydrological regimes. Analysis shows that large-scale diversions like those along the lower Yellow River can cause the development of a convex river bed profile in the long term. Deposition will take place along the whole reach of the river, with an increasing deposition depth from downstream to upstream. This means that the slope of the river bed increases from upstream to downstream. In addition to environmental and ecological impacts, the construction of water diversion projects often involves the relocation of large numbers of people. Thus, there are major social impacts as well. As a result of these types of impacts, and the very large construction costs, water diversion projects are not without controversy in China. In addition to engineering and environmental challenges, water diversion projects provide major institutional and administrative challenges. This is true especially in the case of the South-to-North Diversion Project. The routes for the ERP and MRP traverse several provinces and municipalities, thereby raising trans-jurisdictional water quantity and water quality issues. If the WRP is implemented, such trans-jurisdictional issues will be present there as well. County, township, and village governmental units are also impacted by the water diversion projects and represent constituents with water resource needs. Another issue for China is the impact of climate change. Projected climate scenarios for China in 2050 indicate that North China will be wetter and South China will be dryer than at present. Water security issues in China are most closely related to water availability then attacks on infrastructure.

1.5 Threats Due to Climate Change

This section contains two papers which discuss various aspects of climate change. Climate change is obviously an extremely broad subject and section only addresses a small fraction of the topics that could be addressed in this area. However, they are two important examples of the intersection between climate change and water resources and water supply. One manuscript addresses water quality and water resource availability in Nepal, an area of the world that has abundant water resources and yet has severe water supply problems. Both water resource availability and water quality are definitely threatened by climate change. The second manuscript discusses some of the problems which climate change will cause with regard to water treatment and water treatment technology.

It is clear that global climate change may have a significant impact on the availability and quality of water resources throughout the world. For example, the year round average air temperature of the U.S. has already risen by more than 2°F (1.1°C) over the past 50 years with greater increases projected for the future. The intensity of severe precipitation events has also increased across the U.S. over the past 50 years, and continued increases in both frequency and intensity of precipitation are projected for the future. It is unlikely that climate change impacts, both within the U.S. and across the globe, will be distributed evenly. For example, temperature increases in the past 10 years have generally been greatest in the northern latitudes. Warming was more pronounced at high latitudes, especially in the northern hemisphere and over land. Temperatures in Alaska have increased by approximately twice as much as in the rest of the U.S., with significant impacts on sea ice, ecosystems, and coastal communities. Climate change is expected to adversely affect water quality in streams and other raw water sources and consequently impact the design and operation of existing and future drinking water treatment plants. Increases in precipitation and precipitation intensity will result in increased runoff and stream flows. These increases may result in problematic turbidity levels, increased levels of organic matter, potentially high levels of pathogens and increased levels of pesticides in lakes, rivers, and streams. Climate change may affect both surface water and groundwater quality. Although some areas may experience increases in runoff due to shifts in the spatial and temporal distribution of precipitation, some areas may experience droughts resulting in elevated levels of potentially toxic algae, and high concentrations of organic matters, bacteria, etc (Whitehead et al. 2006, 2009; Interlandi and Crockett 2003; Jacobs et al. 2001).

1.5.1 The Use of Satellite Water Tanks in Developing Countries

Shrestha and Buchberger (Chap. 11, this volume) discuss the technical and economic feasibility of a technique that might materially increase the ability of Nepal's water managers to provide safe drinking water to consumers. Nepal is one of the few countries in Southeast Asia that is uniquely rich in water resources according to a report prepared by Nepal's Water Energy Commission Secretariat. However, availability of water resources in Nepal is increasingly becoming recognized as being sensitive to climate change impacts. Nepal has approximately 6,000 rivers with a drainage area of 191,000 sq. km (73,745.1 sq mi), 74 % of which lie in Nepal alone. If properly developed these rivers could generate hydropower; provide water for irrigation, industrial uses, and supply water for domestic purposes. About 10 % of total precipitation in Nepal falls as snow, and 23 % of Nepal's total land area lie above the permanent snowline of 5,000 m (16,404 ft), with 3.6 % being covered by glaciers. Nepal's economy is largely based on agriculture but Nepalese agriculture is mainly rain fed so that agricultural areas are being badly affected by droughts, flooding, erratic rainfall, and other extreme weather events. Nepal was self-sufficient in food grain production until 1990 but due to drought conditions in 2005/2006, production fell short by 21,553 metric tons (25,863.6 short tons) and by 179,910 metric tons (201,499.2 short tons) in 2006/2007. Only about 72 % of the country's population has access to basic water supply and only 25 % of the population has sanitation facilities. In Nepal, devastating floods may be triggered by:

- Continuous rainfall and cloudbursts (CLOFs).
- Glacial lake outburst floods (GLOFs).
- Landslide dam outburst floods (LDOFs).
- Floods triggered by the failure of infrastructure.
- Sheet flooding or inundation in lowland areas due to obstructions.

Water supply is challenging because water availability, quality, and stream flow are sensitive to changes in temperature and precipitation. Increased demand for water is caused by population growth, changes in the economy, development of new technologies, changes in watershed characteristics, and water management decisions. Nepal suffers, as do many developing countries, from a serious lack of drinking water infrastructure.

The authors propose using community-based satellite tanks to provide reliable continuous supply of suitable quality water at a satisfactory pressure to a residential neighborhood where the existing water distribution system is intermittent. Under this concept, the supply to the network should be sufficient so that the tank does not run empty. The authors have developed a case study based on an existing intermittent water supply network in a neighborhood in Kathmandu, Nepal and they explore the possibility of providing continuous and reliable drinking water using satellite tanks under pressurized conditions. Satellite tanks are designed to serve individual neighborhoods as opposed to large centralized service districts.

Water distribution systems in developed countries are typically designed to provide uninterrupted service to the end users. However, water supply systems in many developing nations are operated in an intermittent fashion due to the lack of proper hydraulic infrastructure and scarce water resources. Approximately 350 million people in South Asia and about 50 million people in Latin America depend on intermittent water supplies which delivers water to the end users for limited hours and only on certain days. However, the service provided is often unsatisfactory due to low network pressure. During non-supply periods, networks may experience negative hydraulic pressures. This may result in the intrusion of contaminants including pathogens which may enter the distribution system through cracks and leaks in the pipeline. The study demonstrates that a suitably sized satellite storage tank in conjunction with a moderate increase in supply can be a viable remedy for water quality and quantity problems that afflict intermittent water systems in developing urban areas. Further, the water quality under the proposed pressurized system is expected to be much better than the quality of the intermittent supply. Using this type of distributed water supply system could greatly enhance water quality and water supply security in developing countries.

1.5.2 Impacts of Climate Change on Water Treatment Design and Operations

According to Clark (Chap. 12, this volume) there is growing awareness that climate change and related global warming will impact every aspect of society, ranging from ecosystems and infrastructure, to public health and economic and national security. It will, no doubt adversely affect the design and operation of such urban critical infrastructure as water and wastewater systems. Climate change is expected to adversely affect water quality in streams and other raw water sources and consequently impact the design and operation of existing and future drinking water treatment plants. Increases in precipitation and precipitation intensity will result in increased runoff and stream flows. These increases may result in problematic turbidity levels, increased levels of organic matter, potentially high levels of pathogens, and increased levels of pesticides in lakes, rivers, and streams. Climate change may affect both surface water and groundwater quality. Although some areas may experience increases in runoff due to shifts in the spatial and temporal distribution of precipitation, some areas may experience droughts resulting in elevated levels of potentially toxic algae, and high concentrations of organic matters, bacteria, etc (Whitehead et al. 2009; Interlandi and Crockett 2003; Jacobs et al. 2001). Some of these changes may have an adverse impact on the ability of drinking water utilities to meet Drinking Water regulations, therefore, requiring treatment systems to make major changes in their operations. Temperature extremes can result in significant challenges within a watershed and distribution system. Increased water temperatures can encourage algal growth, particularly toxic cyanobacteria, which prefer higher temperatures. High temperatures increase the potential for DBP formation in distribution systems and nitrification in chloraminated systems. As air temperatures rise, water temperatures increase and dissolved oxygen in water decreases. Li et al. (2009) conducted a detailed analysis of the effect that climate change may have on water treatment plant (WTP) design and performance, with a special emphasis on the impact of uncertainty in influent water quality and in water demand on WTP performance, design, and operation. For purposes of this analysis the investigators attempted to assess how finished water quality will change in response to changes in influent water quality (pH, TOC, ammonia, UVA, turbidity, bromide, etc.) under possible future climate change conditions. This study was intended to provide insight and guidance to water utility managers as to how to identify key source water quality parameters that can affect regulated water quality parameters such as disinfectant residual, pH, and total trihalomethanes (TTHMs). As part of the study the investigators examined the use of granular activated carbon as a possible treatment technology that is flexible enough to deal with extreme variations is source water quality. A key aspect of the analysis was the use a tool developed by the US EPA called the US Environmental Protection Agency WTP Model (USEPA 2005).

1.6 Integrated Total Water Resources Management

This section contains four chapters that deal with various aspects of integrated total water resources management (ITWRM). One of the biggest concerns for sustaining water resources in the future is the issue of allocating current and future water resources. As access to water becomes increasingly scarce the importance of the decisions about how it is managed, will grow in importance. Finding the balance between human and environmental needs will play a critical role in the sustainability of water resources. In the future water resource managers will be forced to face these allocation issues. The effects of global climate change and the long-term impacts of economic development will make these issues even more difficult to deal with. More effort will be directed toward optimizing the use of water and in minimizing the environmental impact of water use on the natural environment. Optimizing water resource allocation is very difficult because sources of water can cross many national boundaries and it is difficult to assign financial value to many water uses. Half of the world's people live in towns and cities, a figure expected to increase and in areas surrounding urban centers, agriculture must compete with industry and municipal users for safe water supplies, while traditional water sources are becoming polluted with urban wastewater. There may be significant health hazards related to the use of wastewater because it can contain a mixture of pollutants. It is clear that society must address the integrated management of water supply and wastewater systems.

1.6.1 Integrated Control and Detection of Accidental Occurrences

Macan and Macan (Chap. 13, this volume) describe the benefits of communication technologies for use in monitoring water utility performance and illustrate the main components and key features of a hydraulically based performance monitoring system. They discuss an integrated real-time system currently under development and being applied to water systems in Croatia. Although Croatia has access to high quality water only 75 % of the population is served by water supply utilities and only 40 % by sewage treatment facilities. There are 115 local agencies that supply water to consumers with an average population served of 53,000. The smallest utility serves 7,700 people (Ličko-senjska County) and the largest utility serves 260,000 (Zagreb). There are a large number of small utilities which raises the issue of the efficiency and demonstrates the chance for consolidation and for merging utilities to obtain greater economies of scale. Even though water supply systems may have service areas that are close to each other, their integration is complicated by the persistence of the desire to maintain their current state of organizational status. This situation is very similar to the United States. Water supply losses are high and range from 40 to 50 %, and there are water supplies with even larger losses. Croatia has begun working on plans to rationalize and integrate these individual water supplies into regional or even inter-regional water supply systems.

The Central Croatian government employs a policy of periodically passing and updating a Water Master Plan, which defines the main guidelines for the development and safety of public water supply. Through the Water Management Agency (Croatian Waters), the central government is mostly involved in financing of individual water supply development. However, the decisions pertaining to the development are still made independently with individual water supplies. Although there is integrated planning for protecting water supply infrastructure. There are guidelines that provide suggestions that involve the use of video surveillance and physical security for key water supply facilities (pump stations, reservoirs, etc.).

There is also a movement toward more integrated, effective, and reliable techniques that will allow for the more efficient decision making for operating and managing individual water supplies. Such a system is described by Macan and Macan (Chap. 13, this volume) and proposes the integration of automatic meter reading (AMR) and advanced metering infrastructure systems (AMI) systems.

The expansion of mobile telephones and other new communication technologies has opened up new possibilities for development of a low cost, simple, and effective system of real-time telemetry and monitoring of distribution networks. All three components of those systems (telemetry, AMR/AMI, and leak detection) can be integrated into a complete system, are complementary and provide the potential for full control of all hydraulic operational parameters, at all times and in all points of the distribution networks. Such an approach is new, these types of systems are being intensively developed and perfected in several countries (for example in the Republic of Croatia). They have the potential for successfully controlling normal water supply system operations as well as for assisting in the successful control and for maintaining functional safety of water and wastewater systems in emergency and accidental situations.

1.6.2 Planning, Preparing, and Safeguarding Critical Water Infrastructure in Australia

According to Birkett and Mala-Jetmarova (Chap. 14, this volume) in Australia most of the population of 22 million people is concentrated in large populations clusters close to the coast. This includes the cities of Sydney, Melbourne, Brisbane, Perth, Adelaide, Hobart, and Darwin. Most Australians prefer to live close to the coastal fringe, due primarily to the moderating effect of climate, the availability of infrastructure, services, and social support mechanisms. These population clusters have created a heavy reliance on urban services such as water and wastewater critical infrastructure (CI). Although the large Australian cities are not yet categorized as global mega cities, increased reliance and dependency on water and wastewater CI remains a significant issue. This is especially true in light of a history in Australia of natural disasters and terrorist incidents.

The Australian concept of protecting water and wastewater CI is based on an interconnected network that involves the Commonwealth Government through State and Territory Government agencies and business units to the Local Government, private sector and local communities. This is a partnership that encompasses both natural disasters and deliberate human threats with direction and leadership provided by the Australian Government Attorney-General. This network is needed because of a documented history of both natural and manmade disasters. Natural hazards such as bushfires, drought, and floods are part of the national history over the past 200 years. Terrorist incidents have been reported spasmodically in Australia since the 1970s. Considering the fact that Australia has been subjected to a broad history of such incidents, the protection of infrastructure becomes a significant national objective. The development of this approach is being led by the Commonwealth of Victoria and is being observed with interest by other Australian States.

The Australian model for protecting water and wastewater CI is evolving and is striving for continual improvement and effectiveness. Australians have demonstrated national strength in rebuilding and reforming in the face of adversity. In addition, Australians have shown a strong tendency to prepare, and plan and develop procedures and systems designed to act and take action in the event of any incident or emergency. Hopefully, these efforts will result in effective protection of water and wastewater CI and help ensure the continuity of water and wastewater services during any potential future crisis.

1.6.3 Latvian Practices for Protecting Water and Wastewater Infrastructure

Zabašta et al. (Chap. 15, this volume) describe the state of municipal drinking water and wastewater systems in Latvia. They discuss the major problems and issues regarding the monitoring and control of water/sewage systems and provide a comprehensive overview of central and local governmental plans and polices related to water and sewage systems. Laws promulgated by the Republic of Latvia governing nation responses to emergencies in water and wastewater are discussed. According to Zabašta et al. (Chap. 15, this volume), water systems in Latvia have many technical and management problems. Examples of research being conducted on water contamination issues in Latvia are presented and they describe some novel solutions from completed and outgoing projects intended to develop intelligent monitoring and control of water distribution networks. The authors have found, for example, that natural organic molecules are not simply transported, they are deposited through the network. The concentration of organic contaminants in the loose deposits in pipes varies significantly, and its level depends on the pipe material. It was found that total organic carbon concentrations in distribution networks rise with increased water retention time. This indicates that natural organic material is dissolving from the pipe loose deposits and leaching from the pipe walls. The biodegradable organic carbon concentrations decreased with increased water retention time indicating consumption of the substrate by bacteria. The author's overall conclusions were that unidirectional flushing is a very effective method for reduction of turbidity, however, it is more applicable in small communities then in large. From these studies it was found that resuspension potential measurements were better then grab sampling for determining water quality deterioration. The automatic wireless water meter reading system in Talsi is based on a two stage data collection and processing system: the sensors transmit data to the concentrators using short range devices of 868 MHz band and the traffic between concentrators and central server is ensured by a general packet radio service. The system provides opportunities to collect data from difficult to access meters, to utilize existing mobile operator networks and to connect new customers without major investment.

According to Zabašta et al. (Chap. 15, this volume) the increasing dependence on telecommunication and internet service providers due to the growing volume of transmitted data will force Latvian utilities to pay attention to data security and integrity aspects. A hydrodynamic model for network diagnostic application has been created and simulated using an example model of a water distribution network. The authors also suggest some possible future research tasks, based on interviews of local government and water utility representatives.

1.6.4 Austrian Activities in Protecting Critical Water Infrastructure

Möderl et al. (Chap. 16, this volume) give an overview of legal rules and standards concerning risk, safety, and security management in the Austrian water sector, including interdependencies to other sectors such as the energy sector. Further, best practices in management and planning established by Austrian utilities are discussed. Also results from utility performance benchmarks are shown. Research activities of relevant technical institutes in analyzing and protecting critical infrastructure within the frame of EU and national activity plans are also presented. A major part of the chapter deals with the description and linkup between several approaches and tools developed by different institutions for identifying critical control points and minimizing the risks of interfering processes in the water sector. The approaches are presented with focus on urban water systems, but can likewise be used in other sectors, such as the wastewater and storm water management sectors. Several tools and protocols discussed in this manuscript can be used to protect the water supply industry. For example, the spatial vulnerability identification tool (ACHILLES) identifies critical control points spatially and serves as a planning tool to assess system vulnerabilities, using quantitative risk assessment based on hydraulic simulations. The Operational Risk Management Tool and Information System (ORTIS) serves for the qualitative assessment of risks on a municipal scale. The Failure Experience Improvement System (FEIS) provides a functional failure network by means of a qualitative risk assessment as database for the utility sector. The Cyber Attack Information System (CAIS) consists of two tools for both, simulation of information technology infrastructure to analyze cyber attacks, counter measures, and interdependencies as well as for online detection of cyber attacks on water control systems. Condition assessment of large-diameter transmission water mains (ZuHaZu) and Pipe Rehabilitation Management (PiReM) are both pipe rehabilitation planning approaches that can be used to assess the structural condition influenced by deterioration and external hazards like stray current, ground motion, or traffic load. With this information pipe breakage probability can be quantified. An important feature in critical infrastructure protection is the monetary ranking of rehabilitation and safety measures. The results of PiReM can be used for such tasks. The major objective of the Austrian Crisis Management Laboratory (AQUASEC-AUT) is to assist the governmental crisis management by identifying and quantifying chemical contaminants, pathogens, and radiological threats shortly in order to facilitate assessing the risk and to take appropriate measures. ACHILLES can support this objective by analyzing contamination vulnerability maps for the identification of proper sample sites in the network. The interfaces between ORTIS and ACHILLES are developed and tested. Two feasibility studies showed that a linkup of PiReM, FEIS, and ACHILLES improves an integrated risk management approach. The results of this work showed that AQUASEC-AUT, the Cyber Attack Information System (CAIS), the risk of flood and landfill hazard tool (GEDES) and the river flood forecasting network should be included in an integrated safety and security system.

1.7 Summary and Conclusions

Many factors affect water security including direct threats to drinking water supply systems from natural or manmade threats, lack of water resource availability, threats from point- and nonpoint source pollution to water-related ecosystems, and the impact of climate change. In industrialized countries there has been a focus on the vulnerability of water and wastewater systems to manmade threats. This has been particularly true in the United States. However in many parts of the world water security threats are more closely related to the need to balance human and environmental water needs while safeguarding ecosystems and biodiversity. This book is organized in such a way as to reflect this diversity of concerns. Manuscripts from an internationally recognized group of experts are presented to provide specific examples of issues which are related to water security.

The first section of the book addresses the problem of protecting water and wastewater systems from direct physical threats. It characterizes these vulnerabilities including cyber threats and provides several important specific approaches that might either protect against or which may be useful in mitigating the effect of these threats. For example, Janke et al. (Chap. 2, this volume) suggest the importance of real-time modeling as a means of anticipating and responding to deliberate contamination of a drinking water distribution system. Weiss (Chap. 3, this volume) discusses the problems of protecting against the compromising SCADA systems by workers with inside knowledge. He provides some suggestions as to how to protect against such an occurrence. Brill (Chap. 4, this volume) explores the potential of developing algorithims that could evaluate real-time data and make "decisions" as to wheather or not an event is a serious threat to the water quality in a drinking water systems. He provides suggestions and guidance as to how to develop such algorithms. Shen et al. (Chap. 5, this volume) discuss the problem of water demand uncertainty in water distribution systems. Demand uncertainty is a very serious barrier in understanding how contaminants are propagated in water networks and the authors present an algorithmic approach which may assist water managers in understanding propagation patterns. Samuels et al. (Chap. 6, this volume) examine the development of models that can be used to simulate the transport of contaminants resulting from natural and manmade disasters. They illustrate the application of their approach using the destruction of the Fukushima Dai-ichi nuclear power plant after a tsunami and earthquake. Bueken et al. (Chap. 7, of this volume) describe the Dutch drinking water sector, which is a public—private partnership and is characterised as publicly owned, but privately managed. The industry embraces self-regulation in which regulation is initiated by the utility but monitored by the central government but is realistic and well suited to the sector's needs and could be a model for other countries to follow.

The second section of the book addresses the problem of water security as threatened by a lack of water resources and the threats that it can pose to individual and national survival. The section is made up of two chapters. The first chapter by Puño Napoleon (Chap. 8, this volume) describes a long-term conflict over the Puyango-Tumbes watershed which is in both Peru and Ecuador. The two countries had actually gone to war over water resources but have now reached an accord in which the lingering problem of pollution and water resources allocation in the watershed may be improved and could provide an example for other countries to follow. The second chapter by Tularam et al. (Chap. 9, this volume) discusses the serious lack of water resource availability in Asia and some of the political and social issues related to this problem, especially as it affects Australia. Australia also has limitations with regard to water resources but is a very stable and thriving democracy and is very likely to attract significant population growth which in turn is most likely going to put further strains' on its water resources. Tulrama et al (Chap. 9, this volume) propose several solutions to this problem that might be useful for other countries to consider.

The third section deals with the negative impact that the lack of, or the misuse of water resources can have on ecosystems. Chen and Wegner (Chap. 10, this volume) discuss the many challenging water resource issues associated with the construction of large dams and water diversion projects in China. The large-scale water diversions and construction of dams and reservoirs change water flows, nutrient flows, and water chemistry, all of which contribute to alteration and fragmentation of riparian habitats in rivers and streams. Some of the experiences and some of the solutions being considered in China should be very useful to other countries facing similar problems.

The fourth section addresses the serious threat that global climate change will have on water resources and water supply. Shrestha and Buchberger (Chap. 11, this volume) discuss the technical and economic potential for the installation of satellite (or neighborhood) tanks in order to alleviate water quality and water shortage problems in an area of the world with great sensitivity to the impact of global climate change. The approach they suggest has great potential for application in underdeveloped or developing countries. Clark (Chap. 12, this volume) discusses the impact that global climate change will have on water treatment plant design and operation. He suggests the use of Granular Activated Carbon as treatment unit process that has a great deal of flexibility in dealing with some of the extreme water quality variations associated with climate change.

The final and fifth section contains four manuscripts that discuss potential approaches for managing water systems face with security threats that include physical threats, threats to water resources availability, threats to ecosystems due to misuse of water resource management, and threats from global climate change. Birkett and Mala-Jetmarova (Chap. 14, this volume) present a very comprehensive and complete approach to protecting water and wastewater critical infrastructure based on an interconnected network that involves Australia's Commonwealth Governments, State and Territorial Government agencies, the private sector, and local communities. The partnership encompasses both natural disasters and deliberate human threats with direction and leadership provided by the Australian Government Attorney-General. This approach could provide very useful guidance for other countries and regions of the world. Zabašta et al. (Chap. 15, this volume) describe the state of municipal drinking water and wastewater systems in Latvia. They discuss the extensive research they have done and describe some novel solutions from completed and outgoing projects intended to develop intelligent monitoring and control of water distribution networks based on real-time monitoring, Möderl et al. (Chap. 16, this volume) provide an overview of legal rules and standards concerning risk, safety, and security management in the Austrian water sector, including interdependencies with other sectors such as the energy sector. They discuss best practices in management and planning established by Austrian utilities including performance benchmarks and describe one of the most complete water management plans implemented by any developed country today.

References

Bakker K (2012) Water security: research challenges and opportunities. Science 337(6097): 914–915

Bruins HJ (2000) Proactive contingency planning vis-à-vis declining water security in the 21st century. J Conting Crisis Manag 8(2):63–72

Clark RM, Deininger RA (2000) Protecting the nation's critical infrastructure: the vulnerability of U.S. water supply. J Conting Crisis Manag 8(2):73–80

Clark RM, Sethi V (2002) Water conservation practices and policies in the U.S. In: Cabera E, Cobacho R, Lund JR (eds) Regional water system management: water conservation, water supply and system integration. Swets & Zeitlinger B.V., Lisse, The Netherlands, pp 131–148

Clark RM, Geldreich EE, Fox KR, Rice EW, Johnson CH, Goodrich JA, Banik JA, Abdesaken F (1996) Tracking a Salmonella serovar typhimurium outbreak in Gideon, Missouri: Role of Contaminant Propagation Modeling. J Water Suppl Res Technol –Aqua 45(4):171–183

Clark RM, Hakim S, Ostfeld A (2011) Securing water and wastewater systems: an overview. In: Clark RM, Hakim S, Ostfeld A (eds) Handbook of water and wastewater systems protection. Springer, New York, pp 1–25

Fox KR, Lytle DA (1996) Milwaukee's Crypto outbreak: investigation and recommendations. J Am Water Works Assoc 88(9):87–94

Grayman WM, Clark RM, Harding BL, Maslia ML, Aramini J (2004) Reconstructing historical contamination events. In: Mays L (ed) Water supply systems security. McGraw-Hill, New York, pp 10.1–10.55

Interlandi SJ, Crockett CS (2003) Recent water quality trends in the Schuylkill River, Pennsylvania. USA: a preliminary assessment of the relative influences of climate, river discharge and suburban development. Water Resour 37:1737–1748

Jacobs K, Adams DB, Gleick P (2001) Chapter 14, Potential consequences of climate variability and change for the water resources of the United States. In: Meilillo J, Janetos A, Karl T (eds) Climate change impacts on the United States –the potential consequences of climate variability and change. Cambridge University Press, New York

Li Z, Clark RM, Buchberger SG, Yang YJ (2009) Assessing the impact of climate change on drinking water treatment plant design and operation. In: Proceedings of the ASCE/EWRI 2009 annual conference, Kansas city, Kansas, pp 52–60

Liu J, Yang W (2102) Water sustainability for china and beyond. Science 337:649–650

MacKenzie WR, Hoxie NJ, Proctor ME, Gradus MS, Blair KA, Peterson DE et al (1994) A massive outbreak in Milwaukee of Cryptosporidium infection transmitted through the public water supply. N Engl J Med 331:161–167

National Research Council (NRC) (1999) New strategies for American's watersheds, Committee on water science management, National Academy of Sciences, National Academy Press, Washington DC, pp 56–101

Stone R (2012) Transboundry rivers for China and Kazakhstan, no meeting of the minds on water. Science 337:404–407

United States Government Accountability Office (USGAO) (2006) Internet infrastructure: DHS faces challenges in developing a joint public/private recovery plan. GAO-06-672

US Environmental Protection Agency (USEPA) (2005) Water treatment plant model version 2.2 user's manual. Office of Ground Water 16 and Drinking Water, U.S. Environmental Protection Agency, Cincinnati, Ohio

Whitehead PG, Wilby RL, Butterfield D, Wade AJ (2006) Impacts of climate change on in-stream nitrogen in a lowland chalk stream: an appraisal of adaptation strategies. Sci Total Environ 365:260–273

Whitehead PG, Wilby RL, Battarbee RW, Kernan M, Wade AJ (2009) A review of the potential impacts of climate change on surface water quality. Hydrol Sci 54(1):101–123

Chapter 2
Protecting Water Supply Critical Infrastructure: An Overview

Robert Janke, Michael E. Tryby and Robert M. Clark

Technical Terms and Definitions

CANARY	Contamination event detection system
CMMS	Computerized maintenance management systems
CWS	Contamination warning system
DHS	Department of Homeland Security
DSL	Digital subscriber lines
EDS	Event detection system
EPA	U.S. Environmental Protection Agency
GA	Genetic algorithm
GAO	Government Accountability Office
h	Hours
HMI	Human–machine interface
ICS	Industrial control system
IT	Information technology
LIMS	Laboratory information management system
LAN	Local area network
MCMC	Marko chain Monte Carlo
MILP	Mixed integer linear program

R. Janke (✉)
U.S. Environmental Protection Agency, Office of Research and Development, National Homeland Security Research Center, Water Infrastructure and Protection Division, Cincinnati, OH, USA
e-mail: janke.robert@epa.gov

M. E. Tryby
U.S. Environmental Protection Agency, National Risk Management Research Laboratory, Water Supply and Water Resources Division, Cincinnati, OH, USA
e-mail: tryby.michael@epa.gov

R. M. Clark
Environmental Engineering and Public Health Consultant, 9627 Lansford Drive, Cincinnati, OH 45242, USA
e-mail: rmclark@fuse.net

R. M. Clark and S. Hakim (eds.), *Securing Water and Wastewater Systems*, Protecting Critical Infrastructure 2, DOI: 10.1007/978-3-319-01092-2_2,

NFPA	National fire protection association
NTNCWS	Nontransient, noncommunity water supply
ODE	Ordinary differential equation
ORP	Oxidation reduction potential
pH	A measure of acidity
PIN	Possible ingress nodes
PLC	Programmable logic controller
QP	Quadratic programming
RTU	Remote terminal units
S	Seconds
SCADA	Supervisory control and data acquisition
SDWA	Safe drinking water Act
TEVA-SPOT	Threat ensemble vulnerability assessment sensor placement optimization tool
TEVA	Threat ensemble vulnerability assessment
TNCWS	Transient noncommunity water supply
TOC	Total organic carbon
VPN	Virtual private network
VX	An extremely toxic substance that has no known uses except in chemical warfare as a nerve agent
WDS	Water distribution system
WSi	Water security initiative
WS	Water security

2.1 Introduction

Government planners have long been aware that urban water systems are vulnerable to threats and disasters, both manmade and natural, including water shortages and droughts, earthquakes, and storms with high winds and flooding. Since the attacks of September 11, 2001, government planners in the United States have been forced to also consider the vulnerability of the nation's critical infrastructure, including water systems, to terrorism. The Public Health Security and Bioterrorism Preparedness and Response Act of 2002 (U.S. Congress 2002) intensified the focus on water security (WS) research in the United States. Homeland Security Presidential Directive 7 (HSPD-7), signed on December 17, 2003, established a national policy for Federal departments and agencies to identify and prioritize critical infrastructure and to protect them from terrorist attacks. HSPD-7 established the Environmental Protection Agency (EPA) as the lead agency for the Water Sector's critical infrastructure protection activities. Consequently, the EPA developed a Homeland Security Strategy, which is regularly updated (U.S. EPA 2013). The intent of the act was to enhance national security and protect human health and the environment.

Natural threats from water shortages and droughts have led to political, humanitarian, and environmental crises throughout history and in many parts of the world. Drought may affect both developing and developed countries and, according to the United Nation's Office of Foreign Disaster Assistance, no other natural disaster has caused as many displaced persons in the twentieth century. Water played an important role in the Peace Treaty that Israel and Jordan signed on October 26, 1994 and to this point the worst case scenarios have not materialized over water disputes in the Middle East. There is concern, however, that water scarcity might become the basis for future wars.

Unlike droughts, which are described as a creeping phenomenon, the damage associated with earthquakes is concentrated in time and space. In 1906, an earthquake in San Francisco caused numerous pipes to rupture and caused dozens of residents drowning when water from broken pipes flooded the Valencia hotel. It was impossible to control the fires that spread through the area and entire buildings exploded in a huge firestorm during which temperatures were reported to reach 2000 °F (1093.2 °C). In 1995, a major earthquake hit the city of Kobe, Japan. The quake lasted 20 s and 4,069 people died, 14,679 were injured, and 222,127 people were moved into evacuation shelters. There were 67,421 fully collapsed structures of which 6,985 were burned to the ground, and there was a city-wide power failure and a nearly city-wide water supply failure (Clark and Deininger 2001). Also in Japan, on Monday, April 11, 2011, in the Hamadōri region of Fukushima, Japan an earthquake of 9.0 triggered tsunami waves that reached heights of up to 40.5 m (133 ft) in Miyako and traveled up to 10 km (6 mi) inland in the Sendai area. At least 1.5 million households were reported to have lost access to water supplies after the tsunami (Samuels et al. Chap. 6 this volume).

Rapid proliferation of computer systems and telecommunication networks compounds the vulnerability of the nation's critical infrastructure to terrorist attacks (Clark and Deininger 2000). This chapter will discuss the general principles and characteristics of water and wastewater system security and will summarize current research as it relates to system security focusing on intentional threats to water systems.

2.2 U.S. Drinking Water Infrastructure

Most water supply systems in the United States consist of the common elements of a source(s), a treatment facility and a distribution system. Distribution system infrastructure is generally the major asset of a water utility, even though most of the components are either buried or located inconspicuously. Water is transported from its source or sources to various consumers and the system is designed to operate both consistently and economically, and to deliver water in sufficient quantity, of acceptable quality, and at appropriate pressure (Jung et al. 2007). In general, to continuously and reliably move water between a source and a customer, the system would require storage reservoirs or tanks, and a network of pipes, pumps, valves, and other appurtenances. This infrastructure is collectively referred to as the drinking water distribution system (WDS) (Walski et al. 2003).

2.2.1 System Design and Operation

The branch, grid, or loop represents the three basic configurations for most WDSs. A branch system is similar to that of a tree branch with smaller pipes branching off larger pipes throughout the service area. This type of system is most frequently used in rural areas, and the water has only one possible pathway from the source to the consumer. Grid and loop systems are similar, except that a loop system typically contains a larger diameter primary transmission mains that surround the distribution area, contributing water supply within the grid from different directions. Grid and loop systems are the most widely used configurations in large municipal systems and consists of interconnected pipe loops throughout the area to be served. In this type of system, there are several pathways that the water can follow from the source to the consumer. Transmission water mains are typically 20 (7.9 cm) to 24 (9.4 cm) inches in diameter or larger. Dual-service mains that serve both transmission and distribution purposes are normally 12–20 inches (30.48–50.8 cm) in diameter. Distribution mains are usually 6–12 inches (15.25–30.48 cm) in diameter and located in every street. Service lines are typically 1 inch (2.54 cm) in diameter. Single family residences are commonly served by 3/4 inches (19 mm) service lines; While apartment buildings are large residences can have service lines larger than 1 inch (2.54 cm). Specific pipe sizes can vary depending on the extent of the distribution system and the magnitude of demand. Looped systems provide a high degree of reliability should a line break occur, because the break can be isolated with little impact on consumers outside the immediate area (Clark and Tippen 1990; Clark et al. 2004).

Key infrastructure components in a WDS include the following:

- Storage tanks or reservoirs
- Pipe network
- Valves
- Pumps
- Hydrants
- Other appurtenances, e.g., pits, manholes, blow-offs, and meters.

2.2.1.1 Basic Design and Operational Philosophies

A detailed understanding of "how water is used" is critical to understanding WDS design and operation. Almost universally, the manner in which industrial and residential customers use water drives the overall design and operation of a WDS. Generally, water use varies both spatially and temporally. Besides customer consumption, a major function of most distribution systems is to provide adequate standby fire-flow capacity (Fair and Geyer 1971). For this purpose, fire hydrants are installed in areas that are easily accessible to fire fighters and are not obstacles to pedestrians and vehicles. The ready-to-serve requirements for firefighting are governed by the National Fire Protection Association (NFPA), which establishes standards for fire-fighting capacity of distribution systems (NFPA 2003). In order to satisfy this need for adequate standby capacity and pressure (as mentioned earlier), most distribution systems use standpipes, elevated tanks, and large storage reservoirs. Additionally, most large distribution systems are "zoned." Zones are areas or sections of a distribution system of relatively constant elevation. Zones can be used to maintain relatively constant pressures in the system over a range of ground elevations. Sometimes, zone development occurs as a result of the manner in which the system has expanded. Supervisory Control and Data Acquisition (SCADA) systems are key components in operating water distribution networks and have become standard for all medium to large drinking water utilities.

2.2.1.2 SCADA Systems

As with society in general, the use of computer technology in water and waste water technology has become increasingly prevalent. The computer systems for most medium to large water utilities typically include the financial system, the Human Resource system, Laboratory Information Management Systems (LIMS), SCADA systems, and Computerized Maintenance Management Systems (CMMS). The financial, human resources, LIMS, and CMMS are considered to be part of the utilities information technology program and are generally a part of an individual utility or a local governmental information technology (IT) group and are only available 8–10 h a day. SCADA systems are generally run by the utility itself and are available on a 24 h a day, 7 days a week basis. SCADA systems are a computer-controlled type of industrial control system (ICS) that monitors and controls physical industrial processes. SCADA systems historically distinguish themselves from other ICS systems by being integrated into large-scale processes that can include multiple sites and large distances. These processes include industrial, infrastructure, and facility-based processes.

According to Panguluri et al. (2004), a water utility SCADA system usually consists of:

- A human–machine interface (HMI) through which the human operator monitors and controls the process
- A supervisory (computer) system, gathering (acquiring) data on the process and sending commands (control) to the process
- Remote terminal units (RTUs) connecting to sensors in the process, and sending digital data to the supervisory system
- Programmable logic controllers (PLCs), which are more economical, versatile, flexible, and configurable than special-purpose RTUs

- Communication infrastructure connecting the supervisory system to the RTUs
- Various process and analytical instrumentation

Data acquisition begins at the RTU or PLC level, which includes meter readings and equipment status reports that are communicated to SCADA systems as required. Data is then compiled and formatted in such a way that a control room operator using the HMI can make supervisory decisions to adjust or override normal RTU or PLC controls.

A HMI presents process data to a human operator, and the human operator then controls the process through the HMI. HMIs are usually linked to the SCADA system's databases and software programs to provide trending, diagnostic data, and management information such as scheduled maintenance procedures, logistics information, detailed schematics for a particular sensor or machine, and expert-system troubleshooting guides. An important part of most SCADA implementations is alarm processing, i.e., determining when alarms should be activated. The system monitors whether certain alarm conditions are satisfied, to determine when an alarm event has occurred. Once an alarm event has been detected, one or more actions are taken such as the generation of e-mail or text messages to inform management or remote SCADA operators.

The RTU connects to physical equipment. Typically, an RTU converts the electrical signals from the equipment to digital values such as the open/closed status from a switch or a valve, or measurements such as pressure, flow, voltage, or current. By converting and sending these electrical signals out to equipment the RTU can control equipment, such as opening or closing a switch or a valve, or setting the speed of a pump.

The term supervisory station refers to the servers and software responsible for communicating with the field equipment (RTUs, PLCs, etc.), and then to the HMI software running on workstations in the control room, or elsewhere. In smaller SCADA systems, the master station may be composed of a single personal computer (PC). In larger SCADA systems, the master station may include multiple servers, distributed software applications, and disaster recovery sites. To increase the integrity of the system, the multiple servers will often be configured in a dual-redundant or hot-standby formation providing continuous control and monitoring in the event of a server failure.

SCADA systems have traditionally used combinations of radio and direct wired connections (Panguluri et al. 2011). The remote management or monitoring function of a SCADA system is often referred to as telemetry. It is reasonable to consider SCADA as having evolved through three stages. In the first stage, computing was done by mainframe computers. Networks did not exist at the time SCADA was developed. Thus, SCADA systems were independent systems with no connectivity to other systems. Wide area networks were later designed by RTU vendors to communicate with the RTU. In the second stage, processing was distributed across multiple stations that were connected through a local area network

(LAN) and they shared information in real time. Each station was responsible for a particular task thus making the size and cost of each station less than the one used in the first generation. The third stage might be classified as "networked." Due to the usage of standard protocols and the fact that many networked SCADA systems are accessible from the Internet, the systems are potentially vulnerable to remote attack. All three of these stages exist in the water industry today.

2.3 Size and Distribution of U.S. Drinking Water Utilities

Water utilities in the United States vary greatly in size, ownership, and type of operation. The Safe Drinking Water Act (SDWA 1974) defines public water systems as consisting of community water supply systems; transient, noncommunity water supply (TNCWS) systems; and nontransient, noncommunity water supply (NTNCWS) systems. A community water supply system serves year-round residents and ranges in size from those that serve as few as 25 people to those that serve several million. A TNCWS system serves areas such as campgrounds or gas stations where people do not remain for a long period of time. A NTNCWS system serves primarily nonresidential customers but must serve at least 25 of the same people for at least 6 months of the year (such as schools, hospitals, and factories that have their own water supply). There are over 162,000 water systems in the United States that meet the federal definition of a public water system (U.S. EPA 2011). Thirty-three percent (52,838) of these systems are categorized as community water supply systems, 55 % are categorized as TNCWS, and 12 % (19,375) are NTNCWS (U.S. EPA 2011). Overall, public water systems serve 297 million residential and commercial customers. Although the vast majority (98 %) of systems serves less than 10,000 people, almost three quarters of all Americans get their water from community water supplies serving more than 10,000 people (U.S. EPA 2011). Not all water suppliers deliver water directly to consumers; some deliver water to other suppliers. Community water supply systems are defined as "consecutive systems" if they receive their water from another community water supply through one or more interconnections (Fujiwara et al. 1995).

Some utilities rely primarily on surface water supplies while others rely primarily on groundwater. Surface water is the primary source for 22 % of the community water supply systems, while groundwater is used by 78 % of community water supply systems. Of the noncommunity water supply systems (both transient and nontransient), 97 % are served by groundwater. Many systems serve communities using multiple sources of supply such as a combination of groundwater and surface water sources. In a grid/looped system, the mixing of water from different sources can have a detrimental influence on water quality, including taste and odor, in the distribution system (Clark et al. 1988, 1991a, b). Table 2.1 provides a snapshot of the size, and the population served for public water systems in the United States (U.S. EPA 2011).

Table 2.1 Public water system inventory data (U.S. EPA 2011)

		Water system population size category					Totals
		Very small 500 or less	Small 501–3,300	Medium 3,301–10,000	Large 10,001–100,000	Very large >100,000	
Community Water Supply	# Systems	28,3462	13,737	4,936	3,802	419	51,356
	Pop. Served	4,763,672	19,661,787	28,737,564	108,770,014	137,283,104	299,216,141
	% of Systems	55 %	27 %	10 %	7 %	1 %	100 %
	% of Pop.	2 %	7 %	10 %	36 %	46 %	100 %
NTNCWS	# Systems	15,461	2,566	132	18	1	18,178
	Pop. Served	2,164,594	2,674,694	705,320	441,827	203,000	6,189,435
	% of Systems	85 %	14 %	1 %	0 %	0 %	100 %
	% of Pop.	35 %	43 %	11 %	7 %	3 %	100 %
TNCWS	# Systems	80,347	2,726	92	13	1	83,179
	Pop. Served	7,171,054	2,630,931	514,925	334,715	2,000,000	12,651,625
	% of Systems	97 %	3 %	0 %	0 %	0 %	100 %
	% of Pop.	57 %	21 %	4 %	3 %	16 %	100 %
	Total # of systems	124,270	19,029	5,160	3,833	421	152,713

Source U.S. EPA (2011), "Fiscal Year 2011 drinking water and ground water statistics." *CWS* community water supply; *NTNCWS* nontransient, non-community water supply; *TNCWS* transient, noncommunity water supply

2.4 Vulnerable Characteristics of U.S. Water Supply Systems to Intentional Threats

Water systems are vulnerable to a range of intentional threats including physical disruption, contamination, and cyber attack. *Vulnerable* implies the existence of a threat and Haimes and Horowitz (2004) characterize *threat,* in the context of a terrorism scenario, as "a potential adversarial intent to cause harm or damage by adversely changing the states of the system." Willis et al. (2005) expanded the definition of *threat* to include *intent* and *capability* of the perpetrators. Similarly, again in the context of terrorism, *vulnerability* is defined by Haimes and Horowitz (2004) to be the "manifestation of the inherent states of a system (e.g., physical, technical, organizational, and cultural) that can be exploited by an adversary to cause harm or damage." As Haimes and Horowitz (2004) point out "Threats exploit vulnerabilities."

Vulnerable characteristics of water systems include their physical attributes, e.g., reservoirs, tanks, and pump stations. The distribution system itself may be vulnerable to sabotage or intentional contamination. The "trusted insider" is a potential *threat* because he or she has presumably extensive knowledge of the water system and its operation, and, therefore, *capability* (Porco et al. 2006). The largest water systems, i.e., those supporting the largest populations, are believed to be the most *vulnerable* water systems to attack (Copeland 2010).

In addition to physical attributes, a water utility's SCADA could be vulnerable to cyber attack, for example, turning pumps on or off, filling or emptying tanks inappropriately, or causing water hammer events. Cyber attacks could also affect the administrative side of the water system business or operation creating confusion by straining already-strained resources and possibly leading to denial of service for some or possibly leading to compromised water quality (Weiss, Chap. 3 this volume).

An examination of published papers, reports, and studies over the past 10–15 years illustrates the range of threats and vulnerabilities to water systems that have been identified by government agencies, researchers, and commercial sectors of the water community. Some specific threats and vulnerabilities are common in many of the studies examined. For example, contaminant threats are generally identified as the primary threat to water systems. While disruption of water service due to some type of physical destruction is often identified, most studies rank such denial of service or disruption-based attacks below those of contamination, both in terms of magnitude of impact (cost and public health) and the length in time of the disruption. From a vulnerability perspective, many studies cite post treatment storage facilities and the distribution system as being the most vulnerable components (Hickman 1999; Brosnan 1999; Allmann and Carlson 2005; Nuzzo 2006; Porco et al. 2006; Copeland 2010; Tularam and Properjohn 2011).

In the following sections, specific vulnerable characteristic of WDSs are discussed including physical disruption scenarios, intentional contamination, unintentional contamination, and cyber security issues.

2.4.1 Physical Disruption Scenarios

The President's Commission on Critical Infrastructure Protection (PDD 63 1998; PCCIP 1997) identified several features of U.S. drinking water systems that are particularly vulnerable to terrorist attack. For example, community water supplies in the USA are designed to deliver water under pressure and generally supply most of the water for fire-fighting purposes. Loss of water or a substantial loss of pressure could disable fire-fighting capability, interrupt service, and disrupt public confidence (Clark and Deininger 2000). This loss might result from a number of different causes. Many of the major pumps and power sources in water systems have custom-designed equipment and in case of a physical attack it could take months or longer to replace them. Sabotaging pumps that maintain flow and pressure or disabling electric power sources could cause long-term disruption (Clark and Deininger 2001). Many urban water systems are reliant on an aging infrastructure. Temperature variations, large swings in water pressure, vibration from traffic or industrial processes, and accidents often result in broken water mains. Planning for main breaks is usually based on historical experience; however, breaks can be induced by a system-wide hammer effect, which could be caused by opening or closing major control valves too rapidly. This could result in simultaneous main breaks that might exceed the community's capability to respond in a timely manner, causing widespread outages. Recognizing this vulnerability, water systems have been incorporating valves that cannot be opened or closed rapidly. However, many urban systems still have valves that could cause severe water hammer effects. Interrupting the water flow to agricultural and industrial users could have large economic consequences. For example, the California aqueduct, which carries water from northern parts of the state to the Los Angeles/San Diego area, also serves to irrigate the agricultural areas in mid-state. Pumping stations are used to maintain the flow of water. Loss of irrigation water for a growing season, even in years of normal rainfall, would likely result in billions of dollars of loss to California and significant losses to U.S. agricultural exports. Another problem associated with many community water systems is the potential for release of chlorine to the air. Most water systems use gaseous chlorine as a disinfectant, which is normally delivered and stored in railway tank cars. Generally, there is only minimal protection against access to these cars. The release of chlorine gas, whether intentional or unintentional, could injure nearby populations.

2.4.2 *Examples of Unintentional Contamination*

2.4.2.1 Pressure Transients

Pressure transient regimes are inevitable because all systems will, at some time, be started up, switched off, or undergo rapid flow changes such as those caused by hydrant flushing. They will also likely experience the effects of human errors, equipment breakdowns, earthquakes, or other risky disturbances (Boulos et al. 2005, 2006; Wood et al. 2005). LeChevallier et al. (2003) reported the existence of low and negative pressure transients in a number of distribution systems. Gullick et al. (2004) studied intrusion occurrences in live distribution systems and observed 15 surge events that resulted in a negative pressure. Friedman et al. (2004) confirmed that negative pressure transients can occur in the distribution system and that the intruded water can travel downstream from the site of entry. In fact, soil and water samples were collected adjacent to drinking water pipelines and then tested for occurrence of total and fecal coliforms, *Clostridium perfringens*, *Bacillus subtilis*, coliphage, and enteric viruses (Karim et al. 2003). The study found that indicator microorganisms and enteric viruses were detected in more than 50 % of the samples examined.

2.4.2.2 Milwaukee, Wisconsin, USA

In 1993, Milwaukee, Wisconsin, experienced the largest waterborne disease outbreak in documented United States history. The etiological agent was determined to be the *Cryptosporidium* protozoan. In combination with the simultaneous occurrence of frozen ground conditions, recent storms resulted in high levels of surface water runoff while changes in the normal treatment protocols were being introduced were the probable causes of the outbreak. The source of the organism was never officially identified but it was suspected to be caused by the cattle genotype due to runoff from pastures or possibly discharges from a sewage treatment plant outlet two miles upstream in Lake Michigan. Fox and Lytle (1996) and the Centers for Disease Control and Prevention (CDC) showed that this outbreak was caused by *Cryptosporidium* oocysts that passed through the filtration system of one of the city's water-treatment plants. Over the span of approximately 2 weeks, 403,000 of an estimated 1.61 million residents in the Milwaukee area (of which 880,000 were served by the malfunctioning treatment plant) became ill with the stomach cramps, fever, diarrhea, and dehydration caused by the pathogen. At least 104 deaths have been attributed to this outbreak, mostly among elderly and immuno-compromised people, such as AIDS patients (MacKenzie et al. 1994).

2.4.2.3 Cabool, Missouri, USA

Cabool, Missouri, a town of approximately 2,100 people, located in the Southeastern corner of Missouri, experienced a large outbreak of *Escherichia coli* O157:H7 during the winter of 1989–1990 (Geldreich et al. 1992). The waterborne disease outbreak resulted in 243 cases, with 32 hospitalizations and 4 deaths. This was the largest waterborne outbreak of *E. coli* O157:H7 that had been reported in the United States at the time. A precursor model to EPANET WDS modeling software package was applied to examine the movement of water and contaminants in the system. (EPANET is a public sector model that can simulate hydraulic and water quality transport of drinking water networks.) The modeling effort revealed that the pattern of illness occurrence was consistent with water movement patterns in the distribution system assuming two water line breaks. It was concluded, therefore, that some disturbance in the system, possibly the two line breaks and simultaneous meter replacements, allowed contamination to enter the water system. Analysis showed that the simulated contaminant movement covered 85 % of the infected population.

2.4.2.4 Gideon Missouri, USA

In 1993, the town of Gideon, Missouri, located in a rural, agricultural area, suffered an outbreak of salmonellosis that ultimately affected more than 650 people and caused 7 deaths (Hrudey and Hrudey 2004). At the time of the outbreak, Gideon had a population of 1,100. In early November, the town water system had experienced a major taste and odor event. In response, the water system was systematically flushed on November 10. The first cases of acute gastroenteritis were reported on November 29 and diagnosed as *Salmonella typhimurium*. However, the outbreak investigation later revealed that diarrhea cases in Gideon started around November 12 with a peak incidence around November 20. By early December, there was a 250 % increase in absenteeism in the Gideon schools and a 600 % increase in anti-diarrheal medication sales. Over 40 % of nursing home residents suffered from diarrhea and seven people died (Angulo et al. 1997). The U.S. EPA was requested to conduct a field study by the Missouri Department of Health (MDOH) and the CDC (Clark et al. 1996) in early January of 1994. The study utilized water quality modeling to reach the conclusion that the contamination source was bird droppings in the city's largest municipal tank. The tank's hatches had severely deteriorated leaving the surface of the water open to contamination by roosting birds.

2.4.2.5 Walkerton Ontario, Canada

The first documented outbreak of *Escherichia coli* 0157:H7 and *Campylobacter* spp. bacterial gastroenteritis associated with a municipal water supply in Canada occurred in the small rural town of Walkerton, Ontario (population 1261) in May 2000 (Grayman et al. 2004). At the time of the outbreak, the town's drinking water was supplied by three wells (Wells 5, 6, and 7), which fed a common distribution system.

In order to understand the factors that caused the outbreak, a water quality model of the Walkerton WDS was developed. Using a cross-sectional study, it was demonstrated that during the outbreak, residents living in homes connected to the municipal water supply and consuming Walkerton water were 11.7 times more likely to have developed gastroenteritis than those not exposed to Walkerton water.

Modeling of the Walkerton water system required estimations of the following parameters for use in the water quality model:

- Pipe diameter and length, location, age, and composition of all water pipes
- Size, storage capacity, and active volumes of the two stand pipes (water towers) in the system
- Well pump specifications (including pump curves)
- Pipe friction

The results of this study clearly supported the hypothesis that Well 5 was likely the only well involved in the Walkerton *E. coli/Campylobacter* waterborne outbreak. The results also suggested that an extreme rainfall event, which occurred just prior to the peak of the outbreak, may have played a significant role in the propagation of the contaminants. The primary cause of the contamination event, however, was human negligence. The Well 5 chlorinator was not working prior to the outbreak and the responsible operator knew it, but did not report nor correct the problem.

2.4.3 Examples of Intentional Contamination

According to Gleick (2006), attacks on water supply systems have been recorded as long as 4,500 years ago. Hickman (1999) showed that significant harm to public health could be caused by introducing chemical or biological agents into drinking water supplies and the distribution system. Hickman concluded that, "Any adversary with access to basic chemical, petrochemical, pharmaceutical, biotechnological or related industry can produce chemical or biological weapons" (Hickman 1999). Thus, the internet and a small amount of money are sufficient for capability. Hickman identified tanks, reservoirs, and the distribution system as key vulnerabilities. Burrows and Renner (1999) identified a list of biological agents that could be used to efficiently contaminate water supplies. Clark and Deininger (2001) effectively combined the work of Hickman, and Burrows and Renner to highlight how the release of biological organisms into the distribution system could significantly affect public health. Allmann and Carlson (2005) showed how commercially available distribution system modeling tools could be used to study intentional contamination events and demonstrated that service connections and fire hydrants were likely the most vulnerable components of the water system.

The following two case studies are examples of intentional contamination events in a water system. It is noteworthy that in the first example the perpetrators were able to culture the bacterium in their own laboratory. The second example illustrates that a small amount of a pesticide can be strategically placed to cause a significant amount of damage and loss of service.

2.4.3.1 The Dalles, Oregon, USA

In 1984, the Rajneeshee religious cult, using vials of the highly toxic bacterium *S. typhimurium* [*S. enterica* serovar Typhimurium], attempted to contaminate a water supply tank and salad bars in a number of area restaurants in The Dalles, Oregon. Their intent was to cause massive causalities or widespread panic. The attack resulted in a community outbreak of salmonellosis in which at least 751 cases were documented in a county that typically reports fewer than 5 cases per year. It is not clear if the WDS was chlorinated or what role, if any, disinfectant played in possibly mitigating the consequences from the contamination event. The cult apparently cultured the organisms in their own laboratories (Clark and Deininger 2000; Gleick 2006).

2.4.3.2 Pittsburgh, Pennsylvania, USA

In 1980 in Pittsburgh, Pennsylvania, an unknown perpetrator introduced chlordane into the Pittsburgh distribution system. The insecticide was injected at an isolated valve location on a large distribution main feeding, an area of the distribution system of Pittsburgh. This case study has been reported on in several articles but the most comprehensive discussion seems to be by Welter et al. (2009). The contamination event affected an area of the distribution system serving approximately 10,500 people (Welter et al. 2009). It was thought that eight or more gallons of commercial grade chlordane were introduced into the system. The highest measured concentrations of chlordane were 144,000 ug/L and the estimated average concentration across the 2,000 plus customers was estimated to be about 100 ug/L, which was about 50 times the maximum contaminant level (MCL) permitted for chlordane in drinking water (Welter et al. 2009).

The event was first discovered and reported to the utility by customers experiencing taste and odor problems with their tap water (Welter et al. 2009). The utility quickly recognized that there was a water contamination problem due to the number and location of the complaints and, as a result, dispatched personnel to investigate. Utility personnel quickly confirmed (odor was easy to confirm) that there was a contamination event and the likely contaminant was a pesticide. Public health and water utility officials issued a warning through various outlets, i.e., radio, television, and newspaper, to water customers, "do not drink or cook with water until further notice" (Welter et al. 2009). Subsequent sampling and analysis found chlordane concentrations at or above 1 mg/L in many locations (Welter et al. 2009). The utility sought to quickly contain the event, closing valves in order to prevent the contamination from reaching a storage tank. The utility requested and received permission from public health and regulatory officials to initiate hydrant flushing of the pesticide contaminated water to storm sewers in the identified area (Welter et al. 2009). After the contamination was believed to be contained, restoration plans were developed and implemented. Water usage was restored in 1 month, but 9 months of flushing and monitoring were required prior to the release of the water for unrestricted use and some residential appliances and selected pipes had to be replaced (Welter et al. 2009).

The utility and public health officials initially considered shutting down the water system instead of issuing the "do not use for drinking or cooking" order, but the problems associated with no water for sanitation or fire fighting were deemed too critical (Welter et al. 2009). Alternative drinking water was brought in and administered from various locations throughout the contaminated area, especially for residences experiencing high concentrations of chlordane. Additionally, people with sensitive skin were offered the opportunity to bathe nearby but outside the contaminated area. The first action level established was to allow bathing when chlordane concentration dropped sufficiently (below 10 ug/L). Chlordane concentration of 10 ug/L was identified as the odor detection limit for chlordane in heated water. Public health officials allowed drinking and cooking when chlordane concentration dropped below 3 ug/L. However, 3 ug/L was only allowed for 1 month in order to minimize exposure. Additional target action levels were set as the system was flushed and restored, specifically 1 ug/L a month after establishing the 3 ug/L action level, 0.2 ug/L within 2 months, and no greater than the 0.05 ug/L within about 7 months from the start of the event (Welter et al. 2009).

The chlordane incident in Pittsburgh is noteworthy in that extended flushing and intensive monitoring do not tell the whole story. In some cases, customer plumbing was replaced. Such decisions seemed to be based on cost-benefit calculations. Health authorities established progressively lower action levels during the course of the restoration to ensure that customer exposure was minimized. Monitoring continued for months after the system had been restored to unrestricted use (Welter et al. 2009).

2.4.4 Cybersecurity

Growth in the use of the Internet throughout the world has dramatically changed the way that both private and public sectors organizations communicate and conduct business (Clark et al. 2011). Although it was originally developed by the U.S. Department of Defense, the vast majority of the Internet is owned and operated by various entitles in the public and private sectors. It is becoming increasingly recognized that all countries need to prepare for the potential of debilitating Internet disruptions. Therefore in the USA, the Department of Homeland Security (DHS) at the Federal level has been assigned to develop an integrated public/private plan for Internet recovery, should it be impaired. The U.S. Government Accountability Office (GAO) was asked to (1) identify examples of major disruptions to the Internet, (2) identify the primary laws and regulations governing recovery of the Internet in the event of a major disruption, (3) evaluate DHS plans for facilitating recovery from Internet disruptions, and (4) assess challenges to such efforts (U.S. GAO 2006).

GAO found that a major disruption to the Internet could be caused by:

- A cyber incident (such as a software malfunction or a malicious virus)
- A physical incident (such as a natural disaster or an attack that affects key facilities)
- A combination of both cyber and physical incidents.

Recent cyber and physical incidents have, in fact, caused localized or regional disruptions but have not caused a catastrophic Internet failure. The GAO report presents several examples of major interruptions of the Internet, which are summarized briefly in this chapter.

The move from proprietary technologies to more standardized and open software solutions together with the increased number of connections between SCADA systems and office networks has made SCADA systems more vulnerable to attacks (Panguluri et al. 2011). The security of some SCADA-based systems has come into question as they are seen as potentially vulnerable to cyber attacks.

In particular, security researchers are concerned about:

- Lack of concern about security and authentication in the design, deployment, and operation of some existing SCADA networks
- Believing that SCADA systems have the benefit of security through obscurity through the use of specialized protocols and proprietary interfaces
- Believing that SCADA networks are secure because they are physically secured
- Believing that SCADA networks are secure because they are disconnected from the Internet.

There are two distinct threats to a modern SCADA system. First is the threat of unauthorized access to the control software, whether it be human access or changes induced intentionally or unintentionally by virus infections and other software threats residing on the control host machine. Second is the threat of packet access to the network segments hosting SCADA devices and one's ability to control or interrupt critical facility operations. In many cases, there is rudimentary or no security on the actual packet control protocol, so anyone who can send packets to the SCADA device could potentially control it.

The Department of Homeland Security has begun efforts to develop an integrated public/private plan for Internet recovery, but, according to GAO, these efforts are not complete or comprehensive. Specifically, DHS has developed high-level plans for infrastructure protection and incident response. The GAO has provided five examples to illustrate the breadth and depth of both natural and manmade disasters that could have a major effect of electronic communications (U.S. GAO 2006). Clarke and Knake (2010) have explored the potential for cyber attacks from unnamed adversaries on institutions in the United States. They cite an example of a power failure in combination with a programming glitch in a widely used SCADA system; the glitch slowed utility responses to a falling tree, which created a power surge in Ohio. The surge resulted in a power outage that encompassed 8 states, 2 Canadian provinces, and 50 million people. The Cleveland water system was left without electricity causing their pumps to fail and placing the utility in a near crisis. A hacker attack was launched against an electrical system in Brazil with similar results. A more extreme example is the Stuxnet virus that attacks SCADA systems through vulnerability in Microsoft Windows (AWWA Streamlines 2010), which is discussed below.

2.4.4.1 The "Stuxnet" Virus

The "Stuxnet" virus was apparently designed to jump from computer to computer until it found its specific target that, in this case, was Iran's nuclear enrichment program. The virus was apparently successful in finding its targets, which were both of Iran's nuclear enrichment facilities. It entered the operating systems at both facilities and then modified itself when it was discovered. What is especially interesting is that the nuclear facilities in Iran run an "air gap" security system, meaning they have no connections to the Web, making them secure from outside penetration. Stuxnet was apparently designed on the assumption that someone working in the plant would take work home on a flash drive, acquire the worm, and then bring it back to the plant. After defeating the security systems, the worm ordered centrifuges to rotate extremely fast, and then to slow down precipitously damaging the converter, the centrifuges and the bearings, and corrupting the uranium in the tubes. At the same time, it confused Iran's nuclear engineers and left them wondering what was wrong, because computer checks showed no malfunctions in the operating system. It is estimated that this penetration went on for more than a year, leaving the Iranian program in chaos and that the worm grew and adapted throughout the system (Panguluri et al. 2011).

2.4.4.2 Maroochy Shire Council

An attack that threatened public health and safety was carried out in on Maroochy Shire Council's sewage control system in Queensland, Australia (Weiss, Chap. 3 this volume). Shortly after a contractor installed a SCADA system in January 2000, system components began to function erratically. Pumps did not run when needed and alarms were not reported. Sewage flooded a nearby park and contaminated an open surface-water drainage ditch and flowed into a tidal canal. The

SCADA system was directing sewage valves to open when the design protocol should have kept them closed. Monitoring of the system logs revealed the malfunctions was the result of cyber attacks. It was found that the attacks were made by a disgruntled employee of the company that had installed the SCADA system.

2.5 The Threat of Terrorism to Urban Water Systems

As discussed previously, it has become generally accepted that water systems and their customers are vulnerable to terrorist attacks. The President's Commission on Critical Infrastructure (PDD 63 1998) was formed to evaluate the vulnerability of the nation's critical infrastructure to internal and external terrorism and has highlighted this issue. There are a wide range of vulnerabilities associated with municipal water systems including the expansive and spatially distributed infrastructure that can easily be damaged or sabotaged through physical destruction, cyber attack or control, or through the introduction of contamination.

As Beering (2002) points out, "Threats must be analyzed 'in perspective'." The utility must assess its weakest points, and then consider what actions a potential attacker might employ against them. Further, it has been noted that we need to analyze consequences to prioritize responses, identify critical components, harden or secure those that can reasonably be better protected, and develop response plans (Beering 2002; Gleick 2006).

Here we start with a brief discussion on threats to water systems. Broadly, we categorize threats to water systems as either internal or external in origin. Next, we provide some rationale as to why we believe the security emphasis in the water community should be focused on water contamination threats. Next, we discuss water contamination events from the perspective of what is known from selected published papers and reports that have examined the nature and consequences of intentional contamination events. Specifically, we discuss contamination from the perspective of: (1) contaminant quantity, method, and location selection within the water system for contamination injection or release, (2) water contaminants, and (3) magnitude of possible consequences. Finally, we talk about countermeasures that could be employed to defend against possible threats and water system vulnerabilities.

2.5.1 Internal and External Threats

Threats or perpetrators can generally be categorized as either internal or external to the water utility or its community. Porco et al. (2006) suggests the "trusted insider" is perhaps the greatest vulnerability. Copeland (2010) identifies the most likely "vulnerable" water systems to be the relatively small number of water systems serving the largest populated cities in the country.

Internal threats might include disgruntled employees who may or may not be currently employed at the organization. For example, as discussed earlier, in Pittsburgh, some believe the unknown perpetrator was a disgruntled employee (Tucker 2000). Other insider attacks might include a scenario where pipelines from drinking WDS were deliberately cross-connected with a wastewater collection pipeline. Insiders, including current employees, former employees, contractors and vendors, pose a particularly dangerous threat since they have specific knowledge of the utilities' weaknesses.

External threats may range from simple vandals to nation-sponsored terrorist threats. Critical infrastructure is an attractive target for terrorists due to the potential consequences and ripple effects of a successful attack. The distribution components of a water system are especially at risk due to the potentially large number of illness and death that could result from an attack. DHS has issued advisories to water utilities indicating that al-Qaida has shown interest in using cyanide, Botulinum toxin (Botox), *Salmonella typhi* (the causative agent of typhoid fever), and *Bacillus anthracis* (the causative agent of the disease anthrax) to attack U.S. water systems (U.S. DHS 2003). Terrorist organizations such as al-Qaida are not the only external sources with motives to use chemical or biological weapons to attack a water system.

2.5.2 Intentional Water Contamination Events

Hickman (1999); Clark and Deininger (2000, 2001), provided some of the earliest papers raising the awareness of the vulnerability of WDSs to contaminant threats. Hickman (1999), Brosnan (1999), and Clark and Deininger (2000, 2001) have shown that the distributed nature of the distribution system makes it particularly vulnerable to contamination attacks. Clark and Deininger (2000, 2001) specifically highlighted the distribution system as the most vulnerable component of a water system.

Disruption of water service due to some type of physical destruction is often considered in the identification of water threats, but most studies rank such denial of service or disruption based attacks below those of contaminant introduction, both in terms of magnitude of impact (cost or public health consequences) and the length of time of the disruption. Contamination threats represent the greatest risk to water systems and the communities they serve.

Numerous papers have analyzed and reported on the types of contaminant threats that would be of concern to water systems. Prior to the plethora of post-9/11 research studies on the threats and consequences of chemical and biological agents on water systems, Hickman (1999) identified and qualitatively characterized the magnitude of public health impacts that could result given the deliberate introduction of chemical or biological contaminants into a water system to be significant. Hickman (1999) noted from his analysis that "it is not expensive to wage an unsophisticated attack on a community water system."

In 2005, American Water Works Association (AWWA) hosted and led a water utility forum to raise awareness of contamination threats to water systems and identify key research questions that needed to be answered in order to design effective contamination warning systems (CWSs) and response capabilities (Roberson and Morley 2005). In 2007, EPA launched the WS Initiative (WSi), a pilot program to deploy and evaluate CWSs as demonstration projects at four major cities across the country (U.S. EPA 2007). These efforts and others demonstrate the need to focus on water contamination threats to water systems.

In the following sections, the intentional threat is discussed from the perspective of (1) approach, (2) contaminant, and (3) magnitude of potential consequences. Current research work is cited to frame the magnitude of possible public health consequences that could occur given a terrorist attack on an urban water system.

2.5.2.1 Approach

Contamination of a distribution system could occur through contaminant release (e.g., dumping chemicals or pesticides into a water tank) or injection (pressurized back flow of a chemical solution into the distribution system through a service connection). Fire hydrants, tanks, reservoirs, or pump stations are vulnerable to both contaminant release and contaminant injection. Pressurized backflow could theoretically occur anywhere in the distribution system and simply requires a pump with the necessary power to overcome the distribution system line pressure where the injection is to occur.

The amount of material needed to deliberately contaminate a water source (such as a reservoir or aquifer) is large and generally exceeds what an individual or small group of terrorists could easily acquire, produce, or transport. However, contaminants introduced into a tank or directly into the distribution system would be diluted less and would reside in the system for shorter times prior to public exposure and ingestion, thus diminishing the effects of disinfectants and chemical decomposition and oxidation.

A number of researchers have investigated intentional contamination events in WDSs. The objectives of these studies varied from performing threat and consequence studies to developing algorithms and methodologies for designing CWSs. Early work by Hickman (1999), Uber et al. (2004), and Allmann and Carlson (2005) demonstrated the feasibility and shed light on the magnitude of consequences that could result due to intentional contamination of WDSs. Hickman observed that such consequences would be significant. Uber et al. (2004) estimated the consequences that could range from 6 % to above 50 % of the population being exposed to lethal concentrations of a toxic contaminant. Allmann and Carlson (2005) estimated that a single pressure zone (an area of four square miles) could be contaminated at a concentration corresponding to a lethal dose for the chemical agent VX (Allmann and Carlson 2005). However, Davis et al. (2013) describe how less toxic contaminants could be used to contaminant even larger areas at lethal dose concentrations. Grayman et al. (2008) demonstrated the application of hydraulic modeling to a better understanding of possible high-rise building contamination, noting that contamination can originate from outside or from within the building and the extent of contamination is "most sensitive to the operational aspects of the internal water system."

Generally, early studies analyzed a small number of contamination scenarios and later studies have analyzed ensembles or collections of contamination scenarios to provide a statistical analysis of the consequences by injection location. Probabilistic approaches have been applied to the study of contamination events to understand how water usage influences exposure and consequences (Khanal et al. 2006) or better predict the timing of when people drink to better assess dose (Davis and Janke 2008, 2009).

Davis and Janke (2011) and Davis et al. (2010) quantified the consequences from contamination events for a diverse set of 12 real distribution systems. Their modeling and simulation work showed that significant (those similar to worst case) consequences from intentional contamination events would likely only occur at a minority of release or injection locations. These studies also demonstrated that the size of the area exhibiting certain public health consequences was relatively small for less toxic contaminants compared to the size of the area for very toxic contaminants with the relationship being proportional to the quantity of contaminant released or injected.

Grayman et al. (2008) constructed EPANET-based hydraulic models to examine the movement of contaminants within high-rise buildings. Their work showed that contamination movement within residential buildings were very sensitive to the water usage patterns at the fixture level, toilets, faucets, showers, etc. In high-rise buildings, contamination entering the building from the municipal distribution system along with its movement through the high-rise building was found to be most sensitive to the operational practices of the building's water system, i.e., pump operation in filling and draining the building's tanks. Janke et al. (2009) and collaborators later applied the high-rise building model to study consequences and sensor monitoring location performance in two real, but artificially modified system models. These papers along with others illustrated the influence of model detail on estimating consequences.

The nature of the contamination event can be described, generally, by three aspects: (1) type and quantity of the contaminant released as well as the behavior of the contaminant once released into the system, (2) location or locations in the water system where the contaminant is introduced, and (3) the type and distribution of the population downstream of the contaminant introduction and their behavior as the contamination progresses through the water system.

2.5.2.2 Water Threat Contaminants

The President's Commission on Critical Infrastructure Protection (PCCIP 1997) concluded that there is a credible threat to the nation's water supply system from certain known biological agents. Certain chemical agents have also been identified that might constitute a credible threat against water supply systems. The U.S. Army Combined Arms Support Command evaluated 27 agents for the potential for weaponization. Seven of twenty-seven agents are listed as having the potential for being "weaponized" and 14 others are listed as either possible or probable weapons. A number of these organisms are listed as definite or probable threats in water (Clark and Deininger 2000; Burrows and Renner 1999). In addition, newly discovered or emerging pathogens may pose a threat to water supply systems. One such pathogen was isolated during an EPA study (Clark and Deininger 2000) in Peru. Several chemical agents have also been identified that might constitute a credible threat against water supply systems. Although much is known about chemical and biological agents dispersed in air, less is known about these agents in potable water.

Allman and Carlson (2005) conducted a study utilizing a commercial distribution system modeling software program to show how a drinking water system could be impacted by the intentional introduction of chemical contaminants. They examined four highly toxic chemicals such as c-parathion, VX, sodium monofluoroacetate and cyanide, along with a WDS considering water quality models under various scenarios to determine the influence of feed methodology, location, and the contaminant on the effect of contamination. Their results showed that it was possible to accomplish large-scale contamination of a drinking water system through backflow into major network water supply lines.

Most modeling and simulation studies examined do not specify a contaminant but generally refer only to the contaminant as being toxic or harmful, or as being of chemical or biological in nature. Often researchers will specify whether the analysis treats the contaminant as a conservative tracer or considers some form of decay or loss. Most studies to date have generally not considered contaminant decay or loss. Propato and Uber (2004) examined the vulnerability of WDSs to pathogen intrusion to understand how effective a system's disinfectant residual would be. Their findings indicated that disinfectant residual is generally not very effective at reducing the risk of disease from pathogen intrusion.

Davis et al. (2013) consider influence of contaminant decay or loss in their analysis of the consequences of intentional WDS contamination in 12 diverse, real water systems. The extension to EPANET allowing the user to evaluate more sophisticated contaminant interactions in a WDS has been available since 2008 with the release of the multispecies version of EPANET (Shang et al. 2008). The ability to consider the influence of multispecies interactions on estimating consequences was provided with the update of Threat Ensemble Vulnerability Assessment Sensor Placement Optimization Tool (TEVA-SPOT) to include the EPANET-MSX capability in 2011 (U.S. EPA 2013).

Davis and Janke (2011) and Davis et al. (2010) showed that consequences are dependent on the contaminant, where it is released or injected into the distribution system, and the quantity released. The work of Davis and Janke (2011) and Davis et al. (2010) supported the earlier findings of Allman and Carlson (2005). However, the approach used by Davis and Janke (2011) and Davis et al. (2010)

consisted of a flexible approach that was noncontaminant-specific but could be applied to any contaminant for which health effects information was available. Davis and Janke (2011) and Davis et al. (2010) defined "impacts" to be the number of people who receive a dose (mg of some chemical or number of organisms ingested through contaminated tap water) above a certain level. This "impacts"-based approach was extended in Davis et al. (2013).

Davis et al. (2013) expanded on earlier work by examining the consequences from intentional contamination events given contaminant decay or loss as a result of transport in the WDS. A flexible analysis framework for estimating the magnitude of consequences is presented for any system provided the population is specified along with the contaminant and its behavior (decay/loss rate) in the WDS. Specifically, upper bounds on the magnitude of adverse effects are developed for a wide range of water systems, possible contaminants (based on toxicity), and a wide range of contaminant decay/loss rates.

The magnitude of adverse consequences given the release of a contaminant into a WDS is a function of the contaminant: (1) toxicity, (2) quantity released, and (3) behavior in WDS. The behavior of the contaminant is dependent on its interaction with any available disinfectant and naturally occurring biological materials present in WDSs. Adverse health effects are dependent on contaminant solubility and organoleptic properties, which influence exposure and dose.

2.5.2.3 Magnitude of Potential Consequences

Consequences of a water contamination event can be significant. A contamination event in a water system can adversely affect the people, the businesses, and the community it serves due to fear, loss of water service, significant economic costs for decontamination and recovery, and the magnitude of adverse public health effects. Public health consequences can be described and estimated as (1) exposures (i.e., people through their places of residence and business witness contamination in their tap water) (2) doses (i.e., people within the community served by the water system ingest contaminated water or somehow accumulate some measurable quantity of the contaminant or contaminants in their bodies), or (3) health effects, i.e., given some ingested mass of contaminant a health effect can be estimated. Health effects can occur within the short term, i.e., within days or weeks of exposure, or in the long term, i.e., within months or years. Within the short term, health effects could include sickness, incapacitation, or death. In the long term (i.e., Yrs), health effects could include increased cancer risk, although such health effects may be difficult to link to WDS contamination.

Numerous researchers have characterized the magnitude of the consequences that could result from an intentional contamination event in a water system. Most of the research work to describe and estimate the consequences from intentional contamination events have been in the support of WS tools. For instance, Ostfeld et al. (2008), Berry et al. (2005a, b), and Krause et al. (2008) have developed tools to determine where best to place sensor monitoring equipment in support of a

CWS. These researchers, as well as others, used extended period simulation models to predict the consequences of contamination events. Consequences were generally estimated by quantifying (1) tap water contamination concentrations, (2) quantity of pipe experiencing contamination, (3) quantity of contaminant removed at each model node (e.g., gallons of polluted water), (4) population exposed, fraction of population at risk, or number of people who receive a certain dose of contamination, and (5) population sickened or killed, using exposure, dose, and dose response models. Since these studies were intended to develop optimization algorithms, little effort was devoted to accurately estimating public health consequences or infrastructure consequences or even understanding the uncertainties in the process.

A historical evaluation of unintentional contamination events in water systems can provide some insight on the magnitude of adverse public health effects that can occur from contamination events in water systems. Approximately 690,000–1,790,000 *Salmonella typhi* cases, 20,000 hospitalizations, and 400 deaths occur annually in the USA, costing approximately $2.6 billon dollars (US) (Economic Research Service 2008; Scallan et al. 2011). *Salmonella* causes 35 % of all foodborne hospitalizations, 10 % of waterborne disease deaths, and 28 % of foodborne disease deaths (Craun et al. 2006; Scallan et al. 2011). Unintentional *Salmonella* outbreaks can infect large numbers of people. The intentional introduction of *Salmonella* into a WDS could affect a far greater number of people. *Salmonella* was used as a biological terror agent in The Dalles, Oregon in 1984 (Torok et al. 1997).

Salmonella incidence in WDSs and the cost of the associated consequences are difficult to quantify. Incidence, morbidity, mortality, and duration of many historical outbreaks are uncertain (Craun et al. 2006). Current methods fit three categories: (1) incidence models; (2) national illness burden models; and (3) economic impact models. Numerous general or *Salmonella*-specific incidence models exist. Murray et al. (2006) proposed a general susceptible, infected, and recovered population model of the spatial and temporal disease distribution in a WDS. Chandrasekaran (2006) modeled *Salmonella* incidence from contaminated water storage tanks, Danyluk et al. (2006) estimated the risks of consuming raw almonds and Mena et al. (2008) estimated the risks from pipe cross connections. Murray et al. (2006) estimated mortality while the others only estimated incidence.

Herrick et al. (2011) developed a Markov chain Monte Carlo (MCMC) model to estimate illness duration, physician, and emergency room visits, inpatient hospitalizations, mortality, and resultant costs for the Gideon, Missouri *Salmonella* waterborne disease outbreak. Most existing models estimate morbidity, mortality, and cost solely from incidence data but do not estimate illness duration as an independent cost predictor. As a result, such models may underestimate physician visits, hospitalizations, deaths, and associated costs. In the Herrick et al.'s (2011) study, transition probabilities for the Markov analysis were based on a meta-analysis of 53 *Salmonella* studies. His model resulted in an accurate prediction of the public health consequences from the Gideon, Missouri outbreak (Clark et al. 1996).

Predicting the consequences from intentional contamination can be difficult. Most modeling and simulation studies have examined intentional contamination events from the perspective of a single contamination event. Further, actual intentional contamination events have resulted in fairly localized consequences. Little, if any, has been published estimating the consequences from multiple, concurrent contamination events in WDS. Most studies have only varied the location of the contaminant release in the WDS and the various parameters describing the contaminant and how the contaminant is released or injected into the WDS. A few researchers have studied the behavior of potential receptors downstream of the contamination event. Davis and Janke (2011) studied the magnitude of potential consequences, termed "impacts," given a range of dose thresholds or levels (representing a range of contaminant toxicities), location of contaminant release in the WDS, size of population served by the water system, and quantity of contaminant (mass) released. In Davis et al. (2010), the authors examined the nature of the consequences or "impacts" for 12 real and diverse water systems while looking at the sensitivity of the consequences to (1) mass of contaminant injected, (2) time of contaminant injection, (3) duration (hr) of contaminant injection, (4) distribution of population within the WDS model, (5) tap water ingestion pattern, i.e., time of day when people drink and how much. In Davis et al. (2013), consequences were estimated while considering contaminant decay or loss.

In each of these studies, an ensemble of contamination events, described by a contaminant injection at each nonzero demand node in the model, were simulated to determine each location's percentile ranking based on consequences. Generally, only nonzero demand nodes were used as injection locations because they were believed to be most representative of actual service connections. Injection or release of contamination directly into utility facilities, e.g., tanks, was not evaluated.

These studies showed that the magnitude of public health consequences are most influenced by (1) size and nature of the particular WDS, (2) toxicity, quantity of contamination injected, and behavior of the contamination within the WDS, and (3) location of the contamination injection. Given no prior knowledge of the particular WDS, the results further indicated that a random selection of a particular contamination injection location could result in consequences that approach between one-thousandth and one-tenth of the particular system's worst case consequences (Davis et al. 2010), ranging from only a few people to many thousands, which could represent a significant fraction of the population served by the water system. Their work also showed that public health consequences from very toxic contaminants can vary substantially between water systems and is largely a function of the population served. However, for less toxic contaminants, public health consequences can be similar across a wide range and size of water systems. For less toxic contaminants, network population does not significantly influence the magnitude of consequences and increasing the quantity of contaminant injected will have a significant influence on the magnitude of consequences (Davis et al. 2010).

2.6 Countermeasures Against Terrorism

The authors believe that there are several steps that a water utility can take to protect against terrorist threats. These steps will be discussed in terms of physical countermeasures and the CWSs, which include chemical countermeasures and institutional countermeasures.

2.6.1 Physical Countermeasures

Access to a free water surface such as exists in a water reservoir should be eliminated. For example, the ventilation devices in a reservoir must be constructed in such a way as to prevent contamination of the reservoir. The intakes, pumping stations, treatment plants, tanks, and reservoirs should be fenced to secure them against casual vandalism. Beyond that, intrusion alarms should be installed to notify the operator that an individual has entered a restricted area. An immediate response might be to shut down a part of the pumping system until the appropriate authorities determine that there is no threat to the system.

An important extension of the security concept against terrorist attack would be the planning and construction of separate water lines that are fed from a protected water supply-source, which would only be activated during an emergency. Many of the older cities in the United States have separate water lines that have been installed for fire protection in heavily developed downtown areas. These water lines might be upgraded for possible use to supply the population with safe water during emergency conditions. Such proactive planning for WS, including the continuous maintenance and monitoring of chlorine residual in the water, would help to ensure the safety of most water supply systems. Nevertheless, it is of vital importance that system planners and managers be constantly on the alert to prohibit deliberate sabotage of municipal water supply systems.

2.6.2 Contamination Warning System

Among the different threats to a WDS a deliberate chemical or biological contaminant injection is the most difficult to address, both because of the uncertainty of the type of the injected contaminant and its consequences, and the uncertainty of the location and injection time. In principle, a pollutant can be injected at any WDS connection (node) using a pump or a mobile pressurized tank. Although backflow preventers provide an obstacle to such actions, they do not exist at all connections (i.e., generally very rare), they are not always functional at all connections, and they can be overcome.

Online contaminant monitoring systems or simple CWS have been considered for some time (ASCE 2004; AWWA 2004) as a tool to reduce the consequences of a deliberate contamination attack from either a chemical or biological intrusion. A CWS should be designed to detect contamination events and to provide information on the location of the contaminants within the system, including an estimation of the injection characteristics (i.e., contaminant type, injection time and duration, concentration, and injected mass flow rate). Once the type and the characteristics of the contaminant are discovered, a containment strategy can be implemented to minimize the contamination spread throughout the system and to determine which parts of the system need to be contained and/or flushed.

CWSs have been envisioned to include multiple approaches to monitoring. For instance, water quality sensors located throughout the distribution system combined with a public health surveillance system and a customer complaint monitoring program are believed to be capable of detecting a wide range of contaminants in water systems. However, the concept of using a water quality sensor-based CWS has only been piloted within last decade and little experience exists to demonstrate performance. Also, CWSs are expensive to purchase, install, and maintain. To make them a viable option, there is a clear need to maximize the benefits that a CWS can provide beyond those of security.

EPA is piloting the deployment of test CWSs through the Office of Water's Water Security initiative (WSi), formerly called WaterSentinel, at five large drinking water utilities across the nation (U.S. EPA 2007, 2008). (The WSi program was developed by EPA in close partnership with drinking water utilities and other key stakeholders involving the design, deployment, and evaluation of a CWS for drinking water systems.) Design includes the selection of water quality sensors and their strategic placement in the WDS.

The WSi promotes a comprehensive CWS that is capable of detecting a wide range of contaminants, covering a large spatial area of the distribution system, and hopefully providing early detection to mitigate impacts (U.S. EPA 2005, 2007). Components of the comprehensive CWS being piloted through WSi include chemical countermeasures (online water quality monitoring) and institutional countermeasures such as consumer complaint surveillance, public health surveillance, enhanced security monitoring, and routine sampling and analysis. These components are described below (U.S. EPA 2007).

2.6.2.1 Online Water Quality Monitoring

Continuous online monitors for water quality parameters, such as chlorine residual, total organic carbon (TOC), electrical conductivity, pH, temperature, oxidation reduction potential (ORP), and turbidity help to establish expected baselines for these parameters in a given distribution system. Hall et al. (2007) found that free chlorine (in chlorinated water) and TOC were the most useful parameters for observing a changed water quality baseline condition due to a contaminant injection. Event detection systems, such as CANARY (Hart et al. 2007) or Hach

Corporation's Guardian Blue (Kroll 2006), can detect anomalous changes from the baseline to provide an indication of potential contamination. (CANARY is software that can help detect a wide variety of chemical and biological contaminants in drinking water.) The system can also use other monitoring technologies, such as contaminant-specific monitors. The goal is to detect a wide range of possible contaminants.

2.6.2.2 Consumer Complaint Surveillance

Water utilities track consumer complaints regarding unusual taste, odor, or appearance of the water, and they record what steps they took to address these water quality problems. The WSi is developing a process to automate the compilation and tracking of information provided by consumers. Such a system, coupled with anomaly detection software, might be able to rapidly identify unusual trends that indicate a potential contamination incident.

2.6.2.3 Public Health Surveillance

Syndromic surveillance conducted by the public health sector might serve as a warning of a potential drinking water contamination incident. This surveillance includes information such as unusual trends in over-the-counter sales of medication, and reports from emergency medical service logs, 911 call centers, and poison control hotlines. Information from these sources can be integrated into a CWS by developing a reliable and an automated link between the public health sector and drinking water utilities.

2.6.2.4 Enhanced Security Monitoring

Security breaches can be monitored and documented through enhanced security practices that detect anomalous conditions. A tampering event can potentially be detected in progress and thus possibly preventing the introduction of a harmful contaminant into the drinking water system.

2.6.2.5 Routine Sampling and Analysis

Utilities can collect and analyze water samples at a predetermined frequency to establish a baseline for contaminants of concern. This provides a baseline for comparison during the response to detection of a contamination incident. Laboratory staff can engage in regular drills simulating the sampling and analysis of potential contaminants so that they will be better prepared for an actual incident.

Procedures can be periodically reviewed by qualified personnel to ensure that they remain up-to-date and implementable.

Simply placing a collection of monitors and equipment throughout a water system is not enough to effectively detect contamination incidents. To be effective, a CWS must also manage large volumes of data and provide actionable information to decision-makers. Different information streams must be captured, managed, analyzed, and interpreted in time to recognize potential incidents and mitigate the impacts. Each component of a comprehensive CWS provides useful information; however, if the data from these several components were integrated and used to evaluate a potential contamination incident, the credibility of the incident could be established more quickly and reliably than if any single information stream were used.

Many utilities currently implement monitoring and surveillance activities, but few are operating in such a way as to meet the primary objective of a CWS – timely detection of a contamination incident. For example, although many utilities currently track consumer complaint calls, a CWS requires a robust, spatially based system that, when integrated with multiple data streams (from public health surveillance, online water quality monitoring, and enhanced security monitoring), can provide specific, reliable, and timely information for decision-makers to design an effective and timely response. Consequence management plans and advanced laboratory capabilities are also required in order to respond to contamination incidents in a timely and appropriate manner. The utility, public health agencies, local government officials, law enforcement, and emergency responders, and others, must coordinate to develop an effective consequence management plan that ensures appropriate response to detection by different components. An advanced and integrated laboratory infrastructure is needed to support baseline monitoring and analysis of samples collected in response to initial detections. Still, the challenge in applying a CWS is to reliably integrate the multiple information streams in order to decide if a contamination incident has occurred. While the primary purpose of a CWS is to detect contamination incidents, dual-use benefits, such as better monitoring of water age (a surrogate for water quality) under routine circumstances, will likely help to ensure the sustainability of a CWS within a utility.

2.6.3 Cyber Security Countermeasures

Water utility SCADA systems present a major vulnerability to terrorist attacks. As the importance of SCADA systems grow this vulnerability will grow as well (Clark et al. 2011; Weiss, Chap. 3 this volume). DHS established the National Cyber Security Division (NCSD) as a public–private partnership to serve as the flagship for cyber security coordination and preparedness. The NCSD established the U.S. Computer Emergency Readiness Team (U.S. CERT) and the Control Systems Security Program (CSSP).

U.S. CERT provides the operational component of NCSD for advancing cyber security protection across the federal government. Specifically, U.S. CERT is responsible for providing response support and defense against cyber attacks for the U.S. government (nonmilitary). U.S. CERT coordinates with federal agencies, industry, the research community, state and local governments, and others to provide actionable guidance and information on cyber security to the public. CSSP's role is to reduce SCADA and industrial control system risks across all critical infrastructures. The CSSP coordinates activities to reduce the likelihood of cyber attack success and the magnitude of their possible impact against critical infrastructure control systems through risk-mitigation activities. The Cyber Security Evaluation Tool (CSET) is a software tool that assists organizations in protecting their IT assets. CSET guides users through a step-by-step process to assess their network security practices against industry standards. CSET provides a prioritized list of recommendations for improving the cyber security of water utility's SCADA system.

Panguluri et al. (2004) lists 10 vulnerabilities which are common to SCADA system infrastructure as follows:

- Operators are logged-on to the system even when the operator is not present at the workstation, thereby rendering the authentication process useless.
- Easy physical access to the SCADA equipment.
- Unprotected SCADA network access from remote locations via digital subscriber lines (DSL) and/or dial-up modem lines.
- Insecure wireless access points on the network.
- Most SCADA networks are connected directly or indirectly to the internet.
- No firewalls are installed or the firewall configuration is weak or unverified.
- System event logs are not monitored.
- Intrusion detection systems are not used.
- Operating and SCADA system software patches are not routinely applied.
- Network and/or router configuration insecure; passwords are not changed from the manufacturer defaults.

All utilities should periodically review and examine these vulnerabilities. Panguluri et al. (2011) suggest some positive steps that utilities should consider to enhance their cyber security and in particular protect their SCADA systems against cyber threats. It is recommended that utilities take the following six steps to protect cyber system:

- Be proactive in testing software by testing source and binary application code using security scanners with assessments and certifications.
- Employ a variety of network-based intrusion prevention and detection systems combined with network behavior analysis, firewalls, enterprise antivirus, and threat management devices (secure gateways, application firewalls, and managed security services).

- Implement a variety of host-based security measures such as endpoint security, network access control, and system integrity checking tools, application control, and configuration hardening tools.
- Establish vulnerability management by employing penetration testing and ethical hacking techniques, followed by patch and security configuration management and compliance measures.
- Implementing measures such as identity and access management, mobile data protection and storage and backup encryption, content monitoring/data leak prevention, and virtual private networks (VPNs).
- Initiate the use of tools and measures such as log management, event management, media sanitization, mobile device recovery and erasure, security skills development, security awareness training, forensics tools, governance, risk and compliance management tools, and disaster recovery and business continuity planning.

An approach that a utility might consider to effectively increase cyber security is to add network security improvements as a part of the expansion and upgrades to capital projects. If the utility has a current vulnerability assessment that includes network improvement recommendations and the network is well documented, the improvements added can be a step-wise approach to implementing the assessment's recommendations. Without a vulnerability assessment and network documentation, any security improvements implemented may still improve security, but are less likely to be as effective as they might be. In summary, the general technical controls that are very likely to improve network security include (Panguluri et al. 2011):

- Develop secure network topologies.
- Implement logical network separation.
- Effectively employ DMZs (DMZs are separate small buffer networks between a private internal network and an external network).
- Limit physical access.
- Restrict privileges.

2.7 Research into Distribution System Security

Research into drinking WDS security largely began as a result of the events of September 11, 2001. Prior to September 11, 2001, little research was devoted to the improving the security of water systems. While HSPD-7 established a national policy for federal departments and agencies to identify, prioritize, and protect critical from terrorist attacks; HSPD-9, issued on January 30, 2004, directed EPA to "develop robust, comprehensive, and fully coordinated surveillance and monitoring systems… that provide early detection and awareness of disease, pest, or poisonous agents." In 2004 EPA released a research and technical support action plan outlining key research questions along with a list of planned research projects (U.S. EPA 2004).

Since September 2001, research has been focused in several major areas: (1) development of methodologies and tools for assessing the consequences of water contamination events; (2) development of methodologies (algorithms) and tools for the optimal placement of sensors in a distribution system; (3) use of water quality sensors to detect contaminants as the principle components of a contamination event detection system (EDS); and (4) development of methodologies and tools for responding to contamination events, including real-time monitoring and modeling capabilities to provide the necessary foundation for implementing effective response actions. In this section, we briefly examine what has been done in each of these research areas and identify some questions that still need to be answered.

2.7.1 Methodologies and Tools to Assess the Consequence of Contamination Events

Since the early 1990s, various researchers (Clark and Deininger 2000, 2001; Grayman et al. 2004; Hickman 1999) have contemplated the use of distribution system models to better understand contaminant transport in a drinking WDS and the resulting public health exposures and consequences. Nilsson et al. (2005) simulated a deliberate biochemical assault on a municipal drinking WDS to demonstrate an effective method to characterize potential consumer exposure to contaminants and evaluate system vulnerabilities. Linking a novel stochastic water-use simulator (PRPsym, a model for simulating stochastic water demands) with EPANET (Rossman 2000), Nilsson et al. (2005) generated empirical frequency distributions of mass dose loadings at select nodes in a typical WDS. Khanal et al. (2006) extended the work of Nilsson et al. (2005) by examining the sensitivity of network response to the variability in the location, timing, duration, and intensity of a contamination event.

PipelineNet is an EPANET-based software tool to investigate propagation of contamination in a distribution system and the resulting public health consequences (Bahadur et al. 2003). PipelineNet was developed by Science Applications International Corporation (SAIC) through funding from EPA to help protect the Winter Olympics in Salt Lake City, Utah, in 2002 from a possible intentional contamination event. PipelineNet was the first modeling and simulation software tool specifically designed to analyze the consequences from intentional contamination events. In 2004, Uber et al. (2004) used a "systems analysis" approach to assess the consequences from an intentional contamination event. Here "systems analysis" is characterized by an *ensemble* of contamination events, theoretically representing any service connection. Uber et al. (2004) developed an extension to EPANET that allowed the sequential simulation of contamination scenarios using a script approach. The script approach allowed the user to specify a mass injection rate, start and stop time for the contamination event, and the model node name where the injection would take place. The script approach also allowed the sequential running of EPANET simulations. The tool was used to simulate and model contaminant transport in three real, distribution system models (Uber et al. 2004). Using the consequence results from an ensemble of contamination events, a statistical analysis was performed to rank the possible contamination injection locations in the distribution system model based on their ability to cause the greatest downstream consequences.

The work of Uber et al. (2004) led to the development of the Threat Ensemble Vulnerability Assessment (TEVA) Research Program, which resulted in the development of the TEVA-Sensor Placement Optimization Tool (SPOT) (U.S. EPA 2013; Morley et al. 2007; Murray et al. 2004). Without specific intelligence information, it is difficult to predict how a terrorist group might sabotage a water system. Therefore, TEVA-SPOT provides the user with the capability to analyze a large number of possible threat scenarios to help determine the potential magnitude of possible consequences. TEVA-SPOT allows the user to create a threat ensemble, or a set of contamination scenarios, based on varying, for instance, the type of contaminant, the amount and concentration of the contaminant, and the location of the contaminant injection into the distribution system. System vulnerability can then be assessed based on the entire threat ensemble.

TEVA-SPOT incorporates a limited, probabilistic-based framework for analyzing the consequences of contamination events in drinking WDSs. Drinking water consumers can be exposed to contaminants from ingestion, inhalation of volatilized chemicals or particles, and/or dermal exposure. For consequences, TEVA-SPOT (graphical user interface version) provides the following capabilities:

- Simulate and model an *ensemble* of contamination events, e.g., all locations (nodes) in the model, in a distributed, computationally efficient manner making use of a computer's multiple processing cores.
- Estimate consequences based on public health exposures, doses, and health effects (illnesses and fatalities) from ingestion of contaminated tap water. Public health consequences could include injuries, disease, illness, and deaths. TEVA-SPOT provides the probit dose response model with the input parameters of LD_{50} (median lethal dose) and beta slope factor (Holcomb et al. 1999).
- Estimate infrastructure contamination as the length of pipe contaminated or gallons of water contaminated.
- Define the *ensemble* of contamination event locations. Prescribed collections consisting of all nodes, nonzero demand nodes, utility facilities (e.g., tanks and pump stations), user-defined list, and a user-defined random selection of nodes based on number or percentage of nodes from either all nodes, nonzero demand nodes or by pipe diameter. TEVA-SPOT also has the capability of defining the list of contaminant release locations to be those upstream of user-defined critical locations (nodes).

Each scenario in the threat ensemble that is simulated (underlying simulation engine is EPANET) can include an assumption of first order contaminant decay of constituents. In 2011, TEVA-SPOT was upgraded to incorporate the capabilities of EPANET-MSX, i.e., fate and transport modeling of multiple dissolved constituents in distribution systems (Shang et al. 2008). This upgrade permits the modeling of reactions at the pipe wall and in the bulk flow given the specification of constituent reaction kinetics and products, thereby resulting in potentially more accurate estimates of human exposure and health risk. Stochastic modeling is limited to the probabilistic modeling of the timing and volume of tap water ingestion (Davis and Janke 2009). Contaminant water concentration results are collected in binary format, which is used in the optimal placement of sensor monitoring stations.

Given the computational difficulty of analyzing the *threat* ensemble of large or very large WDS models, research continues to investigate ways to minimize the computational burden associated with identifying the extreme high-consequence events. One example is the work of Perelman and Osfeld (2010) who propose an algorithm that would allow the efficient sampling of a subset of possible events in an effort to preferentially identify those of high consequence.

Needed research includes understanding the influence of model detail or skeletonization on estimating consequences. A better understanding is needed as to how operational or seasonal conditions influence the quantification of consequences given a contamination attack. Also, work is needed to better understand and predict how contaminants will interact with pipe walls, biofilms, and disinfectants. How important is the quality of the distribution system model in predicting consequences and the locations of high-consequence events? For example, could infrastructure information, i.e., using GIS information for a particular water system along with other publically available information, be used to create a hydraulic and water quality model that could be analyzed in TEVA-SPOT to "approximately" determine the consequences from an intentional contamination event? Finally, a better understanding and more precise quantification is needed about the influence of post-service connection piping detail for estimating consequences (Grayman et al. (2008), Janke et al. 2009).

2.7.2 Methodologies and Tools for Placement of Sensors

The best approach to mitigate possible consequences from water contamination events involve "early" or advance warning systems (Brosnan 1999). An American Society of Civil Engineer's study in 2004 provided an early, comprehensive discussion of early warning systems design, deployment and operation. This broad-based report discussed the problem of water contamination with respect to (1) rationale for online monitoring and system design basics, (2) identifying detection instruments for potential threat contaminants, (3) selection and placement of instruments, (4) data analysis and use of distribution system models, (5) communication systems requirements, (6) response to contamination events, (7) interfacing with existing surveillance systems, operations, maintenance, and upgrades, and (8) exercising the system (ASCE 2004). Roberson and Morley (2005) helped to focus the discussion of CWS design and implementation in a practical direction. The report emphasizes that detection of contamination in order to provide treatment and response is most likely the best that can be done considering the "a myriad [of] limitations" facing water systems.

Research on methods to mitigate the impacts of contamination incidents converged, by 2006, on the concept of a CWS. The goal of a CWS is to detect contamination incidents early enough to allow for an effective response that minimizes further public health or economic impacts. Janke et al. (2006) showed that a CWS based on real-time monitors could be more effective at reducing public health impacts than sampling-based strategies and that response time was critical to reducing impacts. A CWS is defined as a proactive approach that uses advanced monitoring technologies and enhanced surveillance activities to collect, integrate, analyze, and communicate information to provide a timely warning of potential contamination incidents.

Many different approaches to contamination monitoring have been suggested, including using water quality sensors, composite or grab sampling, and placement of sensors. Since most monitoring programs will be budget constrained, cost is a critical factor in the design, deployment, and operation of a CWS. Since 2003, researchers have published numerous papers on sensor placement in drinking WDSs seeking to maximize the benefit of monitoring while minimizing the cost. The Battle of the Water Sensor Networks study compared 15 different approaches to the problem of sensor placement to support a CWS (Ostfeld et al. 2008). CWS design is typically focused just on the problem of optimizing sensor monitoring station placement within the distribution system. For a good synopsis of the research related to the problem of sensor placement optimization, we refer the reader to Hart and Murray (2010), which includes a thorough review of over ninety papers related to sensor placement for CWS design. In the following paragraphs, key points from Hart and Murray (2010) are provided.

Hart and Murray (2010) outline that sensor placement strategies can be broadly characterized by the technical approach and the type of computational approach used. Hart and Murray (2010) describe the following categories to reflect the important differences in various proposed sensor placement strategies:

- Expert Opinion: These methods rely on the experience and knowledge of experts. As Hart and Murray (2010) indicate, "expert opinion strategies are guided solely by human judgment." Hart and Murray (2010) provide the following references for expert opinion-based approaches: Berry et al. (2005a, b) and Trachtman (2006). These papers consider sensor placements developed by experts with significant knowledge of WDSs. The experts described in these papers did not use a distribution system model to carefully analyze network dynamics. Instead, the experts used their experience to identify locations whose water quality is representative of water throughout the network. Therefore, an advantage of expert opinion-based approaches is that they do not necessarily require a distribution system model.

- Ranking Methods: Another approach is to incorporate user-defined information to rank potential sensor locations (Bahadur et al. 2003; Ghimire and Barkdoll 2006). In this approach, Hart and Murray (2010) indicate that a user provides "preference values for the properties of a 'desirable' sensor location, such as proximity to critical facilities." These user-defined "preferences" can then be used to rank the desirability of particular locations for the placement of monitors. Further, Hart and Murray (2010) suggest that spatial information can then be integrated to ensure good coverage of the network. Generally, ranking-based approaches use a distribution system model.
- Optimization: Sensor placement can be performed with optimization methods that computationally search for a sensor layout that minimizes some objective, such as "contamination risks." Hart and Murray (2010) describe this group of methods as those that use "a computational model to estimate the performance of a sensor configuration." They provide the example, "a model might compute the expected impact of an ensemble of contamination incidents, given sensors placed at strategic locations." Optimization methods typically rely on the use of a detailed distribution system model.

Hart and Murray (2010) identify seven steps common to most of the optimization-based sensor placement strategies: (1) defining the objective or "contamination risk" to minimize consequences (e.g., public health consequences), (2) describing the characteristics of sensors used in the CWS, (3) selecting the performance objective(s), (4) determining the optimization objective, (5) formulating the optimization model, (6) applying an appropriate optimization strategy, and (7) implementing the design. Published sensor placement research studies approach each step with varying degrees of complexity and with different optimization and simulation strategies. Hart and Murray (2010) divide the 90 papers they examined into nine groups according to how the authors addressed each step. In particular, Hart and Murray (2010) use five categories based on (a) use or nonuse of contaminant transport simulations to compute risk ,(b) use of sensor failure model, (c) consideration of multiple design objectives during optimization, (d) type of optimization objective, and (e) whether data uncertainties were modeled. Unfortunately, many smaller water utilities do not possess a sufficiently detailed or accurate distribution system model that would support using an optimization-based approach for sensor placement.

EPA's TEVA-SPOT software is the only open-source program the authors are aware of that assesses the consequences of contamination events and then uses the quantitative consequence results to optimally place sensors in the design of a CWS. TEVA-SPOT is available in two software applications: the command line, toolkit version, and the graphical user interface version. The command line, toolkit version is meant for academic researchers and software developers (U.S. EPA 2013). The graphical user interface version provides an easy-to-use interface and functionality to make use of a computer's multicore processing capabilities for analyzing large WDS models (U.S. EPA 2013). The TEVA-SPOT software applications were developed by the EPA's Threat Ensemble Vulnerability Assessment (TEVA) Research Program composed of researchers from EPA, University of Cincinnati, Argonne National Laboratory, and Sandia National Laboratories.

Utilities can consider a number of possible goals for an online sensor system such as minimizing public exposure to contaminants, the spatial extent of (pipe) contamination, detection time, or costs. Some objectives may conflict with others, making it difficult to identify a single best sensor network design. Quantifying a sensor placement's performance with respect to these goals allows some comparison of competing placements. TEVA-SPOT can optimize with respect to a primary objective, and also consider one or more secondary objectives. TEVA-SPOT provides a regret analysis operation mode to allow the user to analyze multiple sensor placement designs with respect to a range of threats. The regret analysis mode allows the user to determine how well a sensor network design performs when confronted with a threat or objective that is different from that used in its design (Davis et al. 2013). There are many practical constraints and costs faced by water utilities that cannot be easily modeled (Murray et al. 2008, 2009). Designing a CWS is not a matter of performing a simple optimization analysis (Murray et al. 2008, 2009). Instead, the design process is better described by a multiobjective problem that requires informed decision making, using optimization tools to identify possible sensor network designs that work well under different assumptions and for different objectives (Murray et al. 2008, 2009; U.S. EPA 2009). Ultimately, water utilities must weigh the costs and benefits of different designs and understand the significant public health and cost tradeoffs (Murray et al. 2008, 2009; U.S. EPA 2009).

The use of TEVA-SPOT for CWS design is composed of a "modeling process" and a "decision-making process" that employs optimization (Murray et al. 2008; U.S. EPA 2009). The TEVA-SPOT modeling process includes creating or utilizing an EPANET-based network model for hydraulic and water quality analysis, describing sensor characteristics, defining the contamination threats, selecting performance measures, estimating range of utility response times following the detection of a contamination incident, and identifying a set of potential sensor locations (Murray et al. 2008; U.S. EPA 2009). The TEVA-SPOT decision-making process involves applying an optimization method and evaluating sensor placements (Murray et al. 2008; U.S. EPA 2009). The overall process is refined by using TEVA-SPOT to perform regret analyses, analyzing tradeoffs, and comparing preferred designs to account for modeling and data uncertainties (Murray et al. 2008; U.S. EPA 2009; Davis et al. 2013).

Most of the research literature focuses on new or improved sensor placement optimization methods. Little research has been devoted to real utility applications. Some examples of real utility applications include the work by Skadsen et al. (2008) and Davis et al. (2013). Davis et al. (2013) analyze the robustness of sensor placement designs to changed conditions in 11 real and diverse water systems. Their work shows how more robust designs can be achieved by using a high toxicity contaminant, a mass injection rate as high as reasonably feasible, and a design objective that seeks to minimize average consequences (Davis et al. 2013).

2.7.3 Water Quality Sensors and Contamination Event Detection

The use of water quality sensors to detect contamination was conceived as a means to provide broad contaminant coverage in the design of early warning contamination systems. It was recognized early that it was not technically feasible to design an early warning contamination system capable of accurately detecting the plethora of contaminants that could be used to contaminant a drinking water supply/distribution system and to cause public health consequences. Additionally, it was recognized that any technology identified for contamination detection would likely need to be deployed at many locations given the large spatial extent of WDSs. Therefore, any suitable technology would need to be economical for large-scale deployment within a distribution system. As a result, many researchers have focused their efforts on identifying online sensor technologies that could be used to detect anomalous changes in the baseline water quality. Once an anomaly is detected and the water utility operator is alerted, further actions (e.g., grab sampling and analysis) could be undertaken by system personnel to identify and quantify the contaminant whenever possible. Additional discussion of these issues can be found in U.S. EPA (2005).

Research associated with using water quality sensors in contamination event detection systems has been focused in two areas: (1) water quality sensor testing in the laboratory or in laboratory-based distribution system simulators (DSS) to determine water quality sensor response to specific contaminants and (2) statistical algorithm development to decipher anomalous water quality sensor response due to contamination as compared to normal operations. EPA funded the development of the open-source event detection system (EDS), called CANARY, through Sandia National Laboratories. CANARY is available for download at: (https://software.sandia.gov/trac/canary) (U.S. EPA). Hach Corporation developed the Guardian Blue early warning system to detect, alert, and classify a wide variety of threat contaminants in drinking WDSs (Kroll 2006).

Two principal groups have been involved with water quality sensor testing, EPA researchers at the EPA's Test and Evaluation Center in Cincinnati, Ohio and researchers in the sensor industry such as the Hach Corporation in Loveland, Colorado (Kroll 2006). In 2005, EPA published a state of the technology review outlining the technologies and techniques for monitoring and evaluating drinking water quality in the context of early warning systems (U.S. EPA 2005).

Hall et al. (2007) and Hall and Szabo (2010) conducted an in-depth evaluation of how changes in water quality parameters associated with real-time sensors can be used to potentially indicate the presence of contamination. The sensors investigated were off-the-shelf commercial products designed to monitor standard drinking water parameters such as pH, free chlorine, ORP, dissolved oxygen, specific conductance, turbidity, TOC, chloride, ammonia, and nitrate. Sensors were mounted within a re-circulating pipe loop and challenged with contaminants including secondary effluent from a wastewater treatment plant, potassium ferri-cyanide, a malathion insecticidal formulation, a glyphosate herbicidal formulation, nicotine, arsenic trioxide, aldicarb, and *Escherichia coli* K-12 strain with growth media (Hall et al. 2007 and Hall and Szabo (2010). Overall, the sensors that responded to the most contaminants were free chlorine, TOC, ORP, specific conductance, and chloride. It is important to recognize that the characteristics of the re-circulating, pipe loop distribution system used in the investigation likely significantly influenced the results.

Actual distribution system waters are observed to have much greater variability in their water quality parameters than what could easily be tested in a simple pipe loop configuration. However, Hall et al. (2007) and Hall and Szabo (2010) point out that no single water quality sensor responded to all of the contaminants used in the study, yet some sensors responded to a greater number of contaminants than did others. Hall and Szabo (2010) describes water quality sensor test results from a single pass pipe in addition to a recirculating loop. Hall and Szabo (2010) indicate that detecting contamination in a single pass pipe is more challenging. When used in contamination event detection, it is not only the absolute magnitude of the change that is important, but also the magnitude relative to the size and fluctuations in the baseline along with the slope of the change (i.e., to determine whether the changes occur over several hours or several minutes). Thus, the quantitative evaluation makes use of signal-to-noise principles, which is difficult to generalize and is location-specific. The sensors that responded to a larger number of contaminants were specific conductivity, TOC, free chlorine, chloride, and ORP. The chlorine sensors appeared to respond to all the contaminants studied. However, it is important to recognize that some potential contaminants do not react significantly with chlorine. Hall et al. (2007) and Hall and Szabo (2010) indicate that TOC responded to all the organic (carbon-containing) compounds. The TOC monitor, however, has a much higher capital cost when compared with other sensors (Hall et al. 2007, Hall and Szabo (2010). The calibration requirements for the sensors in these systems range from weekly to monthly (Hall et al. 2007, Hall and Szabo (2010). Hall and Szabo (2010) estimate that a multiparameter monitoring station could have reagent and maintenance costs of several hundred dollars per month.

Contamination event detection in a drinking WDS is a case of examining a set of noisy signals in order to detect events having a low probability of occurrence and yet only appear as very subtle deviations from typical background signals. In these situations, the required sensitivity of the monitoring algorithm and overlap in the background and event signal signatures will lead to false alarms in the event detection (Rizak and Hrudey 2006). Testing of EDS methods has focused on baseline water quality monitoring to distinguish between valid alerts (alerts resulting from unusual water quality in the distribution system) and invalid alerts (alerts that are unrelated to unusual water quality) and simulated testing. Simulated testing involves using modeling and simulation to create a contamination event within a SCADA derived background water quality dataset and then determine if the EDS can detect the event.

As a part of the WSi Pilot Program, water quality sensors have been deployed and evaluated at several of the pilot cities as a component of a CWS. Some operational experience has been gained to date from these deployments related to water quality sensor performance and operation as part of an online CWS (Allgeier et al. 2008). Allgeier et al. (2008) reviewed the first year of operation for the Cincinnati Pilot's online water quality CWS. For the Cincinnati Pilot, the CANARY EDS was used to determine whether a given water quality sensor response represented a changed water quality condition and, if so, provide a corresponding alert. Setting an alert threshold sufficiently high will likely not only eliminate the majority of false alarms but also increase the risk of not detecting a contamination event. Allgeier et al. (2008) report that, on average, 3.7 alarms are generated per day across the network of 17 monitoring stations (15 are in the distribution system and 2 are located at the treatment plants). They note that the most common causes of alarms during the baseline operations period consist of "operational changes which resulted in changing water quality and separately, unrelated to operations, sensor errors or malfunctions" (Allgeier et al. 2008). The authors question whether the numbers of false alarms are too high in order for the monitoring system to be sustainable (Allgeier et al. 2008). Later, in 2011, the authors report in more detail on the results of the Cincinnati Pilot's performance using CANARY (Allgeier et al. 2011). In this study, the authors report that 92 % of the alerts were invalid, with 8 % considered valid. Allgeier et al. indicate that valid alerts consisted of (1) unusual plant conditions, (2) process change at the treatment plant, (3) maintenance or repair activities in the distribution system, (4) main breaks, or (5) verified water quality anomaly with unknown cause (Allgeier et al. 2011).

In the 2011 study, Allgeier et al. (2011) also use modeling and simulation based testing to examine the performance of CANARY. Using the Cincinnati Pilot field data as the basis for a modeling study, 1,588 simulated events are added to the water quality signals from the 15 monitoring stations within the distribution system, and run through the CANARY software (Allgeier et al. 2011). The simulated events represented water quality responses to 17 contaminants. Simulated events were created using laboratory data and EPANET-based hydraulic model simulations. Their "total scenario" detection rate, known as "true positives", for the simulated events was 40 %, leaving their false negative rate reported to be 60 % (Allgeier et al. 2011). They note that while only 40 % of the simulated contamination incidents were detected, those scenarios not detected typically represented low consequences (Allgeier et al. 2011).

Most of the research data sets used in the development and testing of EDSs have been relatively stable and do not exhibit significant water quality changes associated with changes in network operations. In cases where the water quality is strongly influenced by changes in utility operations, new approaches are likely needed to recognize the impact of these changes water quality and integrate operational data streams into the online event detection system approach. Potential opportunities to improve event detection in the highly variable water quality conditions of actual systems could include methods to recognize the "recurring patterns in multivariate data streams that are associated with operational changes" (Vugrin et al. 2009) and "direct integration of informative combinations of operational signals to temporarily decrease event detection sensitivity during periods of operational change" (Hart et al. 2010).

To date, EDS analyses have been focused on event detection at each monitoring station independently of observations occurring at other monitoring stations within the distribution network. As utilities continue to add monitoring stations within distribution networks, the concept of "distributed detection," where information from multiple monitoring stations is combined in real time to provide an integrated detection capability, will likely become possible. Koch and McKenna (2011) propose an approach for combining data from multiple locations to reduce false background alarms. Recent development and testing of an approach to distributed detection has shown that integration of EDS results across a network can significantly reduce false positive detections and help to provide better estimates of a contaminant source location (Koch and McKenna 2011). However, determining a contaminant source location after release is an inherently difficult problem and is discussed in the following paragraphs.

Water quality sensors have been demonstrated to be effective in identifying a *change* in water quality conditions. Although it is not necessarily a contamination event, the *change* provides the impetus for additional investigation or perhaps target sampling. What is still needed is ability to connect sensors in a distributed network and integrate the information obtained with a real-time understanding of system hydraulics and water quality to leverage sensor detection information from multiple sites. Leveraging water quality sensor networks for operational and management benefits, i.e., those beyond security, are needed in order to better justify the high capital cost of sensor deployment and operations and maintenance (O&M) for maintenance. Evaluating such connected and distributed networks of water quality sensors for their ability to detect a wide range of contaminant types and their limits of detection is needed. It is probably fair to say, what has been done thus far has largely been illustrative, i.e., more rigorous deployment and testing is needed.

2.7.4 Responding to Contamination Events

Transport of contamination in a drinking WDS to unsuspecting customers can be quick, generally as fast as a few hours, depending on the system. The resulting public health and economic consequences as discussed can be significant. The response time of a water utility and its community to a water contamination event are dependent on the capability of their CWS to identify the contamination event quickly and then implement the necessary procedures to minimize public health consequences and the spread of contamination. As discussed in the previous section, considerable uncertainties and needed research underlie the performance needed of a CWS to detect a contamination event in sufficient time for the utility and community to properly respond. Even with a rapid notification of the contamination event, an accurate and timely understanding of contaminant transport is needed to properly execute response actions that are specific to the contamination event and that effectively aid in reducing public health and economic consequences. For instance, it is easy to recognize that commensurate with effective and timely contamination event notification is the need for the near immediate identification of the source location of the contamination. With an accurate and timely identification of the contamination source location, response tools could, for instance, dictate which valves should be immediately closed to contain the contamination, or which fire hydrants should be opened to flush quickly and efficiently rid the system of contamination. The underlying critical component needed to support these needs, i.e., accurate and timely CWS-based detection, contamination source identification, and real-time response, is a continuous, real-time understanding of system operations or a real-time model.

Research related to the development of methodologies and tools for responding to contamination events have largely been focused in four broad areas. First, methods to identify the contamination event and initiate response actions quickly. The problem is being able to distinguish a contamination event from normal operations. Second, research to design monitoring networks and methods to identify the location of the source of the contamination. Third, research to develop algorithmic methods and tools to allow the optimal implementation of containment (e.g., valve closure) and recovery actions (e.g., fire hydrant flushing) to reduce public health consequences and the extent of contamination. Fourth, research to link an infrastructure model with SCADA data to provide a continuous, real-time understanding, i.e., model, of system operations.

The effectiveness of any response strategy depends largely on the length of time needed to deploy the necessary actions to stop public health consequences. Response delay time can be defined as the time period from the first uncertain detection to the cessation of any additional public health consequences. Numerous researchers have examined the influence of response time on the magnitude of public health consequences (Janke et al. 2006; Skadsen et al. 2008; Murray et al. 2008) and showed that the effectiveness of a CWS to reduce public health exposures can decrease by 50 % or more when response delay increases from 12 to 48 h. Murray et al. (2008) show that the performance of a CWS can decrease substantially (as much as 70 %) when the response delay time is increased from 6 to 24 h. It is hard to imagine how response could be initiated in sufficient time without a continuous, real-time understanding of system flows and chemical (water quality) transport.

Bristow and Brumbelow (2006) examine the "temporal and procedural space" between the detection of anomalous water quality event to the response decision(s) which result in the cessation of individuals ingesting contaminated tap water. This includes the "process by which decision-makers realize and affirm contamination and activate the initial phases of an emergency response plan." Bristow and

Brumbelow (2006) show that the cumulative time required to detect the contamination event, perform emergency response, and address the "compliance process" can take a considerable amount of time, generally on the order of days (Bristow and Brumbelow 2006). They also show that the "first three phases of the response process—transmission of water quality parameters, verification of water contamination, and drafting of warning messages—are the most significant sources of delay" (Bristow and Brumbelow 2006).

Many important aspects of a utility's response during contamination events could be derived from a system's water quality monitoring. In a WDS contamination scenario, water quality monitoring results would provide crucial information such as confirmation of a contamination event, the nature of the event, and the extent of contamination, all of which are critical when rapidly planning and executing a mitigating response. Water quality monitoring data could also be used to determine the source of the contamination attack, which is also known as a contamination source inversion problem. Inverse problems are computationally difficult to solve by their nature. Their solution can be computationally demanding, making them difficult to solve in a reasonable amount of time. Observational data needed for their solution is generally in short supply making it difficult to uniquely identify sources of contamination in the network. If the data is of poor quality and the solution procedure is sensitive to error and noise it may be impossible to solve the problem altogether.

Inverse problems are difficult to solve for several reasons including rank deficiency and ill-posedness. "Ill-posedness" means there is no unique solution or stability of solutions. Rank deficiency refers to data that is insufficient to meet the information necessary for the mathematical model to generate a unique or usable solution. Solution uniqueness and stability problems can result in a computational tractability problem, which means the problem cannot be solved in a reasonable amount of time. Some of these issues are a function of monitoring design which dictates the amount and quality of information available to formulate and solve the problem. Model and measurement errors can be an important factor contributing to identification uncertainties. Generally, source inversion problems are underdetermined, i.e., not solvable, because the data is too limited and there are more unknown variables than data or observations. This leads to an inverse problem description where there are an infinite number of solutions and inherent nonuniqueness.

Work on source inversion started after research on monitoring system design was well under way. This early work resulted in the development of novel solution procedures not regularly applied to inverse problems. Preis and Ostfeld (2006) developed model tree linear programming method for solution of the source inversion problem. Guan et al. (2006) developed a solution for the source identification problem using a predictor corrector algorithm.

Solutions techniques that use simulation models that advance forward through time represent a conventional approach for solving inverse problems. Laird et al. (2005), however, formulated the source inversion problem by developing an origin tracking algorithm that facilitated the embedding of ordinary differential equations

(ODEs) describing water quality transport as constraints directly within a quadratic programming (QP) problem. The approach was highly efficient and scalable; however, the solutions identified were frequently not unique. Employing a somewhat more conventional approach DiCristo and Leopardi (2008) developed a two step procedure. First, a transport pathway analysis is performed that identifies a feasible subset of potential contamination source nodes. Then using the reduced set of feasible sources a more compact discrete linear inverse problem was solved.

Solution nonuniqueness describes a condition when there are many solutions to an inverse problem that are indistinguishable from one another. This makes the problem difficult to solve because the true solution can not be identified. Solution nonuniqueness occurs when the potential sources in a network outnumber the data observations available for their identification. Laird et al. (2006) extended their previous work to address solution nonuniqueness. They develop a two phase solution approach where the QP is solved then the solution subspace identified is searched using mixed integer QP for distinct injections giving rise to the solution of the original NLP. Extending Laird's work, Wong et al. (2010) and Mann et al. (2012) formulated the source inversion problem considering discrete (positive/negative) grab samples. The problem was solved as a mixed integer linear program (MILP) problem. The procedure developed is adaptive; in that, it can accommodate additional sampling cycles to improve the accuracy of the identification. An origin tracking algorithm is utilized in the formulation of the MILP to effectively reduce problem size.

Solving inverse problems using simulation models running in reverse time is typically not possible because the numerical solutions become unstable. Numerical results become unstable when errors and noise become amplified within the solution algorithm. Generally, conventional numerical solution techniques do not work in reverse time. One of the interesting features of WDSs is the ability to develop stable techniques for solution in reverse time. Starting from the point of observation a contaminant can be transported back in time flowing past potential source locations along the way. DeSanctis et al. (2009) uses a particle back tracking algorithm and binary sensor data to identify potential sources along contaminant transport pathways assuming known network hydraulics. Neupauer et al. (2009) developed a backwards particle tracking method in a probabilistic framework. The algorithm developed treats observation nodes as instantaneous sources of probability which are transported backwards in time to obtain probability density functions (PDFs) of possible prior times when a particle was at an up gradient node. Back tracking time PDFs are conditioned with sensor data to determine likely source node locations and contaminant release times. The example assumes steady state hydraulics, but the method described is flexible and can be expanded to support dynamic hydraulics as well as other sources of uncertainty. Tao et al. (2012) developed a probabilistic treatment of DeSanctis backtracking based solution procedure using consumer complaints in place of binary sensor observations.

Formulating inversion problems using a probabilistic framework is another approach for addressing nonuniqueness by assigning each potential contaminant source a likelihood of being the true source. Propato et al. (2010) formulated the source inversion problem using a linear description of the input output dynamics of a WDS. The problem is then solved using minimum relative entropy, an entropic-based Bayesian inversion technique. (Liu et al. 2011a) used logistical regression to formulate the source inversion problem. The method required a large number of source realizations to be simulated offline to estimate likelihood model coefficients. The method is flexible as measurement errors and uncertainty can be incorporated into the coefficient calculations. The solutions, however, were better suited to enumerating the set of nonunique source locations than for identifying the true source location.

The global search characteristics of evolutionary algorithms are well suited to solution of nonunique source identification problems. Evolutionary algorithms are coupled with forward simulation models for objective and constraint evaluation. A major shortcoming of this approach, however, is computational tractability. Hundreds of thousands, if not millions, of forward model evaluations may be required to solve a typical inverse problem. Preis and Ostfeld (2007) formulated the source characterization problem as a least squares minimization problem solved using a standard genetic algorithm (GA). A sensitivity analysis was performed to characterize the robustness and accuracy of the proposed solution procedure. The authors noted the substantial computational cost of the procedure. Building on their previous work, Preis and Ostfeld (2008) described a technique for storing water quality transport simulation results in an offline matrix structure to speed up objective function evaluation. The flexibility of their solution approach allowed them to explore the affect of imperfect sensors on solution performance and response time.

Some researchers exploring evolutionary algorithms resorted to distributed computing approaches to improve computational tractability. Zechman and Ranjithan (2009) solved the source identification problem using a hybrid procedure that combined evolutionary strategies with gene encoding techniques found in genetic programming. A binary tree data structure is used to encode variable length contaminant mass loading schedules. Mutation operators modify contaminant release characteristics such as location, start time, release schedule, and schedule duration over the course of the search. The evolutionary strategy, a type of evolutionary algorithm, and encoding chosen allow the solution procedure to flexibly adapt to the uncertainties associated with the source characterization problem. Kumar et al. (2012) formulated the source inversion problem using low resolution sensor data and solved it using an evolutionary algorithm that simultaneously searches objective space for solutions that best explain observed data and decision space to characterize solution nonuniqueness.

A common assumption of early work on the source inversion problem was that of known hydraulics. This is a significant and limiting assumption reflecting the best understanding of the problem available at the time. That assumption can be relaxed somewhat by describing the uncertainties associated with the hydraulic state of the system. Vankayala et al. (2009) used Gaussian and auto regressive models to describe demand uncertainties. The source identification problem is formulated using an approach to minimize the maximum prediction error objective and then solve the resulting problem with stochastic search procedures. A stochastic demand GA, a different type of evolutionary algorithm, was used to solve the problem for each stochastic demand realization. Many realizations are simulated to build up a statistically valid characterization of the results. This approach can be contrasted with a Noisy GA, which uses a different demand realization for each generation of the GA search and computes fitness of an individual as an average over the set of realizations simulated. Both techniques performed well, but the noisy GA identified the contaminant source with a higher probability. Wang and Harrison (2013) formulated the source characterization with stochastic demands as a Bayesian probabilistic inverse problem. A unique Marko chain Monte Carlo (MCMC) algorithm was developed to address the discrete nature of the PDF associated with transport processes in the WDS.

Rather than assume that system hydraulics or a statistical description of them is available, they can become another inversion problem where system demands are estimated given pressure, flow, and boundary measurements. Preis and Ostfeld (2011) formulated a coupled inverse problem that seeks to simultaneously characterize hydraulic regime and contaminant releases given limited and low resolution data. They extended their previous work by considering hydraulic uncertainty and low resolution sensor data. The approach they described uses a Monte Carlo technique to characterize nodal demands to generate a hydraulics regime that best reflects limited hydraulic observations. The formulation considers a fixed set of observations gathered over an arbitrary length of time.

Operational source inversion requires a solution in real time. (Liu et al. 2010b) formulated the source inversion problem using streaming data from a fixed set of sensors located in the WDS. The adaptive dynamic optimization procedure they developed considers time varying streams of sensed data. Multiple populations are used to simultaneously search objective space as observation information is updated over time and search decision space for maximally different explanations of the observations. Liu et al. (2012) extended their hybrid search algorithm using logistical regression as a pre-screening step to reduce the search space by eliminating unlikely sources from the problem. A local search technique (pattern move) was also incorporated as a selection operator into the evolutionary strategy-based dynamic optimization procedure to improve the algorithms convergence characteristics.

Haxton (formerly Baranowski) (Baranowski et al. 2008) performed a case study analysis at Ann Arbor to identify and evaluate potential response actions following a contamination event. Working closely with the water utility representatives, the authors identified "practical bounds" to determine feasible response strategies and evaluated them using modeling and simulation. A hydraulic response tool was used to identify valve closure locations and hydrant flushing locations and rates to best minimize the extent of pipe contamination. The case study assumed the source or location of the contamination event was known. Valve closure consisted of isolating the tank directly downstream of the contamination event. While optimal hydrant flushing somewhat reduced the extent of pipe contamination, isolation of the tank proved to be most effective at reducing the spread of contamination (Baranowski et al. 2008). The analysis relied on the identification of the contamination source location. Without the identification of the contamination source location, it would not have been possible to isolate the tank.

Haxton et al. (2012) use an updated modeling and simulation approach to again identify and examine effective hydrant flushing locations but now to minimize public health impacts in addition to the extent of pipe contamination used in the earlier case study at Ann Arbor. Using a comparatively simple distribution system model, Haxton et al. (2012) compare algorithm-based optimized response actions against enumeration-based response actions in the development of a suite of WS tools. Together the studies point to the difficulties (computationally difficult for small models possibly insurmountable for large models) and problems (may spread rather than reduce contamination) with an a priori selection of the "best" hydrant flushing and valve closure locations based modeling and simulation studies (Haxton et al. 2012).

Effectively monitoring for, detecting, and responding to contamination events, unintentional or intentional, will require an integration of infrastructure modeling, data, and information on system operations. The fusion of operational (SCADA) data with infrastructure-aware predictive models is not new to the water community. WDS models are being used increasingly in the planning and decision-making process for drinking water utilities in the USA. These models have and are being used for predicting water quality, sensor location and for assessing the impacts of disaster. In addition, many medium to large unities in the USA recognize the need to incorporate SCADA data systems into infrastructure model and use them both more effectively in their day-to-day operations (Janke et al. 2011). However, institutional constraints, e.g., lack of resources, organization structure, and historical (outdated) operational procedures, hamper the adoption of real-time modeling capabilities at many water utilities in the USA. Typically utility operations are organizationally separated from engineering and modeling activities (Janke et al. 2011). Operations personnel often do not believe in model predictions and generally operate the WDS in a procedural manner without regard to optimization. Enormous quantities of SCADA data are continuously collected at many water utilities but never actually used. WDSs are generally operated without the employment of real-time analytical (e.g., hydraulic or water quality) data to optimally manage distribution system operations, i.e., to reduce energy costs, to manage disinfectant residual, or to identify pipe leaks and water losses. Engineering and modeling staff has a difficult time convincing operations staff to believe in and use their predictive models when they themselves are often not convinced. Convincing either side to explore distribution system optimization will require the fusion of SCADA data with up-to-date infrastructure models.

The need for real-time modeling and operational control is not new. Commercial companies purporting to offer real-time modeling capabilities have been around since early 2000s. Generally, commercial real-time modeling applications follow one of two approaches: (1) "software as a service" provider or (2) off-the-shelf real-time modeling application, which is typically an infrastructure and GIS centered software tool based off the particular water utilities existing application. Generally, the "software-to-service" providers do not offer a ready, off-the-shelf application for water utilities to install to incorporate real-time modeling into their decision making and management activities. The commercial off-the-shelf real-time modeling applications by design offer such tools, but, generally, the products are new and untested. There are few documented real-time modeling case studies in the literature and fewer still that have definitively substantiated the benefits and value of real-time modeling.

Hatchett et al. (2011) discuss integration of data with a network hydraulic model. Specifically, data available from interconnected and open information architectures, such as from a SCADA system, are fused within a real-time hydraulic modeling frame work. The authors describe efforts at field-testing such a system with a partnering water utility. It is based on a "Real-Time Extension" to EPANET (so-named EPANET-RTX). The software libraries seek to connect a model's controls, demands, and boundary conditions to real-time SCADA data, and provide a visual output of the model's predictions along with statistical accuracy metrics. In addition to being able to analyze error statistics and data time series, the hydraulic and water quality model is adjustable to the hydraulic model's parameters dynamically and for exporting historical scenarios (as EPANET input files) for offline analysis.

Hatchett et al. (2011) describe the steps taken to field-validate critical model details and implement real-time simulation. The research seeks to document the utility's experience and provides recommendations for future development and deployment of the software tools (Janke et al. 2011). The authors believe that the development of the EPANET-RTX platform and pilot-scale installation is a first step in the path to creating an extensible and open-source framework for real-time hydraulic and water quality modeling (Hatchett et al. 2011; Janke et al. 2011). Clearly, many technical issues remain which provide the impetus for further research. Some of the major technical issues include: (1) demand estimation for reliable forecasting (2) data assimilation (e.g., SCADA data cleanup and filtering and modeling data processing), and (3) information technology communication and security (i.e., hardware, software, and network connectivity issues and concerns). Advances in infrastructure hydraulic monitoring technologies (e.g., flow monitoring) and automated meter reading should provide assistance in addressing these issues.

2.8 Summary and Conclusions

A reliable source of clean, potable water is fundamental to a community's health and viability. Water is the most essential commodity, one that is used every day, by every person in the USA and around the world. Protecting water supply and infrastructure is central to maintaining the water commodity for a community and its survival. Most water supply systems in the USA consist of a water source(s), treatment facility, and distribution system. The distribution system infrastructure is typically the biggest asset and liability of a water utility. WDS infrastructure is necessarily complex in part because of the required redundancy needed to ensure clean and reliable tap water for every customer every day. WDSs are also complex because of the manner in which the infrastructure have been built, i.e., in parallel with cities and communities growing and expanding to meet the needs of the people they support. Water supply infrastructure in the USA is one of the oldest infrastructures. With this complex and aging infrastructure are numerous potential threats and risks.

Water systems are vulnerable to unintentional and intentional threats. Unintentional threats can occur from natural causes (e.g., droughts, floods, and earth quakes), accidents, or equipment failures, e.g., pipe breaks. Accidents or equipment failures can lead to utility disruptions and customer loss of service or even result in water contamination causing public health exposures, illness, disease, or even death. Cases of accidental contamination in water systems are numerous with illnesses sometimes reaching the many thousands and deaths numbering in double figures.

Intentional threats can include physical acts of sabotage, cyber attack of information or SCADA systems, or contamination. Water systems are vulnerable to such intentional threats due to their physical size, number of physical attributes (e.g., reservoirs, tanks, and pump stations), and sheer number of open access points for sabotage or contamination entry. Here the authors examined numerous published papers, reports, and studies indicating that post treatment storage facilities and the distribution system represent the most vulnerable components of a water system.

Haimes and Horowitz's (2004) definition of intentional threat is an "adversarial intent to cause harm or damage" and as modified by Willis et al. (2005) to include "intent" and "capability" of the perpetrators. Internal and external threats are discussed, with the observation that the "trusted insider" threat may represent the greatest concern because he or she may not only have the intent but also the knowledge and capability. A synopsis of research studies over the past 10–15 years is reviewed to discuss and describe intentional contamination threats in terms of approach, type of contaminant, and magnitude of possible consequences. The authors also describe possible countermeasures (i.e., physical, CWS, and cyber) which could be implemented by water utilities and communities to help protect and respond to physical and contamination threats.

Due to the magnitude of public health and economic consequences which could result from a contamination event, the consideration of online CWSs has become the focus of the water community. EPA began the WS Initiative pilot utility program to field test a five component conceptual model for a CWS at five large water utilities in the U.S. customer complaint surveillance, public health surveillance, and enhanced security monitoring are important components of the proposed CWS architecture. In terms of the earliest detection and concomitant response, the online contamination monitoring component promises the best opportunity to minimize the consequences of intentional contamination. However, while many utilities implement some form of monitoring and surveillance activities few operate in a manner to meet the primary objective of a CWS – timely detection of the contamination incident. The reasons for this are many, e.g., technical difficulties, immature technologies, lack of resources, and institutional constraints. Effective online contamination monitoring (i.e., to ensure timely detection of contamination) must be integrated with routine monitoring and routine monitoring must be integrated with operations.

The authors review the state of research related to (1) development of methodologies and tools for assessing the consequences of water contamination events; (2) development of methodologies and tools for the optimal placement of sensors in a distribution system; (3) use of water quality sensors to detect contamination and function as part of an event detection system (EDS); and (4) development of methodologies and tools for responding to contamination events. Effectively monitoring for detecting and responding to contamination events, unintentional or intentional, will require an integration of real-time analytical (SCADA) data with infrastructure-aware predictive models.

Water systems, especially the distribution system, are vulnerable to unintentional and intentional threats. Water contamination threats likely represent the greatest threat to the water utilities and communities they serve, due to the possible significant public health, including loss of life, and economic consequences which could occur. Online contamination monitoring that is integrated with real-time operational control is likely the only approach which can promise early detection and potentially effective response. Effective response is, however, dependent on the timely identification of the contamination source location.

Acknowledgments The authors gratefully acknowledge the efforts and contributions of the peer reviewers and technical editor Marti Sinclair [contract GS35F4594G, TO1533) who helped to significantly improve the material presented here. The views expressed here are those of the authors and do not necessarily reflect the views or policies of EPA.

References

Allgeier SC, Lee Y, Umberg K, Tyree M (2008) Operational experience with water quality event detection during the Cincinnati water security initiative pilot. In: Proceedings of the AWWA water quality technology conference, Cincinnati, OH, 16–20 Nov, 2008. American Water Works Association, Denver

Allgeier SC, Umberg K, Johnson R, Sahni N (2011) Detection of distribution system contamination incidents using online water quality monitoring. In: Proceedings of the AWWA water quality technology conference, Phoenix, AZ, 13–17 Nov, 2011. American Water Works Association, Denver

Allmann TP, Carlson KH (2005) Extended summary: modeling intentional distribution system contamination and detection. J Am Water Works Ass 97(1):58

American Society of Civil Engineers (ASCE) (2004) Interim voluntary guidelines for designing an online contaminant monitoring system. American Society of Civil Engineers, Reston

American Water Works Association (AWWA) (2004) Interim voluntary security guidance for water utilities. American Water Works Association, Denver

American Water Works Association (AWWA) (2010) Virus exploits USB vulnerability to reach SCDA systems. Streamlines (e-newsletter), July 27, 2010, Vol. 2, No. 19

Angulo FJ, Tippen S, Sharp DJ, Payne BJ, Collier C, Hill JE, Barrett TJ, Clark RM, Geldreich EE, Donnell HD Jr, Swerdlow DL (1997) A community waterborne outbreak of salmonellosis and the effectiveness of a boil water order. Am J Public Health 87(4):580–584

Bahadur R, Samuels WB, Pickus J (2003) Case study for a distribution system emergency response tool. American Water Works Association Research Foundation, Denver, CO

Baranowski T, Janke R, Murray R, Bahl S, Sanford L, Steglitz B and Skadsen J (2008) Case study analysis to identify and evaluate potential response initiatives in a drinking water distribution system following a contamination event. In: Proceedings, 2008 Borchardt conference, Ann Arbor, Michigan, 27–28 Feb, 2008

Beering PS (2002) Threats on tap: understanding the terrorist threat to water. J Water-ASCE 128(3):163–167

Berry J, Fleischer L, Hart WE, Phillips CA, Watson JP (2005a) Sensor placement in municipal water networks. J Water Res Pl-ASCE 131(3):237–243

Berry J, Hart WE, Phillips CA, Uber JG, Walski TM (2005b) Water quality sensor placement in water networks with budget constraints. In: Proceedings, world water and environmental resources congress. American Society of Civil Engineers, Reston

Boulos PF, Karney BW, Wood DJ, Lingireddy S (2005) Hydraulic transient guidelines for protecting water distribution systems. J Am Water Works Ass 97(5):111–124

Boulos PF, Lansey KE, Karney BW (2006) Comprehensive water distribution systems analysis handbook for engineers and planners. 2nd edn. MWH Soft Pub, Broomfield. 660 pp

Bristow E, Brumbelow K (2006) Delay between sensing and response in water contamination events. J Infrastruct Syst 12(2):87–95

Brosnan TM (ed) (1999) Early warning monitoring to detect hazardous events in water supplies. In: An ILSI (International Life Sciences Institute) Risk Science Institute workshop report. ILSI Press, Washington

Burrows WD, Renner SE (1999) Biological warfare agents as threats to potable water. Environ Health Perspect 107(12):975–984

Chandrasekaran L (2006) Predicting disease incidence due to contaminant intrusion in a water distribution system. Master's thesis, University of Cincinnati, Cincinnati. 178 pp

Clark RM (2011) U.S. water and wastewater critical infrastructure. In: Clark RM, Hakim S, Ostfeld A (eds) The handbook for securing water and wastewater systems. Springer, New York

Clark RM, Deininger RA (2000) Protecting the nation's critical infrastructure: the vulnerability of U.S. water supply systems. JCCM 8(2):73–80

Clark RM, Deininger RA (2001) Minimizing the vulnerability of water supplies to natural and terrorist threats. In: Proceedings of the American Water Works Association's IMTech conference, Atlanta, GA, 8–11 April, 2001, pp 1–20

Clark RM, Tippen DL (1990) Water supply. In: Corbitt RA (ed) Standard handbook of environmental engineering. McGraw-Hill, New York, pp 5.173–5.220

Clark RM, Grayman WM, Males RM (1988) Contaminant propagation in distribution systems. J Environ Eng-ASCE 114(2):929–943

Clark RM, Grayman WM, Goodrich JA (1991a) Water quality modeling: its regulatory implications. In: Proceedings, AwwaRF/EPA conference on water quality modeling in distribution systems, 4–5 Feb, 1991, Cincinnati, OH. AWWA Research Foundation, Denver

Clark RM, Grayman WM, Goodrich JA, Deininger RA, Hess AF (1991b) Field testing of distribution water quality models. J Am Water Works Ass 83(7):67–75

Clark RM, Geldreich EE, Fox KR, Rice EW, Johnson CH, Goodrich JA, Barnik JA, Abdesaken F (1996) Tracking a *Salmonella* serovar Typhimurium outbreak in Gideon, Missouri: role of contaminant propagation modeling. J Water Supply Res T 45(4):171–183

Clark RM, Grayman WM, Buchberger SG, Lee Y, Hartman DJ (2004) Drinking water distribution systems: an overview. In: Mays LW (ed) Water supply systems security. McGraw-Hill, New York, pp 4.1– 4.49

Clark RM, Hakim S, Ostfeld A (2011) Securing water and wastewater systems: an overview. In: Clark RM, Hakim S, Ostfeld A (ed) Handbook of water and wastewater systems protection. Springer, New York. pp 1–25

Clarke RA, Knake RK (2010) Cyber war: the next threat to national security and what to do about it. HarperCollins Publishers, New York

U.S. Congress (2002) Public health security and bioterrorism preparedness and response act of 2002: Public law 107–188. 107th Congress, June 12, 2002

Copeland C (2010) Terrorism and security issues facing the water infrastructure sector. Congressional Research Service, 7-5700, www.crs.gov, RL32189, 16 Mar, 2010

Craun MF, Craun GF, Calderon RL, Beach MJ (2006) Waterborne outbreaks reported in the United States. J Water Health 4(Supplement 2):19–30

Danyluk MD, Harris LJ, Schaffner DW (2006) Monte Carlo simulations assessing the risk of salmonellosis from consumption of almonds. J Food Protect 69(7):1594–1599

Davis MJ, Janke R (2008) Importance of exposure model in estimating impacts when a water distribution system is contaminated. J Water Resour Plann Manage 134(5):449–456

Davis MJ, Janke R (2009) Development of a probabilistic timing model for the ingestion of tap water. J Water Resour Plann Manage 135(5):397–405

Davis MJ, Janke R (2011) Patterns in potential impacts associated with contamination events in water distribution systems. J Water Res Pl-ASCE 137(1):1–9

Davis MJ, Janke R, Taxon TN (2010) Assessing potential impacts associated with contamination events in water distribution systems: a sensitivity analysis. EPA/600/R-10/061. U.S. Environmental Protection Agency, Cincinnati

Davis MJ, Janke R, Magnuson M (2013) A framework for estimating the adverse health effects of contamination events in water distribution systems and its application. Accepted Risk Analysis

Davis MJ, Janke R, Phillips C (2013a) Robustness of designs for drinking-water contamination warning systems under uncertain conditions. In preparation

DeSanctis A, Shang F, Uber JG (2009) Real-time identification of possible contamination sources using network backtracking methods. J Water Res Pl-ASCE 136(4):444–453

DiCristo C, Leopardi A (2008) Pollution source identification of accidental contamination in water distribution networks. J Water Res Pl-ASCE 134(2):197–202

Economic Research Service (2008) Foodborne illness cost calculator: *Salmonella*. United States Department of Agriculture

Fair GM, Geyer JC (1971) Water supply and waste-water disposal. Wiley, New York

Fox KR, Lytle DA (1996) Milwaukee's crypto outbreak: investigation and recommendations. J Am Water Works Ass 88(9):87–94

Friedman M, Radder L, Harrison S, Howie D, Britton M, Boyd G, Wang H, Gullick R, LeChevallier M, Wood D, Funk J (2004) Verification and control of pressure transients and intrusion in distribution systems. AWWA Research Foundation, Denver

Fujiwara M, Manwaring JM, Clark RM (1995) Drinking water in Japan and the United States: conference objectives. In: Clark RM, Clark DA (eds) Drinking water quality management. Technomic Publishing Company Inc, Lancaster

Geldreich EE, Fox KR, Goodrich JA, Rice EW, Clark RM, Swerdlow DL (1992) Searching for a water supply connection in the Cabool, Missouri disease outbreak of *Escherichia coli* 0157:H7. Water Res 26(8):1127–1137

Ghimire SR, Barkdoll BD (2006) A heuristic method for water quality sensor location in a municipal water distribution system: mass-released based approach. In: Proceedings, 8th annual water distribution systems analysis symposium, Cincinnati, OH, 27–30 Aug, 2006. American Society of Civil Engineer, Reston

Gleick PH (2006) Water and terrorism. Water Policy 8:481–503

Grayman WM, Clark RM, Harding BL, Maslia ML, Aramini J (2004) Reconstructing historical contamination events. In: Mays L (ed) Water supply systems security. McGraw-Hill, New York. pp 10.1–10.55

Grayman WM, Buchberger S, Samuels W (2008) Hydraulic models of buildings for use in contamination studies. In: Van Zyl JE, Ilemobade AA, Jacobs HE (eds) Proceedings, 10th annual water distribution system analysis conference, 17–20 Aug, 2008, Kruger National Park, South Africa. American Society of Civil Engineers, Reston

Guan J, Aral MM, Maslia ML, Grayman WM (2006) Identification of contaminant sources in water distribution systems using simulation–optimization method: case study. J Water Res Pl-ASCE 132(4):252–262

Gullick RW, LeChevallier MW, Svindland RC, Friedman MJ (2004) Occurrence of transient low and negative pressures in distribution systems. J Am Water Works Ass 96(11):52–66

Haimes YY, Horowitz BM (2004) Adaptive two-player hierarchical holographic modeling game for counterterrorism intelligence analysis. J Homel Secur Emerg 1(3): article 302

Hall J, Szabo J (2010) On-line water quality monitoring in drinking water distribution systems: A summary report of USEPA research and best practices. JAWWA 102(8):20–22

Hall J, Zaffiro AD, Marx RB, Kefauver PC, Krishnan ER, Haught RC, Herrmann JG (2007) On-line water quality parameters as indicators of distribution system contamination. J Am Water Works Ass 99(11):66–77

Hart WE, Murray R (2010) Review of sensor placement strategies for contamination warning systems in drinking water distribution systems. J Water Res Pl-ASCE 136(6):611–619

Hart D, McKenna SA, Klise K, Cruz V, Wilson M (2007) CANARY: a water quality event detection algorithm development tool. In: Proceedings, world environmental and water resources congress, Tampa, FL, 15–19 May, 2007. American Society of Civil Engineers, Reston

Hart D, McKenna SA, Murray R, Haxton T (2010) Combining water quality and operational data for improved event detection. In: Proceedings, 12th annual conference on water distribution systems analysis, Tucson, AZ, 12–15 Sept, 2010. American Society of Civil Engineers: Reston

Hatchett S, Uber J, Boccelli D, Haxton T, Janke R, Kramer A, Matracia A, Panguluri S (2011) Real- time distribution system modeling: development, application, and insights. In: Savic DA, Kapelan Z and Butler D (eds) Proceedings of the eleventh international conference on computing and control for the water industry, Exeter, UK. Centre for Water Systems, University of Exeter, Exeter, Sept 5–7, 2011

Haxton T, Murray R and Klise K (2012) Examining the application of modeling tools to identify effective flushing locations. In: Proceedings, world environmental and water resources congress 2012, Albuquerque, New Mexico, 20–24 May, 2012, pp 3071–3081. American Society of Civil Engineers, Reston

Herrick RL, Buchberger SG, Clark RM, Kupferle M, Murray R, Succop P (2011) A Markov model to estimate *Salmonella* morbidity mortality, illness, duration, and cost. Health Econ 21(10):1169–1182. Published online in Wiley Online Library (wileyonlinelibrary.com). DOI: 10.1002/hec.1779

Hickman DC (1999) A chemical and biological warfare threat: USAF water systems at risk. Future Warfare Series No. 3. Air University, U.S. Air Force Counterproliferation Center, Maxwell AFB, Alabama, p. 36. Accessed 23 April, 2013. http://www.au.af.mil/au/awc/awcgate/cpc-pubs/hickman.htm

Holcomb DL, Smith MA, Ware GO, Hung YC, Brackett RE, Doyle MP (1999) Comparison of six dose-response models for use with food-borne pathogens. Risk Anal 19(6):1091–1100

Hrudey SE, Hrudey EJ (2004) Safe drinking water: lessons from recent outbreaks in affluent nations. IWA Publishing, London

Janke R, Murray R, Uber J, Taxon T (2006) Comparison of physical sampling and real-time monitoring strategies for designing a contamination warning system in a drinking water distribution system. J Water Resour Plann Manage 132(4):310–313

Janke R, Haxton T, Grayman W, Bahadur R, Murray R, Samuels W, Taxon T (2009) Sensor network design and performance in water systems dominated by multi-story buildings. In: Proceedings, world environmental and water resources congress, Kansas City, MO, 17–19 May, 2009. American Society for Civil Engineers, Reston

Janke R, Morley K, Uber J, Haxton T (2011) Real-time modeling for water distribution system operation: integrating security developed technologies with normal operations. In: Proceedings, American water works association, water security conference and distribution systems symposium, Nashville, TN, 11–14 Sept, 2011. American Water Works Association, Denver

Jung BS, Karney BW, Boulos PF, Wood DJ (2007) The need for comprehensive transient analysis of distribution systems. J Am Water Works Ass 99(1):112–123

Karim MR, Abbaszadegan M, LeChevallier MW (2003) Potential for pathogen intrusion during pressure transients. J Am Water Works Ass 95(5):134–146

Khanal N, Buchberger SG, McKenna SA (2006) Distribution system contamination events: exposure, influence, and sensitivity 2006. J Water Res Pl-ASCE 132(4):283–292

Koch MW, McKenna SA (2011) Distributed sensor fusion in water quality event detection. J Water Res Pl-ASCE 137(1):10–19

Krause A, Leskovec J, Guestrin C, VanBriesen JM, Faloutsos C (2008) Efficient sensor placement optimization for securing large water distribution networks. J Water Resour Plann Manage 134(6):516–526

Kroll DJ (2006) Securing our water supply: protecting a vulnerable resource. PennWell Corp, Tulsa. ISBN 10: 1593700695

Kumar J, Brill ED, Mahinthakumar G, Ranjithan S (2012) Contaminant source characterization in water distribution systems using binary signals. J Hydroinform 14(3):585–602

Laird CD, Biegler LT, van der Bloeman Waanders BGV, Bartlett RA (2005) Contaminant source determination for water works. J Water Res Pl-ASCE 131(2):125–134

Laird CD, Biegler LT, van der Bloeman Waanders BGV (2006) Mixed-integer approach for obtaining unique solutions in source inversion of water networks. J Water Res Pl-ASCE 132(4):242–251

LeChevallier MW, Gullick RW, Karim MR, Friedman M, Funk JE (2003) The potential for health risks from intrusion of contaminants into distribution systems from pressure transients. J Water Health 1(1):3–14

Liu L, Ranjithan S, Mahinthakumar G (2011a) Contamination source identification in water distribution systems using an adaptive dynamic optimization procedure. J Water Res Pl-ASCE 137(2):183–192

Liu L, Sankarasubramanian A, Ranjithan SR (2011b) Logistic regression analysis to estimate contaminant sources in water distribution systems. J Hydroinform 13(3):545–557

Liu L, Zechman EM, Mahinthakumar G, Ranjithan SR (2012) Identifying contaminant sources for water distribution systems using a hybrid method. Civil Eng Environ Syst 29(2):123–136

MacKenzie WR, Hoxie NJ, Proctor ME, Gradus MS, Blair KA, Peterson DE et al (1994) A massive outbreak in Milwaukee of *Cryptosporidium* infection transmitted through the public water supply. New Engl J Med 331:161–167

Mann AV, McKenna SA, Hart WE, Laird CD (2012) Real-time inversion in large-scale water networks using discrete measurements. Comput Chem Eng 37:143–151

Mena KD, Mota LC, Meckes MC, Green CF, Hurd WW, Gibbs SG (2008) Quantitative microbial risk assessment of a drinking water—wastewater cross-connection simulation. J Environ Eng Sci 7(5):525–530

Morley K, Janke R, Murray R, Fox K (2007) Drinking water contamination—warning systems: water utilities driving water security research. J Am Water Works Ass 99(6):40–46

Murray R, Janke R, Uber J (2004) The threat ensemble vulnerability assessment program for drinking water distribution system security. In: Proceedings, ASCE-EWRI world water and environmental resources congress, Salt Lake City, UT, June 27–July 1, 2004

Murray R, Uber J, Janke R (2006) Model for estimating acute health impacts from consumption of contaminated drinking water. J Water Res Pl-ASCE 132(4):293–299

Murray R, Janke R, Hart WE, Berry JW, Taxon T, Uber J (2008) Sensor network design of contamination warning systems: a decision framework. J Am Water Works Ass 100(11):97–109

Murray R, Hart WE, Phillips CA, Berry J, Boman EG, Carr RD, Riesen LA, Watson JP, Haxton T, Herrmann JG, Janke R, Gray G, Taxon T, Uber JG, Morley KM (2009) US Environmental

Protection Agency uses operations research to reduce contamination risks in drinking water. Interfaces 39(1):57–68
National Fire Protection Association (NFPA) (2003) Fire protection handbook, 19th edn. In: Cote AE (ed). National Fire Protection Association, Quincy
Neupauer RM, Records MK, Ashwood WH (2009) Backward probabilistic modeling to identify contaminant sources in water distribution systems. J Water Res Pl-ASCE 136(5):587–591
Nilsson KA, Buchberqer SG, Clark RM (2005) Simulating exposures to deliberate intrusions into water distribution systems. J Water Res Pl-ASCE 131(3):228–236
Nuzzo JB (2006) The biological threat to U.S. water supplies: toward a national water security policy. Biosecur Bioterr 4(2):147–159
Ostfeld A, Uber JG, Salomons E, Berry JW, Hart WE, Phillips CA, Watson JP, Dorini G, Jonkergouw P, Kapelan Z, di Pierro F, Khu ST, Savic D, Eliades D, Polycarpou M, Ghimire SR, Barkdoll BD, Gueli R, Huang JJ, McBean EA, James W, Krause A, Leskovec J, Isovitsch S, Xu JH, Guestrin C, VanBriesen J, Small M, Fischbeck P, Preis A, Propato M, Piller O, Trachtman GB, Wu ZY, Walski T (2008) The battle of the water sensor networks (BWSN): a design challenge for engineers and algorithms. J Water Res Pl-ASCE 134(6):556–568
Panguluri S, William PR Jr, Clark RM (2004) Cyber threats and IT/SCADAsystem vulnerability. In: Mays LW (ed) Water supply systems security. Mc Graw-Hill, New York. pp 5.1–5.18
Panguluri S, William PR Jr, Ellis P (2011) Cyber security: protecting water and wastewater infrastructure. In: Clark Robert M, Hakim Simon, Ostfeld Avi (eds) Handbook of water and wastewater systems protection. Springer-Science, New York, pp 285–318
Perelman L, Ostfeld A (2010) Extreme impact contamination events sampling for water distribution systems security. J Water Resour Plann Manage 136(1):80–87
Porco JW, Elwell B, McFee J (2006) Domestic municipal end-to-end water security architecture study. In: 8th annual water distribution systems analysis symposium, Cincinnati, Ohio, 27–30 Aug, 2006. American Society of Civil Engineering, Reston
Preis A, Ostfeld A (2006) Contamination source identification in water systems: a hybrid model trees–linear programming scheme. J Water Res Pl-ASCE 132(4):263–273
Preis A, Ostfeld A (2007) A contamination source identification model for water distribution system security. Eng Optimiz 39(8):941–947
Preis A, Ostfeld A (2008) Genetic algorithm for contaminant source characterization using imperfect sensors. Civil Eng Environ Syst 25(1):29–39
Preis A, Ostfeld A (2011) Hydraulic uncertainty inclusion in water distribution systems contamination source identification. Urban Water J 8(5):267–277
President, Presidential Decision Directive (PDD) 63 (1998) Protecting America's critical infrastructure. Presidential decision directive 63. The William J. Clinton Presidential Library, directives. Accessed 17 April, 2013. http://www.clintonlibrary.gov/pdd.html
President's Commission on Critical Infrastructure Protection (PCCIP), Critical Foundations: protecting America's Infrastructures (Washington, DC: October 1997). Report can be accessed at: http://www.fas.org/sgp/library/pccip.pdf
Propato M, Uber J (2004) Vulnerability of water distribution systems to pathogen intrusion: how effective is a disinfectant residual? Environ Sci Technol 38(13):3713–3722
Propato M, Sarrazy F, Tryby M (2010) Linear algebra and minimum relative entropy to investigate contamination events in drinking water systems. J Water Res Pl-ASCE 136(4):483–492
Rizak SN, Hrudey SE (2006) Misinterpretation of drinking water quality monitoring data with implications for risk management. Environ Sci Technol 40(17):5244–5250
Roberson JA, Morley KM (2005) Contamination warning systems for water: an approach for providing actionable information to decision-makers. American Water Works Association, Denver
Rossman LA (2000) EPANET version 2 users manual. EPA/600/R-00/057. Drinking Water Research Division, U.S. Environmental Protection Agency, Cincinnati
Safe Drinking Water Act (SDWA) (1974) U.S. Public Law 93-523, 93rd Cong., 16 Dec, 1974

Scallan E, Hoekstra RM, Angulo FJ, Tauxe RV, Widdowson M, Roy SL, Jones JL, Griffin PM (2011) Foodborne illness acquired in the United States–major pathogens. Emerg Infect Dis 17(1):7–15

Shang F, Uber JG, Rossman LA (2008) Modeling reaction and transport of multiple species in water distribution systems. Environ Sci Technol 42:808–814

Skadsen J, Janke R, Grayman W, Samuels W, TenBroek M, Steglitz B, Bahl S (2008) Distribution system on-line monitoring for detecting contamination and water quality changes. J Am Water Works Ass 100(7):81–94

Tao T, Huang H, Xin K, Liu S (2012) Identification of contamination source in water distribution network based on consumer complaints. J Cent South Univ 19(6):1600–1609

Török TJ, Tauxe RV, Wise RP, Livengood JR, Sokolow R, Mauvais S, Birkness KA, Skeels MR, Horan JM, Foster LR (1997) A large community outbreak of salmonellosis caused by intentional contamination of restaurant salad bars. JAMMA 278(5):389–395

Trachtman GB (2006) A "strawman" common sense approach for water quality sensor site selection. In: Proceedings, 8th annual water distribution systems analysis symposium, Cincinnati, OH, 27–30 Aug, 2006. American Society of Civil Engineers, Reston

Tucker JB (2000) Lessons from case studies. In: Tucker JB (ed) Toxic terror: assessing terrorist use of chemical and biological weapons. MIT Press, Cambridge. pp 250–251

Tularam GA, Properjohn M (2011) An Investigation into modern water distribution network security: risk and implications. Secur J 24(4):283–301

U. S. Environmental Protection Agency (U.S. EPA) (2013). Models, tools, and applications. Available at: http://www.epa.gov/nhsrc/toolsandapps.html. Accessed on 5 Feb, 2013

U. S. Environmental Protection Agency (U.S. EPA) (2004) Water security research and technical support action plan. EPA/600/R-04/063. U.S. Environmental Protection Agency, Office of Research and Development and Office of Water, Washington

U. S. Environmental Protection Agency (U.S. EPA) (2007) Water security initiative: Interim guidance on planning for contamination warning system deployment. EPA 817-R-07-002. U.S. Environmental Protection Agency, Office of Water, Cincinnati

U. S. Environmental Protection Agency (U.S. EPA) (2008) Water security initiative Cincinnati pilot post-implementation system status: covering the pilot period: December 2005 through December 2007. EPA-817-R-08-004. U.S. Environmental Protection Agency, Office of Water, Office of Ground Water and Drinking Water, Washington

U. S. Government Accountability Office (U.S. GAO) (2006) Internet infrastructure: DHS faces challenges in developing a joint public/private recovery plan. GAO-06-672 (June, 2006) U.S. Government Accountability Office, Washington

Uber J, Murray R, Janke R (2004) Use of systems analysis to assess and minimize water security risks. J Contemp Water Res Edu 129(1):34–40

U.S Department of Homeland Security (USDHS) (2003) Advisory: potential Al Qaeda threats to U.S. water supply. June 23, 2003

U.S. Environmental Protection Agency (U.S. EPA) (2005) Technologies and techniques for early warning systems to monitor and evaluate drinking water quality: a state-of-the-art review. EPA/600/R-05/156. U.S. Environmental Protection Agency, Washington

U.S. Environmental Protection Agency (U.S. EPA) (2009) Sensor network design for drinking water contamination warning systems: a compendium of research results and case studies using TEVA-SPOT. EPA/600/R-09/141. Office of Research and Development, National Homeland Security Research Center, U.S. EPA, Cincinnati

U.S. Environmental Protection Agency (U.S. EPA) (2011) Fiscal Year 2011 ground water and drinking water statistics. EPA 816-R-13-003. U.S. Environmental Protection Agency, Office of Water, Washington

U.S. Environmental Protection Agency (U.S. EPA) (2013a) Homeland security and EPA's strategic plan. www.epa.gov/homelandsecurityportal/strategicplan.htm. Accessed April 23, 2013

Vankayala P, Sankarasubramanian A, Ranjithan SR, Mahinthakumar G (2009) Contaminant source identification in water distribution networks under conditions of demand uncertainty. Environ Forensics 10(3):253–263

Vugrin E, McKenna SA, Hart D (2009) Trajectory clustering approach for reducing water quality event false alarms. In: Proceedings, world environmental and water resources congress, Kansas City, Missouri, 17–21 May, 2009. American Society of Civil Engineers, Reston

Walski TM, Chase DV, Savic DA, Grayman WM, Beckwith S, Koelle E (2003) Advanced water distribution modeling and management. Haestad Press, Waterbury, pp 1–4

Wang H, Harrison KW (2013) Bayesian update method for contaminant source characterization in water distribution systems. J Water Res Pl-ASCE 139(1):13–22

Weiss J. Industrial Control System (ICS) Cyber Security for Water and Waste Water Systems to appear in securing water and waste water systems: global experiences.In: Clark RM and Hakim S (ed). Springer, New York, 10013

Welter G, Lechevallier M, Cotruvo J, Moser R, Spangler S (2009) Guidance for decontamination of water system infrastructure. In: Water Research Foundation, Report No. 2981. 2009

Willis HH, Morral AR, Kelly TK, Medby JJ (2005) Estimating terrorism risk. RAND Corporation, Center for Terrorism Risk Management Policy

Wong AV, McKenna SA, Hart WE, Laird CD (2010) Real-time inversion and response planning in large-scale networks. Comp Aided Chem Eng 28:1027–1032

Wood DJ, Lingireddy S, Boulos PF (2005) Pressure wave analysis of transient flow in pipe distribution systems. MWH Soft Pub, Pasadena

Zechman EM, Ranjithan SR (2009) Evolutionary computation-based methods for characterizing contaminant sources in a water distribution system. J Water Res Pl-ASCE 135(5):334–343

Chapter 3
Industrial Control System (ICS) Cyber Security for Water and Wastewater Systems

Joe Weiss

Technical Terms and Definitions

CIA	Confidentiality, integrity, and availability
COTS	Commercial off-the-shelf
DCS	Distributed control systems
DOE	Department of energy
ICS	Industrial control system
IP	Internet protocol
ISA	International society of automation
IT	Information technology
MHI	Human–machine interface
NT	NT is a microsoft windows product
PDD	Presidential decision directive
PLC	Programmable logic controllers
RTOS	Real-time operating systems
RTU	Remote terminal units
SCADA	Supervisory control and data acquisition
TCP	Transmission control protocol

3.1 Background

Industrial Control Systems (ICSs) operate industrial infrastructures worldwide including water/wastewater, electric power, oil/gas, pipelines, chemicals, mining, pharmaceuticals, transportation, and manufacturing. ICSs measure, control, and provide a view of the process (once only the domain of the operator). These types of systems are commonly utilized throughout the global industrial infrastructures.

J. Weiss (✉)
Applied Control Solutions, LLC, 10029 Oakleaf Place, Cupertino, CA 95014, USA
e-mail: joe.weiss@realtimeacs.com

R. M. Clark and S. Hakim (eds.), *Securing Water and Wastewater Systems*,
Protecting Critical Infrastructure 2, DOI: 10.1007/978-3-319-01092-2_3,

The commonality of ICSs and their architecture enabled the International Society of Automation (ISA) to establish one general process industry committee for cyber security—S99 (ISA 1999).

Security is like a three-legged stool consisting of physical security, Information Technology (IT) security, and ICS security. Physical security is generally well understood and often addressed by experts coming from the military or law enforcement. IT security generally deals with traditional commercial off-the-shelf (COTS) hardware and software and connections to the Internet with experts coming from IT and the military. There is little doubt that IT security is necessary and that systems are continuously being probed, tested, and hacked. The third leg, ICS security, is much less understood, has little expertise, is often not considered critical. Those working in this area are generally either from the IT security community with little knowledge of ICSs or ICS experts knowledgeable in the operation of systems, not security.

ICS cyber security awareness was very low and its perceived importance even lower. Generally, cyber security is viewed as a corporate IT issue with little direct impact on water/wastewater operations. Moreover, it is viewed as a hindrance to ICS technology advancements. From a security perspective, ICSs were generally isolated networks and the concept of "security by obscurity" is still alive and well.

The fundamental reason for securing ICSs is to maintain the mission of the systems be it to provide clean water and clean-up dirty water, produce or deliver electricity, make or distribute gasoline, etc. I do not believe it is possible to fully electronically secure ICSs. However, we can make them more secure and also minimize the possibilities of unintentional incidents that have already cost hundreds of millions of dollars as well as a number of lives.

Cyber security is generally viewed in the context of traditional business Information Technology (IT) systems and Defense computer systems. ICSs are frequently not viewed as "computers" nor are they often considered to be susceptible to cyber security threats. Consequently, cyber security is taught within Computer Science Departments focusing on traditional IT concepts. Control system theory and applications for ICSs are addressed in the various engineering disciplines but do not address computer security.

3.2 What are ICSs

ICSs include Supervisory Control and Data Acquisition (SCADA), Distributed Control Systems (DCS), Programmable Logic Controllers (PLC), Remote Terminal Units (RTU), and field instrumentation. ICSs are commonly utilized throughout the global industrial infrastructures. Almost all university engineering programs offer courses in control system theory.

ICS networks and workstations including the human–machine interface (HMI) are generally IT-like systems and may be susceptible to standard IT vulnerabilities and threats. Consequently, they can utilize IT security technologies and traditional

IT education and training apply. The field instrumentation and controllers generally do not utilize commercial off-the-shelf operating systems and are computer resource constrained. They often use proprietary real time operating systems (RTOS) or embedded processors. These systems have different operating requirements and can be impacted by cyber vulnerabilities typical of IT systems and also cyber vulnerabilities unique to ICSs.

Figure 3.1 illustrates the different aspects of an ICS from the Windows-based workstation with built-in security to the field elements with generally little to no security.

ICSs continue to be upgraded with advanced communication capabilities and networked to improve process efficiency, productivity, regulatory compliance, and safety. This networking can be within a facility or even between facilities continents apart. When an ICS does not operate properly, it can result in impacts ranging from minor to catastrophic. Consequently, there is a critical need to ensure that electronic impacts do not cause, or enable, misoperation of ICSs.

3.3 Securing ICS and IT Systems

Securing systems consist of physical security, IT security, and ICS security. Physical security is generally well understood and often addressed by experts coming from the military or law enforcement. IT security generally deals with traditional commercial off-the-shelf (COTS) hardware and software and connections to the Internet with experts coming from IT and the military. There is little

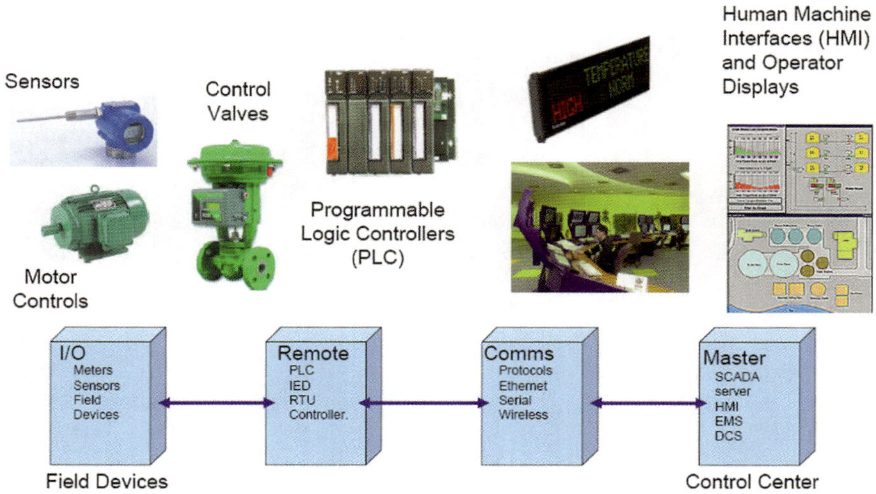

Fig. 3.1 Control system basics

doubt that IT security is necessary and that systems are continuously being probed, tested, and hacked. The third aspect unique to the industrial community, ICS security, is much less understood, has little expertise, and is often not considered critical. Those working in this area are generally either from the IT security community with little knowledge of ICSs or ICS experts knowledgeable in the operation of systems, not security.

The Triad of Confidentiality, Integrity, and Availability (CIA) effectively defines the technologies needed for securing systems. In the IT domain, cyber attacks often focus on acquisition of proprietary information. Consequently, the CIA triad results in Confidentiality being the most important attribute which dictates that encryption is required. However, in the ICS domain, cyber attacks tend to focus on destabilization of assets. Moreover, most ICS cyber incidents are unintentional and often occur because of a lack of effective message integrity and/or appropriate ICS security policies. Consequently, Integrity and Availability are much more important than Confidentiality which lessens the importance of encryption and significantly raises the importance of authentication and message integrity. ICS security research and education should focus on technologies that address Integrity and Availability.

ICS security is an engineering problem requiring engineering solutions. Resilience and robustness are the critical factors in the survivability of compromised ICSs. ICS security requires a balanced approach to technology design, product development and testing, development and application of appropriate ICS policies and procedures, analysis of intentional and unintentional security threats, and proactive management of communications across view, command and control, monitoring, and safety. It is a life cycle process beginning with conceptual design through retirement of the systems.

3.4 Differences Between IT and ICS Systems

Cyber security is generally viewed in the context of traditional business IT systems and Defense systems. IT systems are "best effort" in that they get the task complete when they get the task completed. Unlike IT systems, ICSs are not general-purpose systems and components but designed for specific applications. The ICS design criteria was performance and safety, not security. ICS systems are "deterministic" in that they must do it now and cannot wait for later as that will be too late. ICSs often are not viewed as "computers" or susceptible to cyber security threats.

Legacy ICSs were not designed to be secured, easily updated, or with efficient security troubleshooting, self-diagnostics, and network logging. The ICS community has the knowledge-base to understand what physical parameters are required to perform a root-cause analysis of an incident. Consequently, the ICS community has developed the detailed forensics for physical parameters—temperature, pressure, level, flow, motor speed, current, voltage, etc. However, the legacy/field device portions of an ICS have minimal to no cyber forensics. This area is ripe for research and development to determine what specific types of forensics are needed and how they would be performed in the least noninvasive manner possible.

Reliable and timely communications are critical for maintaining the operations of ICSs. ICSs are deterministic systems with hard real time requirements meaning communications must occur within a prescribed period of time. There was some signal validation, no authentication, no encryption, and adequate speed (i.e., some latency was acceptable). TCP/IP is not deterministic and consequently, TCP/IP is only used for non-process or non-safety critical communications. Since there is a movement to utilize TCP/IP protocols from RTUs or PLCs to SCADA or DCS, there is another common look and feel between the IT community and the ICS community. The use of TCP/IP and Windows was a natural progression to the Internet. There are now many instances where ICSs are connected directly to the Internet using TCP/IP through Windows or other commercial off-the-shelf operating systems. An ongoing research effort called Project Shine has identified more than 1,000,000 ICS devices directly connected to the Internet. In the future, TCP/IP will be used for safety applications. This really needs to be done with great care.

In the IT community, software security and secure software are often discussed in the context of software assurance. Software assurance is broader than software security as it encompasses the additional disciplines of software safety and reliability. Software assurance aims to provide justifiable confidence the software is free of vulnerabilities, that it functions in the intended manner and that the intended manner does not compromise the security and other required properties of the software, its environment, or the information it handles. Software assurance also aims to provide justifiable confidence that the software will remain dependable under all circumstances. These include the presence of unintentional faults in the software and its environment; exposure of the operational software to accidental events that threaten its dependability; and exposure of the software to intentional threats to its dependability in development and operation. Software assurance addresses trustworthiness, predictable execution, and conformance where trustworthiness means no exploitable vulnerabilities exist whether intentional or unintentional; predictable execution provides confidence will function as designed; and conformance means the software and products conform to applicable standards and requirements. To date, the ICS community has not actively and formally applied these principles.

Certain mainstream IT security technologies can adversely affect the operation of ICSs or result in operator confusion. Examples include using port scanning tools that result in components freezing up or worse. Block encryption can slow down control system operation resulting in a Denial of Service. Locking up a system after a specified number of password failures can have devastating consequences if it occurs during off-normal conditions when operators are under great stress.

The current state of the IT world insures a high degree of intelligence and processing capability on the part of the various devices within an IT system. The standard implementation provides centralized control points for authentication and authorization of IT activities. The lifetime of the equipment in an IT network, typically, ranges from 3 to 7 years before anticipated replacement and often does not need to be in constant operation. By the very nature of the devices and their intended function, ICS devices may be 15–20 years old, perhaps older, before anticipated replacement. Since security was not an initial design consideration, ICS devices do not have excess computing capacity for what would have been considered unwanted or unneeded applications.

Of considerable importance is intra- and intersystems communication in both the IT and ICS realms. ICS systems are intended to operate at all times, whether connected to other systems or not. This independence makes the ICS very flexible, indeed. The age of the equipment makes it difficult to authenticate communications properly. Not just between servers, but between servers and devices, devices and devices, workstations and devices, and devices and people. The older technologies do not have the ability, by want of adequate operating systems, to access centralized authentication processes. By want of the ability of the ICS network to be broken into very small chunks, the use of centralized authentication is impractical, using the technologies of today. In an IT network, the authentication rules take place in the background and are hidden, for the most part, from the end user. In an ICS network, the authentication rules take place in the foreground and require interaction with the end user, causing delay and frustration.

Patching or upgrading an ICS has many pitfalls. Patches need to be verified to determine the patch is really the same as the one that was sent and to determine that the patch really fixes a bug and would not adversely affect the system performance. This is not as easy as it seems. The field device must be taken out of service which may require stopping the process being controlled. This in turn may cost many thousands of dollars and impact thousands of people. An important issue is how to protect unpatchable, unsecurable workstations such as those still running NT Service Pack 4, Windows 95, and Windows 97. Many of these older workstations were designed as part of plant equipment and control system packages and cannot be replaced without replacing the large mechanical or electrical systems that accompany the workstations. Additionally, many Window patches in the ICS world are not standard Microsoft patches but have been modified by the ICS supplier. Implementing a generic Microsoft patch can potentially do more harm than the virus or worm against which it was meant to defend. As an example, in 2003 when the Slammer worm was in the wild, one DCS supplier sent a letter to their customers stating that the generic Microsoft patch should not be installed as it would shut down the DCS. Another example was a water utility that patched a system at a water treatment plant with a patch from the operating system vendor. Following the patch, they were able to start pumps, but were unable to stop them!

The perceived distinctions between IT systems and ICS are starting to blur with grave consequences. Table 3.1 provides a comparison between key characteristics of business IT and ICSs. These differences can have very dramatic impacts on ICS operation and education.

Table 3.1 Comparison of typical IT and ICS characteristics (Weiss 2010a)

Attribute	IT	ICS
Confidentiality (Privacy)	High	Low
Message integrity	Low medium	Very high
Availability	Medium	Very high
Authentication	Medium-high	High
Time criticality	Delays tolerated	Critical
Security Skills/ awareness	Usually good	Usually poor
Security education	Good	Usually poor
Engineering education	Usually none	Yes
Certification	CISSP (Certified Information Systems Security Professional)	Professional engineer (PE)
Life cycle	3–5 years	15–25 years
Forensics	Available	Minimal
Impacts	Business impacts	Business impacts, safety, environmental

3.5 Current Status

Figure 3.1 characterizes the relationship of the different types of special technical skills and certifications needed for ICS cyber security expertise, and the relative quantities of each at work in the industry today. Most people now becoming involved with ICS cyber security typically come from a mainstream IT security background and not an ICS background. This trend is certainly being accelerated by the Smart Grid initiatives, where the apparent lines between IT and ICS are blurring. Many of the entities responsible for control system cyber security, industry, equipment suppliers, and government personnel do not fully appreciate the difficulties created by this trend.

As can be seen in Fig. 3.2, IT encompasses a large realm, but does not include control system processes. The arrows indicate that most people coming into the ICS security domain (from academia and the work force) are coming from the IT domain. This needs to change. It does not take "rocket science" to compromise an ICS; however, it does take "rocket science" to be able to protect an ICS and still have it be able to perform its functions. Being able to do that is what constitutes an ICS security expert. Arguably, there are few people worldwide who fit into the tiny dot called ICS cyber security.

There are many explanations for this imbalance. First, there are simply more trained IT security personnel than ICS security personnel. There is now significant money in securing critical infrastructure. There is an old adage: "to a carpenter, everything looks like a nail." As ICS systems get more of an IT look, IT views them as IT systems and wants to apply their expertise to them. Second, there is often little funding for training ICS personnel in cyber security as Operations often does not view this area as under their purview. Is there any question as to why there are so many more IT-trained security personnel than ICS-trained? The timing is ripe for the academic community to address the need to educate more ICS technologists, researchers, and experts.

There is a convergence of highly integrated industrial automation combined with ICSs sharing more and more constructs with IT. ICS cyber security is an emerging and interdisciplinary field encompassing computer science, networking, public policy, and engineering control system theory and applications. The lack of ICS cyber security understanding in the industrial community is also often reflected in the academic community. Lectures at the National Defense University, the Naval Postgraduate School, the University of Washington, the University of Illinois, and Mississippi State University, have focused on the need for interdisciplinary focus and coordination (Weiss 2010a). Unfortunately, today's computer science curriculum often does not address the unique aspects of ICSs. Correspondingly, the electrical engineering power systems focus, chemical engineering, mechanical engineering, nuclear engineering, and industrial engineering curricula do not address computer security. Consequently, there is a need to form joint interdisciplinary programs for ICS security.

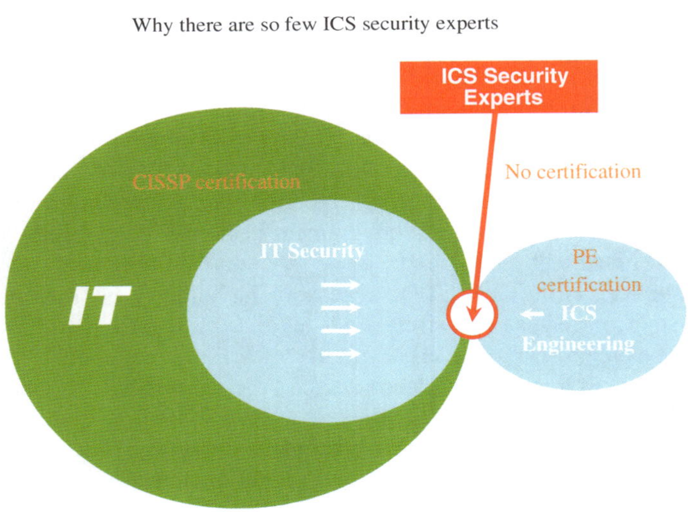

Fig. 3.2 Relative availability of ICS cyber security expertise (Weiss 2010b)

3.6 Why Care, and More Specifically, Why Should Water and Wastewater Care?

More than 300 control system cyber incidents globally have been documented in many industries (Weiss 2010a). These cases include water/wastewater, electric power, chemicals, pipelines, manufacturing, transportation, etc. Although most of these incidents are not malicious, they have still caused considerable concerns. Moreover, most of the unintentional cases could have been done maliciously. Impacts range from trivial to significant environmental discharges to significant equipment damage to major electric outages to deaths. In the water/wastewater industry, more then 20 cases have been documented to date (Weiss 2010b). Impacts range from trivial to significant environmental discharges to loss of drinking water to pumping contaminated water into the drinking water system to broken water mains. Moreover, the impacts have come from very large water systems to very small, rural water systems.

Experience as well as the 2011 McAfee report [In The Cross Fire…, 2011), has shown that the water and wastewater industries are generally believed to lag most other industries in securing their control systems. At least three have been malicious attacks. Consequently, there is a need to educate the water and wastewater industries on how to secure their control systems.

3.7 Examples of Water and Wastewater Control System Cyber Incidents

Two examples are worth considering—a malicious attack and an unintentional incident—both caused physical problems (Weiss 2010b).

3.7.1 Maroochy Wastewater Wireless SCADA Attack

Vitek Boden, a man in his late 1940 s, worked for Hunter Watertech, an Australian firm that installed SCADA radio-controlled sewage equipment. Hunter Watertech installed radio-controlled sewage equipment for the Maroochy Shire Council in Queensland, Australia. On at least 46 occasions from February 28 to April 23, 2000, Boden issued radio commands to the sewage equipment he (probably) helped install. Boden caused millions of liters of raw sewage to spill out into local parks, rivers, and even the grounds of a Hyatt Regency hotel. “Marine life died, the creek water turned black, and the stench was unbearable for residents,” said a representative of the Australian Environmental Protection Agency. Boden got caught when a policeman pulled him over for a traffic violation after one of his attacks. A judge sentenced him to 2 years in jail and ordered him to reimburse the one major cleanup mentioned above. Boden's attack became the first widely known example of someone maliciously breaking into a control system (Fig. 3.3).

3.7.2 Maroochy Shire Site Description

Maroochy Shire has 880 km of gravity sewers treating an average of 35 million liters/day. Maroochy Water Services Sewerage SCADA System consists of 142 Sewage Pumping Stations with two Monitoring Computers utilizing three Radio Frequencies. Hunter Watertech Pty Ltd installed the "PDS Compact 500" computer device at each pumping station capable of receiving instructions from a central control center, transmitting alarm signals, and other data to the central computer and providing messages to stop and start the pumps at the pumping station. Communications between pumping stations and between a pumping station and the central computer were by means of a dedicated analog two-way radio system operating through repeater stations. Each repeater station transmitted on a different frequency.

3.7.3 The Attack

The offences occurred between February 9, 2000 and April 23, 2000. Vitek Boden accessed computers controlling the Maroochy Shire Council's sewerage system, altering electronic data in particular sewerage pumping stations causing malfunctions in their operations. Vitek Boden had been employed by Hunter Watertech as its site supervisor on the Maroochy SCADA project for about 2 years until resigning in December 1999. At about the time of his resignation he approached

Fig. 3.3 Maroochy SCADA system

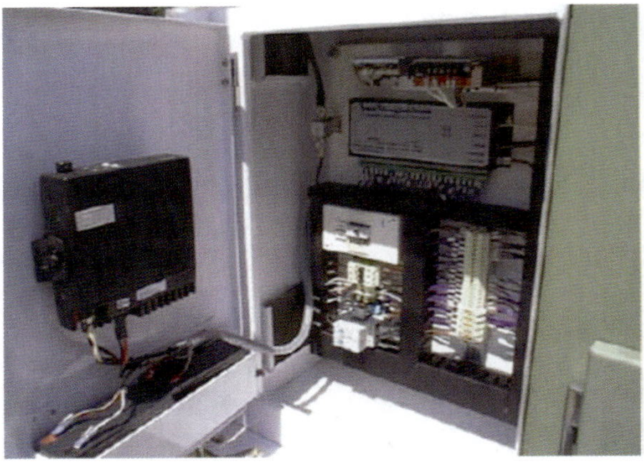

the Council seeking employment. He was told to enquire again at a later date. He made another approach to the Council for employment in January 2000 and was told that he would not be employed. The sewerage system then experienced a series of faults:

- Pumps were not running when they should have been,
- Alarms were not reporting to the central computer,
- A loss of communication between the central computer and various pumping stations.

An employee of Hunter Watertech was appointed to look into the problem. He began monitoring and recording all signals, messages, and traffic on the radio network. As a result of his investigations he concluded that many of the problems being experienced with the system resulted from human intervention rather than equipment failure. Other technical experts shared his opinion. Further, the evidence revealed that the problems associated with the attack ceased when Vitek Boden was arrested. On an occasion during the investigation, the investigator determined that pumping station 14 appeared to be the source of the messages corrupting the system. He physically checked the pumping station and ascertained that it was working properly and bore no signs of having been physically tampered with. He concluded that the source of the false messages was a PDS Compact 500 computer with an address of 14 and he changed the identification number of pumping station 14–3 so that any legitimate messages from that station could be identified as coming from station 3. Conversely, any messages coming from a station identifying itself as 14 would be known to be bogus.

On March 16, 2000, when malfunctions occurred in the system, the investigator communicated over the network with a bogus pump station 14 was sending messages to corrupt the system. He was temporarily successful in altering his program to exclude the bogus messages but then had his computer shut out of the network for a short period. The intruder was now using PDS identification number 1 to send messages. Further problems then occurred as a result of a person gaining remote computer access to the system and altering data so that whatever function should have occurred at affected pumping stations did not occur or occurred in a different way. The central computer was unable to exercise proper control and, at great inconvenience and expense, technicians had to be mobilized throughout the system to correct faults at affected pumping stations. On one occasion, a pumping station overflowed causing raw sewerage to escape.

On April 23, 2000, an intruder, by means of electronic messages, disabled alarms at four pumping stations using the identification of pumping station 4. The intrusions began just after 7:30 pm and concluded just after 9:00 pm. By this time Vitek Boden had fallen under suspicion and was under surveillance. Police officers located a vehicle driven by him. When Boden's vehicle was pulled over and searched at around 10:00 pm, a PDS Compact 500 computer, later identified in evidence as the property of Hunter Watertech, was found, as was a laptop computer. On examination it was found that the software to enable the laptop to communicate with the PDS system through the PDS computer had been reinstalled in the laptop on February 29, 2000. The PDS Compact computer had been programmed to identify itself as pump station 4—the identification used by the intruder in accessing the Council sewerage system earlier that night. The software program installed in the laptop was one developed by Hunter Watertech for its use in changing configurations in the PDS computers. There was evidence that this program was required to enable a computer to access the Council's sewerage system and had no other practical use. Once it was demonstrated that the malfunctions resulted from human intervention, the existence of other problems became of limited significance, the investigator was adamant that the malfunctions in the system could only have been caused by unauthorized human intervention. Boden sought to establish that some of the electronic messages that gave rise to the charges could have been caused by system malfunction or by error on the part of Council employees. One of his arguments in this regard showed three sets of identical messages on the same day from addresses 000, 099, and 004. The Crown contended that only the message emanating from address 004 was initiated by Boden. Boden pointed to the other messages as evidence that defective messages of the nature of those relied on by the Crown may have been caused other than by human intervention. Another witness, an engineer specializing in computer engineering who, for a time, was Hunter Watertech's project engineer on the installation of the computerized sewerage system said that all three messages were generated by the PDS configuration program used on the PDS Compact computers. His opinion was that the messages, other than the ones from address 004, were generated by persons attempting to rectify the result of the alleged unauthorized intervention. He also gave evidence that 000 and 099 messages were not causing damage to the computer system. The investigator gave evidence some days later than the computer engineer and thus had more opportunity to consider the possible explanations for the 000 and 099 messages. His evidence was that these messages occurred over several days and resulted from the actions of maintenance staff who were either employees of Hunter Watertech or Council employees under direction of the former. He ruled out the possibility of mechanical error. He said that the 004 messages were definitely generated by a person different from the one who generated the other messages.

3.7.4 Timeline

1997–December 1999—Vitek Boden employed by Hunter Watertech as site supervisor

December 3, 1999—Boden resigns from Hunter Watertech

Early December 1999—Boden approached City Council seeking employment

Early January 2000—Boden reapproached City Council and was turned down

February 9–April 23, 2000—SCADA system experiences series of faults

March 16, 2000—Hunter Watertech investigator tried to troubleshoot system

April 19, 2000—Log indicates system program had been run at least 31 times

April 23, 2000—Boden disabled alarms at four pumping stations using the identification of pumping station 4. The intrusions began just after 7:30 pm and concluded just after 9 pm

April 23, 2000—Boden pulled over by police with computer equipment in car

October 31, 2001—Boden convicted in trial—sentenced to 2 years

March 21, 2002—Appeal rejected

3.7.5 Attack Summary

- Vitek Boden was an insider who was never an employee of the organization he attacked. He was an employee of a contractor that supplied IT/control system technology to the Maroochy Shire Council. With his knowledge he was the "ultimate insider."
- The service contract was deficient or inadequate concerning Watertech's responsibilities. Management, technical, and operational cyber security controls were inadequate. Personnel security controls that applied to its employees such as background investigations and protection from disgruntled employees were inadequate.
- A number of anomalous events occurred before recognition that the incidents were intentional. As a skillful adversary, Boden was able to disguise his actions. Extensive digital forensics over a period of time were required to determine that a deliberate attack was underway
- There were no existing cyber security policies or procedures.
- There were no cyber security defenses.

3.7.6 Observations

As reported by Slay and Miller, Robert Stringfellow was the civil engineer in charge of the water supply and sewage systems at Maroochy Water Services during the time of the breach and has presented his analysis in closed forums. Stringfellow observed:

- At first it was easier to blame installation errors for the problems.
- Upon reinstalling all the software and checking the system, pump station settings kept changing beyond the ability of the system to do this automatically.
- Conclusion: an external malicious entity was using wireless equipment to access the SCADA system.

Stringfellow's analysis of the incident made several important points:

- It is very difficult to protect against insider attacks.

- Radio communications commonly used in SCADA systems are generally insecure or are improperly configured.
- SCADA devices and software should be secured to the extent possible using physical and logical controls.
- It is often the case that security controls are not implemented or are not used properly.
- SCADA systems must record all device accesses and commands, especially those involving connections to or from remote sites; this requires fairly sophisticated logging mechanisms.

Stringfellow also recommended the use of antivirus and firewall protection along with appropriate use of encryption. He emphasized a need for upgradeable SCADA systems (from a security perspective), proper staff training, and security auditing and control.

3.7.7 Possible Solutions

This case revolves around a disgruntled insider who was never an employee of the organization he attacked. Some of the issues raised by analysis of this case are just being addressed by cyber security practitioners. Some are unresolved with no solution in sight.

3.7.8 Policy and Procedures

Every organization should have ICS cyber security policies and procedures that are formally documented and audited.

3.7.9 Personnel Security

Since Boden was never a Maroochy Council employee, direct hiring controls by the Maroochy Council were technically not applicable. However, is it prudent to have a key personnel clause to protect the client from unilateral changes in key personnel by the contractor. Determining which controls to apply to on-site contractor personnel may not be easy since it depends on the role played by the individual.

3.7.10 System and Services Acquisition

Hunter Watertech supplied hardware, software, and services to the Maroochy Shire Council. The cyber security responsibilities of the contractor organization (e.g., Hunter Watertech) and the contractor's employees (e.g., Boden) should be included in the contract between the organizations. Almost all of the controls applicable to direct employees are also applicable to contractor employees, but the exact details may vary.

3.7.11 Awareness and Training

Personnel need to be trained in preventing, recognizing, or responding to ICS cyber-related incidents.

3.7.12 Contingency Planning

The analysis indicates that there were no plans to deal with an emergency or system disruption. Effective contingency planning, execution, and testing are essential to mitigate the risk of system and service unavailability.

3.7.13 Incident Response

Response to the sewerage discharge was ad hoc. Considerable time elapsed during troubleshooting before malicious intent was considered. An incident response capability is necessary for rapidly detecting incidents, minimizing loss and destruction, mitigating the weaknesses that were exploited, restoring computing services, and apprehending malefactors.

3.7.14 Information Protection

Cryptography is employed as a protective mechanism for many objectives. A second line of defense is encryption protection of information in stolen and lost devices.

3.7.15 Malicious Activities

Boden's malicious activities included:

- Stealing equipment
- Issuing radio commands
- Falsifying network address
- Sending false data and instructions
- Disabling alarms.

3.7.16 Access Control

Access Control is the process of granting or denying specific requests for obtaining and using information and related information processing services. This is one of the fundamental controls on any IT system. Most access controls are based on the identity of the person, process, or device involved. Therefore, Identification and Authentication are intimately tied with access control. Access controls need to be applied appropriate to the communications environment.

3.7.17 Identification and Authentication

Identification is the process of determining the identity of a user, process, or device, and authentication is the process of verifying the putative or claimed identity, often as a prerequisite to allowing access to resources in an information system.

3.7.18 Portable Device Protection

Part of defense-in-depth is recognizing that when some protections fail, there should be more controls to help protect the organization. Theft of portable electronic equipment occurs, often for the value of the equipment rather than the information stored. Boden's attack was an exception to this generality.

3.7.19 System Monitoring

Another part of defense-in-depth is determining that something is going wrong. The cause may be an intruder, often masquerading as an authorized user, or a malfunction.

3.7.20 Conclusions

The 2000 Maroochy Shire cyber event is important because it provides a public record of an intentional, targeted attack by a knowledgeable person on an ICS. The attack by an insider who is not an employee demonstrates several critical physical, administrative, and supply chain vulnerabilities of industrial control systems. The key issue is the treatment of vulnerabilities coming from suppliers or others outside the organization. Contractor and subcontractor personnel are often overlooked as potential attack sources. The technical issues demonstrate the difficulty in identifying a control system cyber attack and retaking control of a "hijacked" system. Once alerted to this type of attack, ICS owners and operators may have adequate controls to protect their assets. However, a determined, knowledgeable adversary such as Vitek Boden could potentially defeat these controls.

Observations. This was arguably the first public case of a malicious attack on an ICS. It was done by someone with insider knowledge who was not a traditional insider. It also demonstrated that ICS cyber attacks are not always immediately obvious as a cyber attack.

3.7.21 Nonmalicious ICS Cyber Incident

A water utility was upgrading the operator workstation from Windows NT4 Service Pack 4 to Service Pack P6a. The Ethernet driver from the operator workstation to the PLC was via an Ethernet Card. Everything appeared to be working well. All validation tests were signed off. Following the patch, the utility was able to start pumps at the water treatment plant, but unable to stop them! The problem was caused by interactions between products. There were no noticeable warnings from the operator workstation.

3.7.21.1 Observations

System interactions are critical with ICSs and are very difficult to test in a non-production environment. This is another case of meeting IT security requirements and yet the system failed with lack of control system cyber forensics.

3.8 What Should be Done

3.8.1 General Recommendations

- Develop a clear understanding of ICS cyber security (Weiss 2010b). This includes developing a clear understanding of the associated impacts on system reliability and safety on the part of industry, government, and private citizens.
- Define "cyber" threats in the broadest possible terms including intentional, unintentional, natural, and other electronic threats such as Electro Magnetic Pulse (EMP) and electronic warfare against wireless devices. ICS cyber security threats are more than botnets and malware.
- Change the culture in critical industrial infrastructure such that security is considered in the same context as performance and safety (not as critical, but important to consider).
- Get IT and operations to work together.
- Establish a means for vetting ICS experts rather than using traditional security clearances or IT certifications.

3.8.2 Administrative and Procedural

- Get senior management support as the process fails without it. Identify division of responsibilities and reporting structure all the way to the Board of Directors as cyber security is a corporate risk. IT staff have the ears of senior management and the Board of Directors. I believe that the IT audit staff can be an asset in securing ICSs if approached in a teaming manner. However, this is not an IT task and should not be housed within the IT organization. Establish a credible budget to accomplish what will be identified. Recognize cyber security is critical to safe and reliable ICS operation which translates directly to the bottom line. Incorporate security into executive and employee performance measures. Ensure appropriate awareness and training is regularly provided to all personnel.
- Identify all affected stakeholders and their interactions. This is not an easy process as there are many subtle relationships. These relations extend beyond Facility Operations and even beyond the overall organization to encompass vendors, contractors, regulators, first responders, and even the public.
- Mandate *effective* cyber security requirements so this is not a compliance exercise. An effective, living program is critical. Take ICS cyber security as seriously as you take enterprise cyber security. Secure based on electronic connectivity, not just the size of the facilities or equipment.
- Determine what you really have. The hardware, software, and firmware that affect cyber security are often not identified in any formal system diagrams or vendor documentation. Identify all control system hardware and communication infrastructure. Document what you have and what you have done. Establish a living Configuration Management/Configuration Control Program that includes

the ICS as well as cyber security-specific software (e.g., patch versions), hardware (e.g., network interface cards), and firmware.
- Determine what you really need from the ICSs in terms of functions, features, and communication. This requires obtaining input from throughout the organization because cyber security will affect any new systems.
- Determine what you want to do and do it. Again, this is not as easy as it seems. This requires understanding of what features need to have security enabled.
- What risks are present? Traditional risk approaches of addressing probability and consequence need to be modified. Probability should be 1 (it will happen) and consequences should be based on "design basis threat" (worst case). Risk assessments require a cost-benefit trade-off between performance/safety and security. This would involve assessing the risk of both security and performance features. It also addresses externalities including external personnel (e.g., vendors and contractors), economic conditions, and social concerns. Questions to be answered include how risks can be mitigated structurally, technically, and administratively.
- Develop *ICS-specific* policies and procedures. This is what I consider to be one of the most important tasks in the entire process. Recognize that complexity significantly adds security overhead and potential performance/safety impacts. Work with IT to make sure that the ICS policies and procedures are not inconsistent. But first and foremost, they must be developed for the specific equipment to be secured and in the way they are expected to be operated.
- Make your equipment suppliers and contractors partners in securing your systems. Require detailed documentation of what has been provided and how it has been tested and secured. Do appropriate background checks to assure key vendor and contractor personnel are competent and vetted—notice the use of the word vetted not "cleared." (Do not forget about your own critical employees.) Work closely with your vendors and contractors to make sure that they are doing what *YOU* want.
- Consider lifecycle issues. ICSs can be cyber vulnerable from initial design until they are retired or destroyed.
- Consider system recovery issues following an incident or attack. (see later recommendations).

References

In the Crossfire: Critical infrastructure in the age of cyber war, P.1, http://www.mcafee.com/us/resources/reports/rp-in-crossfire-critical-infrastructure-cyber-war.pdf

ISA S99, industrial automation and control system security, http://www.isa.org/MSTemplate.cfm?MicrositeID=988&CommitteeID=6821

Presidential Decision Directive (PDD)-63 (1998) http://www.fas.org/irp/offdocs/pdd/pdd-63.htm

Weiss J (2010a) The need for interdisciplinary programs for cyber security of industrial control systems. WorldComp, Las Vegas

Weiss J (2010b) Protecting industrial control systems from electronic threats. Momentum Press

Chapter 4
Implementing Machine Learning Algorithms for Water Quality Event Detection: Theory and Practice

Eyal Brill

Technical Terms and Abbreviations

AWWA	American Water Works Associations
DT	Decision tree
EDS	Event detection system
LR	Logistic regression
ML	Machine learning
NN	Neural network
USEPA	Environmental Protection Agency
WQE	Water quality event
WQED	Water quality event detection

4.1 Introduction

For several reasons, water quality event detection (WQED) technologies have received great attention over the past several years. First, as a result of the attack against the World Trade Center in New York, NY, USA on September 11, 2001, many countries are more aware of the vulnerability of public infrastructure, including water systems, as a target for intentional sabotage (Berman and Gavious 2007). More and more resources are being invested in the protection of air traffic, communication systems, and other possible terrorist targets. In addition, it has become clear that potential damage can also result from harming critical infrastructure, such as electric or water and waste water systems. Harming water infrastructure has the potential for threatening the health and welfare of large numbers of people (even more in the case of an airplane such as was used in New York); thus, it could present a "favorite" target for terrorist attacks.

E. Brill (✉)
Faculty of Technology Management (MOT), Holon Institute of Technology, Holon, Israel
e-mail: eyalb@hit.ac.il

R. M. Clark and S. Hakim (eds.), *Securing Water and Wastewater Systems*, Protecting Critical Infrastructure 2, DOI: 10.1007/978-3-319-01092-2_4,

Second, in many countries water systems have suffered for long years from a suboptimal level of investment and from poor repair, replacement and rehabilitation policies. Old and unmaintained equipment and the loss of professional staff have, in many situations, generated accidents and other unfortunate events. Moreover, in many water facilities the average age of engineers is increasing, a fact which also generates a growing problem related to a loss or lack of knowledge (Mann and Runge 2010).

These two forces have generated a growing need for automatic water systems management and control, in general, and for WQED, in particular. This growing market and the need for WQED systems have motivated several companies to develop WQED systems. Examples of such systems are given in Appendix A.

In 2005, the United States Environmental Protection Agency (USEPA) published guidelines for Environmental Detection Systems (EDSs). These guidelines focus on both the physical protection of water facilities, as well as statistical methods for detecting water quality abnormalities. It is hypothesized that statistical models can help detect water quality problems, similar to banking systems that detect fraud.

During 2008, the USEPA initiated a project called the "Water Security Initiative" in which private companies were invited to test their EDSs using datasets that were compiled by the EPA. Project results indicate that machine learning methods were a critical tool in this task. Most of the systems (mentioned above) tested in this project used machine learning methodology.[1]

The Water Quality Event Detection (WQED) topic has also been addressed in the literature. A partial list includes studies by Skadsen (2008), Story et al. (2011), Yang et al. (2009), Chang et al. (2012), and Helbling and VanBriesen (2008). A set of commercial reports with many useful pieces of information can be found at the following EPA website: http://www.epa.gov/nhsrc/pubs.html (see EPA 2005).

Global changes have had an impact on Israeli water systems. In Israel (and even before Israel was established) water systems were owned by the state under the principle that land rights do not create water rights. Water was allocated to agricultural usage by administrative quotas managed by the state. Water for urban usage historically was priced using tier pricing by the state owned company which distributed water to municipalities and which performed maintenance of urban water networks and billing for urban water use.

During the last 10 years this arrangement has totally changed. First, corporate-based water utilities were established and were given full responsibility for water allocation and distribution to the end users. These corporations operate as non-profit organizations and are managed by shareholders in order to maximize the benefit to water users. The managers of these organizations have personal responsibility for the quality of water delivered to the end users. Even more, if contamination events occur, that put user's health at risk, managers may lose their

[1] Results from this project have been presented at the annual American Water Works Association conference in 2011, but the official report has not yet been published.

jobs. Therefore, these managers have a personal interest in the efficient detection of WQEs.

Today, 55 water utilities distribute almost 2 billion m^3 of water. These utilities are also responsible for sewage treatment. 75 % of the sewage generated is reused by agriculture after proper processing.

The main goal of this manuscript is to focus on important managerial considerations when implementing machine learning (ML) for WQED. The major questions are:

- What concepts can be used and what drives their efficiency?
- How the concepts of efficiency and reliability can be measured?

The current manuscript begins with an introduction and then describes the properties of water systems that are related to WQED. Section 4.3 gives a short description of machine learning, how it is used for WQED, and what type of techniques and options exist. Section 4.4 describes how to use the supervised learning concept for WQED. Section 4.5 discuses the same issues for unsupervised learning, while Section 4.6 addresses the question of measuring ML quality and how water system properties influence the validity of the concepts' efficiency measurements. Section 4.7 provides a numerical example for implementing ML algorithms for WQED, and the final section presents a summary and conclusions.

4.2 The Nature of Water Systems

Water system data are characterized in many cases by variability and noise. Data properties vary from site to site and change on a daily basis—from day to night, winter to summer—and as one water source of water is replaced by another (for example, underground water is replaced by either surface water or desalinated water).

The spatial nature of water systems, where water monitoring stations are located over large areas and in separate locations, has also generated a need for transferring water quality data over distances to a control system in a central location. This transmission which, in many cases, is based on cellular or radio communication also sometimes "contributes" to the noisy data due to the poor quality of communication lines.

Another issue is the equipment which is commonly used for quality measurement and which may also, in some cases, be the source of data problems. For example, some equipment requires constant maintenance and monitoring, which is costly and often not feasible or is impractical. Nevertheless, water utilities are obligated to monitor water quality and prevent a situation in which low-quality water is supplied to the consumers. A general definition of a WQE is one in which a short-term change in water quality, at a given site, may indicate possible contamination, making it unusable.

A water quality event (WQE) may occur as a result of intentional or unintentional contamination, due to an equipment breakdown or an operational malfunction or error. Another cause of water quality events (WQEs) may be an operational change. When operators switch from one operational mode to another, water quality measurements may be unstable before becoming stable once again. This instability may be miss-interpreted as an abnormal situation or a WQE.

Another issue relates to instrument calibration, which must be conducted periodically. Such calibration sometimes changes the basic data references and may be perceived as an abnormal event.

Since water frequently travels through a distribution system at a high velocity, it is most important to detect such adverse events as-soon-as-possible in order to limit their potential damage. Due to the nature of water systems, delayed detection of even a few hours may affect thousands of users. On the other hand, oversensitivity in regard to the detection system may generate many false alarms as a result of misidentification, and may also cause economic damage and frustration.

It is clear that, given the above description, a fixed set of global rules or limits may not be sufficient for the process of detecting a WQE. Furthermore, water quality which is considered safe in one location may not be suitable in another location. For example, in some cases a water site with a steady turbidity measurement of 0.6 is considered safe, while in another situation a sudden change of turbidity in a site with normal measurements of 0.2–0.5 indicates an event that should be examined by an expert. Thus, a global set of rules and regulations regarding statistical limitations are not very feasible.

The task of identifying a water quality event (WQE) is complicated and requires human intelligence and wisdom. Even more, it requires constant study of the water properties that change over time, and the continual updating of the detecting mechanism.

In answer to this challenge, and similar to the situation in many other industries, attempts have been made in recent years to cope with the above problem by practicing machine learning technology. Such technology, which has made great advancements in the fields of speech recognition and fraud detection, seems to also show promise in the WQED arena.

4.3 Machine Learning in a Nutshell

Machine learning is a mathematical technique that allows computers to evolve behaviors based on empirical data. Practicing machine learning technology for the purpose of WQED may occur through either supervised or unsupervised learning. In both concepts, a mathematical procedure is used to build a model which is a relationship between inputs and outputs. Inputs are quality or physical measurements of water. Output is a categorical variable which indicates if the situation is a WQE, or not. The mathematical procedure which is used to create these relationships is defined as an algorithm. The procedure of using the algorithm to construct the model is called in-machine learning "training the model".

With regard to supervised learning concepts, a model is trained with a dataset that contains both normal data and abnormal events and abnormal events are tagged as such. After learning, the algorithm is tested with a different dataset, which also contains both normal and abnormal events. The role of the trained model is to generate an alert in the case of abnormal events only. Results are then summarized in order to express the model's ability to detect these events (taking into account True Positive, False Positive, and False Negative).

With regard to the unsupervised learning concept, it is assumed that information about True Positive does not exist or is very rare. Thus, the training algorithm is fed with data which indicates normal behavior and abnormal behavior, but no prior classification is made with regard to the data. It is assumed that, for the most part, the process/system behaves normally because a system that generates bad events most of the time is not likely to happen. In other words, abnormal behavior is assumed to be rare. The algorithm's role, then, is to use this dataset to understand/discover the system's normal behavior pattern. This is a cornerstone in the unsupervised concept. Abnormal is rare!

Once a solid baseline has been established, the model should be capable of detecting abnormal events. The problem is that not every rare event is a bad event. The challenge is to differentiate between "normal-abnormal" and "abnormal–abnormal", i.e., between situations in which rare combinations occur, but the reason relates to a transition, equipment malfunction or other non-hazardous situation, rather than a real threat. It is important to note that when an event occurs in the unsupervised model, a human classification is made. Thus, over time the unsupervised model also learns from human intelligence and wisdom. However, unlike the supervised model, this is done "post facto".

Another topology of machine learning refers to parametric or non-parametric modeling. With regard to parametric modeling, a functional form of the model is assumed. The role of the algorithm is to find optimal values for these parameters in order to optimize detection.

With regard to the non-parametric approach, a non-functional form is assumed. The algorithm has several criteria; the data is analyzed and classified based on these criteria.

The third characteristic of machine learning algorithms, which is also related to WQED, is the difference between lazy algorithms and eager algorithms. Lazy algorithms postpone any processing until a query for classification is requested. This enables local learning and a more accurate answer, but it also requires more response time. It also requires storing all relevant data until an actual query is requested.

Eager algorithms perform learning processes on the data ahead of time. The algorithms generate a mathematical model, which encapsulates the relations between inputs and outputs, i.e., between input values and True Event probabilities. When a query is presented, the answer is based on the model, rather than the raw data that was used to generate the model. As will be shown in the next section, water system properties have a crucial effect on the efficiency of machine learning algorithms and one should carefully consider which algorithms are the most suitable in each situation.

4.4 Supervised Learning Approach and WQED

As was explained previously, a supervised learning approach is based on the idea that an expert is able to classify historical results into good events and bad events, and that bad events have occurred in the past with such a level of frequency that they can be used to build a statistical model. Using a supervised concept allows us to learn from these examples about the pattern of bad events, and enables the extraction of rules about the relationship between inputs and outputs in bad situations. The extracted rules (the model) can then be used to automatically classify new samples, i.e., when the rules about the mathematical relation between inputs and outputs in the case of a bad event is known, given a new set of inputs, it is possible to predict whether the output will be bad or good. Several algorithms can be used in order to implement supervised learning for the WQED. Examples are: Logistic regression (LR), Neural Networks (NN), and Decision Trees (DT).

With regard to LR, a dependent categorical variable's probability (a variable that can take on a limited number of categories) is predicted by a set of independent variables using a linear relation. Whenever the "bad situation" probability is higher than a given value, an alarm is generated. In the case of WQED, we refer to a binary logistic regression in which the outcome may be "Good" or "Bad". In order for the LR to be valid, several preassumptions should exist in the data. First, it is assumed that inputs are not interrelated, i.e., the correlation between inputs should be low. This assumption is not always true in the case of water. Water quality measurements do have strong correlations with one another; for example, the pH is correlated with conductivity and free chlorine is also correlated with conductivity. This may have adverse effects on LR efficiency. Thus, when using LR, one should verify that violating these prerequisites do not damage the model's validity.

Another property of LR is the assumption that no lag relations exist among variables. This means that the values from previous timestamps are assumed not to be related to present variables' values. It is also known that this does not always exist in a real life dataset, which should also be considered when using LR. Last but not least, it is also important to note that, in the long run, water properties do not behave in a linear manner. This also puts some limits on the efficiency of linear-based models.

An additional method, which is widely used for modeling the relation between inputs and outputs, and which can be implemented for WQED, is that of Neural Networks (NNs). The NN is based on a mathematical model that mimics the behavior of neurons in a biological nerve system. The NN is based on an input layer, output layer, and a hidden layer, which create the relation between input and output layers. The secret of the NN is in the weights of the connections between layers and how they affect the output. During the learning process, these weights are optimized to generate the proper answer. Without going too deeply into the nature of this technique and its performance, the main issue with this methodology is the fact that the coefficients of the resulting model are not "understandable" for the common engineer. In most cases, only NN experts can understand the precise meaning of each of the NN model's coefficients. Following the need to integrate engineers in the process of modeling WQEs, NN is also considered a less preferable solution. Moreover, since water quality modeling also involves non-numerical and non-ordinal inputs, it should be noted that construction of NN with such variables makes the modeling process very complicated.

Another method which is also used for supervised learning and can be practiced in the case of WQED is the Decision Tree (DT). The idea behind this methodology is as follows: Assume the value of an output has some tolerance (i.e., alarm, non-alarm, and the gray area in-between). Given a set of records with both values, i.e., some not tagged as alarms and some tagged as alarms, one can gain information by splitting the dataset into two sub-datasets. The majority of output values differ substantially among the two groups. In the best case, the groups are totally different. One group is tagged only as an alarm and the other as a non-alarm. Splitting the dataset into two groups is achieved by selecting a deviator—an input value which differs among the two groups. This technique was used by Quinlan (1986, 1993) in building the ID3 and the C4.5 algorithms and by Brieman et al. (1984) in building the CART and regression tree algorithms. The process of dividing the dataset into smaller groups continues, as long as the resulting groups have significant information beyond their aggregate form. The deviators create a set of rules that define each of the final sub-groups.

The DT algorithm has an additional advantage when used for WQED, due to the nature of variables that can be found in this domain. In the case of WQEs, modeling input variables are a combination of floats, integers, Boolean, and non-ordinal values. For example, water quality measurements, such as turbidity and free chlorine, are floats. Operational variables, such as the number of operating pumps, are integers. Operating day of the week, hour of the day, or month of the year, are ordinal variables. All these variables are different and were found to be relevant in the case of water quality. The DT is an easy way to combine variables of different types into a single model.

The methodology of building a model is based on splitting the original dataset into two sub-datasets. Traditionally, two-thirds of the dataset is used as a learning set, while the rest is used as a testing set. For each record in the learning set, an expert has to define a binary output indicating whether or not this record expresses a quality event. After tagging the learning set, the algorithm is used in training the model.

To summarize this section, we conclude that supervised learning is the preferred method when:

- Historical records about true events exist (enough to generate a statistical model),

- An expert is available for data classification,
- No non-ordinal variables exist (if NN or LR are used).

4.5 Unsupervised Learning Approach and WQED

The main idea behind unsupervised learning is the assumption that abnormal situations are rare. This is similar to the abnormal situation where someone uses someone else's credit card without being authorized to do so. The algorithm must learn the frequency of each data combination. Combinations which occur below a predefined rate are suspect, and should be examined. Once a new/rare combination occurs, an alert is generated. It is then the role of the user to classify this combination. If the combination is tagged as "good", no additional alarm will be generated for this combination or a similar combination in the future. If the combination is tagged as "bad", hitting a similar combination in the future will yield an alarm, depending on the question of how similar the new record is to the known bad situation.

The question is: How to generate the predefined known combination? And, when such a definition exists, how to detect whether or not a new combination is similar enough to the existing one?

The methodology for solving these issues is to use the clustering method. Clustering is an old statistical technique with many variations. In general, it is a statistical procedure that enables the assigning of single records into groups that share similarities. Each group is defined as a cluster. Each cluster has a center of gravity (centroid), which is the focal point shared by all of the similar records. Clustering may be either density-based or distance-based. Density-based clustering uses some form of kernel function in order to define the map shape. This method is parametric and suitable for "eager" algorithms. Distance-based clustering, on the other hand, uses the actual distance between a point and its neighbors in order to define similarities. This method is likely to be non-parametric and is suitable for lazy algorithms.

Once a clustering map (CM) has been generated (with either of the methods), it is possible to test each new record according to its location on the map. If the new records are "close enough" to one of the known cluster centers, and if this cluster is not tagged as a bad cluster, then it is safe to define this new record as "safe" or "good". If the new record is too far from any known cluster, it is suspect and considered "bad", after which a human expert is called to classify it.

Similar to the decision tree, a learning set is needed as a starting point. However, this is not for classification purposes, but rather for frequency learning purposes. After the frequency map is stable enough, the model can begin being used for detection. The result of such an evaluation is, once again: alarm or non-alarm. Given the actual classification of each record, it is once again possible to express model quality. The next section discusses how model quality should be evaluated.

Hence, we conclude that practicing WQED with unsupervised learning can be performed through using the clustering method. This method is preferable when:

- Historical records about true events do not exist or are rare,
- No available expert exists to conduct ongoing data classification,
- Non-ordinal variables exist as part of the variables set.

One should note, also, that the lazy algorithm requires the system to have access to an original dataset. This is since actual calculation is done only when the query is asked. Thus, unsupervised lazy algorithms require more on-line storage of historical data.

4.6 Measuring WQED Model Quality

A crucial question relates to the quality of the model constructed by each of the above algorithms. Several methods exist for this purpose. The current manuscript focuses on two methods: The receiver operating characteristic (ROC) curve and the confusion matrix with Kappa measurement.

The term ROC (receiver operating characteristic) refers to a curve (see Fig. 4.1), where the horizontal axis is the number of false alarms and the vertical axis is the number of true alarms. In order to create an ROC curve, the model is tested by using different levels of sensitivity. In the case of the supervised concept, the level of sensitivity may be the minimal number of records with bad classification that triggers alarms. In the case of unsupervised learning, the level of sensitivity may be the frequency below which alarm is raised. At each level of sensitivity, while performing the test for a given dataset, the model produces some

Fig. 4.1 ROC curve

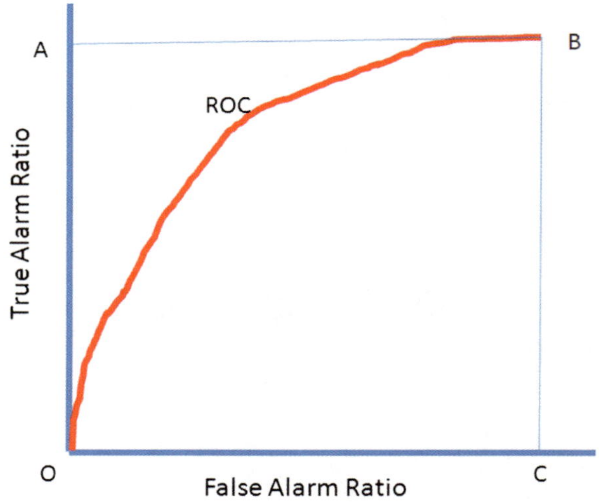

false alarms, while also detecting true alarms. Naturally, in order to detect more true alarms, the model misses some and produces false alarms. Each point in Fig. 4.1 graph represents a test for the same dataset with different levels of sensitivity. The quality of the model is measured by the area under the ROC curve, relative to the area denoted by OABC in Fig. 4.1. This value ranges from 0 to 1. The bigger the value is, the better the model.

Producing the ROC curve requires that the tester set the model's sensitivity level at different levels. Each algorithm must examine which sets the model is sensitive to, and determine the pros and cons of using this device. In general, it is preferable to change the sensitivity level by using a measurement that is related to the output probability, rather than the input measurement. This allows to the user to examine different types of models that use different inputs to produce a similar output.

Note that the ROC curve includes only the true positive and false positive. It does not include the effect of the false negative events and does not give these events a different weight, i.e., cases in which the model missed true alarms. In the case of water quality, false negative are a big issue. The fact that the weight of false negative errors is not explicitly included in the measuring method of the model quality is one of this method's disadvantages.

A special note should be made with regard to this issue, concerning the difference between supervised and unsupervised concept. In order for a supervised method to "miss" an event, all "that is needed" is for an expert to miss classifying an event as a "bad" event. This is simply human error and can easily happen. With regard to the unsupervised method, it is assumed that bad events are rare. Thus, missing the event is not a question of human classification, but is rather a question of the level of rareness that triggers the alert. Thus, in most cases, the unsupervised method will miss less true alarms, but will probably generate more false alarms. Thus, for the same datasets, ROC will, in most cases, yield results that are in favor of the supervised algorithms if good experts exist and are in favor of the unsupervised method, and vice versa.

Another method which may be used to measure model quality and takes into account all four possible situations is the confusion matrix, presented in Fig. 4.2.

The rows in the matrix present the model prediction; the columns in the matrix present the true results. Each of the cells in the matrix represents True Positive (TP), False Positive (FP), True Negative (TN), and False Negative (FN), respectively. For example, the upper left cell of the matrix represents a situation in which

		Actual situation	
Model Prediction		Alarm	No alarm
	Alarm	True Positive	False Positive
	No alarm	False Negative	True Negative

Fig. 4.2 Confusion matrix

the model declared a record as an alarm, and the situation actually reflected bad quality. Thus, this cell is denoted as True Positive (TP). The numbers in the confusion matrix are obtained by classifying all records in the test set and comparing this prediction with the actual results.

Using this matrix, one can calculate the Cohen (1960) Kappa statistics. These statistics measure the level of agreement between the model and reality. The value of the Kappa is between a value of 1 (total agreement) and -1 (total disagreement). In practice, a value of Kappa above 0.6 is considered good.

Much like the ROC curve, the Kappa statistics also have some drawbacks. In some cases, the number of true negatives relative to other situation is high and has a critical influence on the Kappa sensitivity. In the case where most of the records are true negative, this method may not be very efficient.

4.7 Numerical Example

The current section illustrates a simple numerical example of how two algorithms (one supervised and one unsupervised) are implemented in order to classify a WQE. Figure 4.3 shows the raw data measurements, which include four different measurements: Free chlorine, Conductivity, TOC, and Turbidity.

The dataset was divided into two parts. The first part, the learning set, located on the left side of the vertical line, includes 10,216 records, each of which is a 5-min average. Using these records, a water expert identified 317 min in which water measurements were considered abnormal. All records are marked with circles. The rest of the dataset, located on the right side of the vertical line, was used as a test set. This dataset includes 7,354 records, out of which 62 were considered to contain abnormal conditions, also based on an expert's classification.

In order to implement the detecting algorithm, the Weka software package,[2] version 3.6.6 was used.

4.7.1 Supervised Concept Implementation

The supervised algorithm for building a decision tree was the REPTree, which is a version of the C4.5 algorithm. The algorithm yielded the following tree, as can be seen in Fig. 4.4.

The first deviating variable was the TOC. All records with TOC > 2.59 are classified as abnormal. The rest of the records were tested against the value of the free chlorine. If this value was above 1.17, the record was considered as normal. If the value was below, the record was tested against the turbidity. If the turbidity value was above 0.55, the record was marked as abnormal; otherwise, as normal. The confusion matrix generated by the algorithm is presented in Fig. 4.5.

[2] http://www.cs.waikato.ac.nz/ml/weka/.

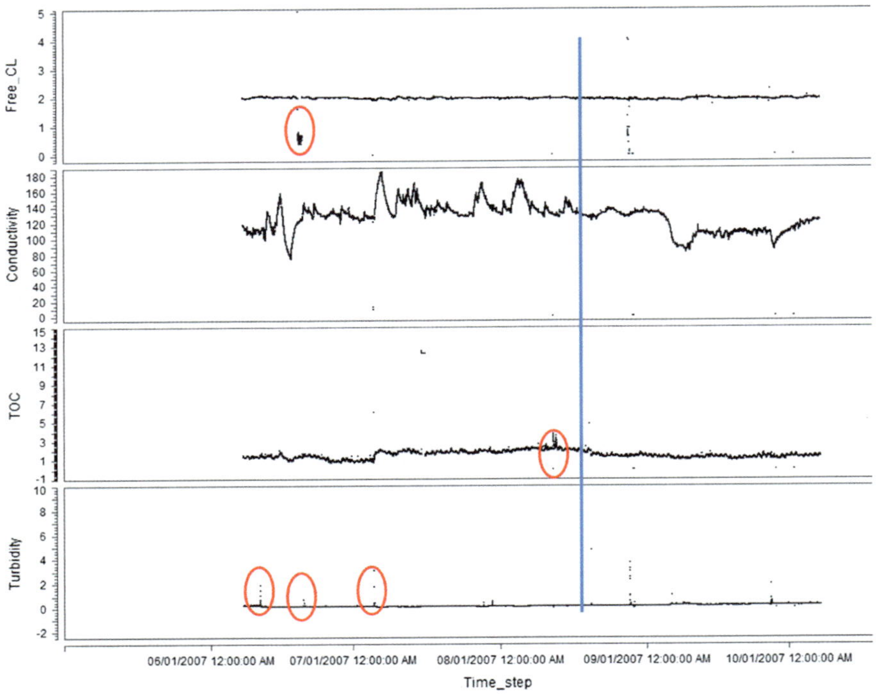

Fig. 4.3 Water quality data-set

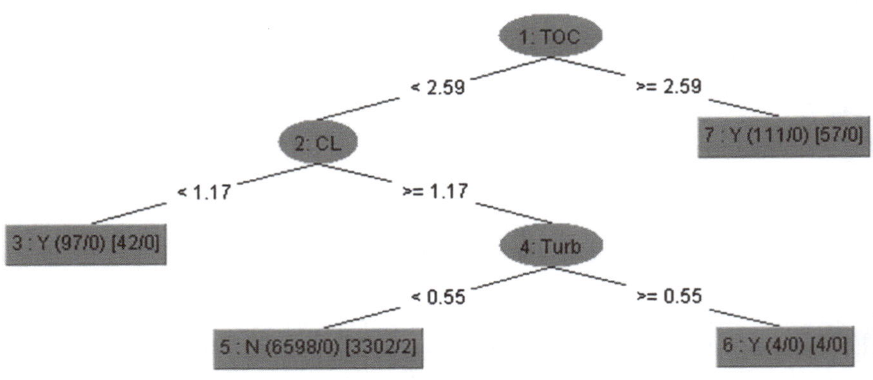

Fig. 4.4 Decision tree

The ROC value for this model was 0.984; the Kappa value was 0.9597. Thus, it is evident that the algorithm performs very well, but mainly since an expert was available to classify the predefined dataset from which the algorithm can learn.

	Data	
Model	True Negative : 7288	False Positive: 3
	False Negative : 2	True Positive: 60

Fig. 4.5 Decision tree confusion matrix

4.7.2 Unsupervised Concept Implementation

What happens if this is not the case—if an expert is unavailable to classify water quality data?

In this case, the unsupervised methodology may be used. In order to examine this situation, the same learning set (without expert classification) was processed to generate clusters. Then, the test set was examined with the same algorithm in an attempt to allocate each record to one of the predefined clusters. If the distance of the new record to the nearest cluster was above a given distance, an alarm was generated. This distance was then determined as the sensitivity parameter, which is related to the probability that this record is rare and thus should be further examined.

Figure 4.6 shows the clusters table. The table includes the center of each cluster and the number of members in each cluster (last column on the right, entitled "Count").

The cluster table can also be visualized using a PCA algorithm. This algorithm transforms the four results dimensions into a 3D linear combination. This 3D picture is illustrated in Fig. 4.7. The different clusters are presented with different

Id	Free CL	Conductivity	TOC	Turbidity	Count
1	1.99	93	1.15	1.84	5
2	1.99	121	1.16	0.15	3675
3	0	121	32.79	0.29	1
4	1.91	126	32.79	0.20	9
5	5.00	121	1.45	0.14	1
6	1.98	137	12.22	0.12	117
7	0.61	123	1.38	0.14	137
8	0	0	0	0	2
9	0	125	32.79	96.71	2
10	1.95	141	1.89	0.11	6266

Fig. 4.6 Clusters list for unsupervised algorithm

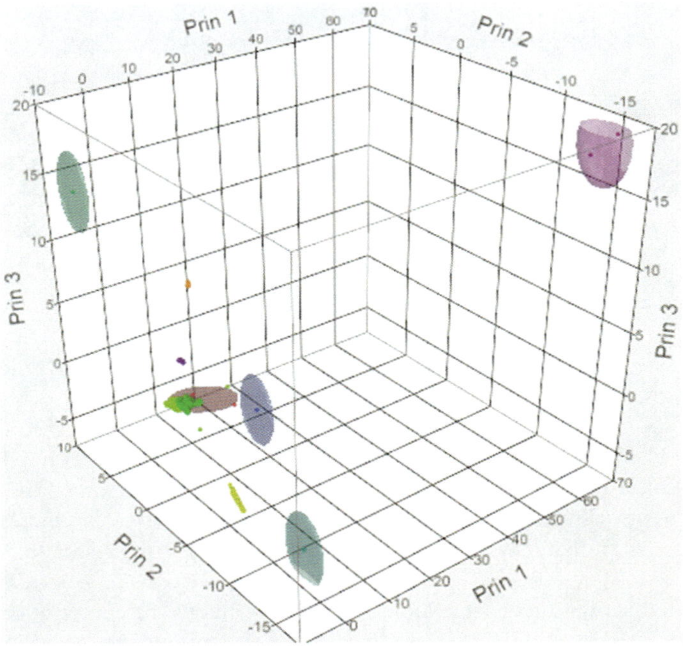

Fig. 4.7 Clusters' 3D picture in PCA

shapes in the PCA cube. Any record that is not located near one of the shown clusters will trigger an alert and should be examined by an expert.

Given this clusters map, the same test dataset was evaluated. The evaluation process allocated each record in the test set into one of the existing clusters. Figure 4.8 shows the result of the evaluation.

It can be seen that the unsupervised algorithm's performance was weaker than that of the supervised algorithm, as it generated more alarms (totally 68 while the supervised generated 65). These alarms were generated because records showed up in the test set in locations with low density. Obviously, once these alarms are classified as non-alarms, they will not trigger alarms in the future. However, without expert classification any cluster below density of 1 % triggered alarms.

4.8 Concluding Remarks

This manuscript examines the implementation of machine learning for water quality event detection (WQED), and shows that several algorithms are likely to be used for this purpose. The main issues that should be considered are whether history about true events exists, and if these events have been classified by experts.

Fig. 4.8 Test set results

Cluster	Assigned	Alarms
1	0	No
2	5,294	No
3	0	No
4	2	Yes
5	5	Yes
6	0	No
7	40	Yes
8	16	Yes
9	5	Yes
10	1,996	No

If this is the case, then supervised learning is a better way in which to proceed. However, in the case of water systems, since automatic data recording is relatively new, and since many water utilities do not have a classified history of water events, another methodology should be considered. This alternative methodology is based on unsupervised learning. According to this methodology, that which is rare should be further examined. Thus, the algorithm must learn the frequency of historical combinations in the data. Once this frequency is known, any rare combination which did not occur or occurred with low frequency should generate an alarm, and be classified by an expert for future reference.

As can be seen from the numerical example, in the presence of an expert, supervised learning performs better. However, the existence of an expert should be carefully examined. Sometimes the expert may not be aware of new types of contaminations or abnormalities. Moreover, experts are not always in agreement with one another. The advantage of the unsupervised learning method is its independence and ability to detect a possible or potential wrong event without an expert. This method may be accompanied at its first steps by a higher rate of false alarms, but as they say: "There's no such thing as a free lunch". As can be seen from the numerical example, the unsupervised test dataset generated 68 alarms overall but did not miss any true alarm. This methodology was carried out without prior information or classification about water quality history; all that was needed was combination distribution. In some cases, this methodology may be the only possible way to detect WQEs.

A.1 4.9 Appendix A

Following is a list of websites for common EDS systems

Gardian Blue by Hach: http://hachhst.com/products/cityguard-virtual-command-center

Canary by Sandia Labs: https://share.sandia.gov/news/resources/news_releases/canary/

Moni::tool by S::can: http://www.s-can.at/

BlueBox by WhiteWater: http://www.w-water.com/qualitysecurity/Product.aspx?id=64

References

Berman O, Gavious A (2007) Location of terror response facilities: a game between state and terrorist. Eur J Oper Res 177(2007):1113–1133

Breiman L, Friedman JH, Olshen RA, Stone CJ (1984) Classification and regression trees. Chapman & Hall/CRC, New York

Chang N-B, Pongsanone NP, Ernest A (2012) A rule-based decision support system for sensor deployment in small drinking water networks. J Cleaner Prod 29–30:28–37

Cohen J (1960) A coefficient of agreement for nominal scales. Educ Psychol Measur 20(1):37–46

EPA (2005) WaterSentinel online water quality monitoring as an indicator of drinking water contamination. EPA 817-D-05-002

Helbling DE, VanBriesen JM (2008) Continuous monitoring of residual chlorine concentrations in response to controlled microbial intrusions in a laboratory-scale distribution system. Water Res 42(12):3162–3172

Mann J, Runge J (2010) State of the industry report 2010: how water professionals are meeting ongoing challenges and economic uncertainty. J AWWA 102:10

Quinlan JR (1986) Induction of decision trees. Mach Learn 1:81–106

Quinlan JR (1993) C4.5: programs for machine learning. Morgan Kaufmann Publishers, San Francisco

Skadsen J (2008) Distribution system on-line monitoring for detecting contamination and water quality changes. J AWWA 100:7

Story MS, Van der Gaag B, Burns B (2011) Advances in on-line drinking water quality monitoring and early warning systems. Water Res 45:741–747

Yang YJ, Haught RC, Goodrich JA (2009) Real-time contaminant detection and classification in a drinking water pipe using conventional water quality sensors: techniques and experimental results. J Environ Manage 90(8):2494–2506

Chapter 5
Sensor Placement Under Nodal Demand Uncertainty for Water Distribution Systems

Hailiang Shen, Edward McBean and Yi Wang

Technical Terms and Definitions

CWS	Contaminant warning system
e-NSGA-II	Epsilon-nondominated sorting genetic algorithm-II
EPANET	A software program developed by the United States Environmental Protection Agency designed to model the hydraulic behavior and transport on constituents in water
F1	Time delay
*F*2	Sensor detection redundancy
NSGA-II	Nondominated sorting genetic algorithm-II
PF	Pattern factor
PINs	Possible intrusion nodes
SHARCNET	Shared hierarchical academic research computing network
SQL	Structural query language
TOC	Total organic carbon
WDS	Water distribution systems

H. Shen (✉)
Water Resources Engineer, Conestoga-Rovers & Associates, Waterloo, ON N2V 1C2, Canada
e-mail: hshen@craworld.com

E. McBean
School of Engineering, Canada Research Chair in Water Supply Security, University of Guelph, Guelph, ON N1G 2W1, Canada
e-mail: emcbean@uoguelph.ca

Y. Wang
School of Engineering, University of Guelph, Guelph, ON N1G 2W1, Canada
e-mail: ywang09@uoguelph.ca

R. M. Clark and S. Hakim (eds.), *Securing Water and Wastewater Systems*, Protecting Critical Infrastructure 2, DOI: 10.1007/978-3-319-01092-2_5,

5.1 Introduction

The governing requirements for security for water supply systems in Canada are provincially mandated, not federally regulated. The national strategy works on the basis of collaborative efforts from federal, provincial, territorial, and critical infrastructure sectors to provide the infrastructure resiliency.

Approximately 89 % of the Canadian population obtains their water from a public-serving water distribution system (WDS). Of the remaining, 11 % are in vast majority from private wells, with less than 1 % obtaining their water by truck.

The considerable majority (typically 90 % of the 89 % referred to above) receive their water from municipally owned distribution systems (approximately 1,500 in number), with the remainder of the consumers receiving their water from many small distribution systems. The large, municipally owned systems are highly regulated at the individual provincial levels. These systems are owned and operated under provincial regulatory systems and require permits for construction, source water protection, and certification of operators. On the other hand, the small systems (numbering in the thousands, many with fewer than 15 connections) are much less stringently regulated, and therefore have much greater potential for public health issues.

The potential for contaminant intrusion into water distribution systems (WDSs), as a result of either deliberate injection and/or an accidental event is attracting increasing concerns for public health and security. Accidental events such as cross-connection/backflow, pipe leakage/breakage, as well as the possibility of deliberate injections (where the type and patterns of contaminants will never be known), have the potential to contaminate water within the water distribution systems (WDSs). Accidental contaminants may be mixed agents, chemical (e.g., diesel fuel), viral, bacteria, protozoa, parasitic, and some unidentified, while intentional ones may include Ricin, as an example. Some examples of events include Bailey, CO 1981 (antifreeze from solar heating system at a school), Plantation, FL 1987 (carbon dioxide from soda machine at a theater), and Stratford, ON 2005 (detergent from a car wash). On this premise, issues associated with placement of sensors as part of a contaminant warning system (CWS) are relevant. A CWS consists of: (i) online water quality sensors, which monitor surrogate parameters, e.g., chlorine, TOC, turbidity, temperature, and pH, (ii) an event detection algorithm, which determines the existence of contamination and its level, (iii) a contaminant source identification algorithm, which identifies the locations of possible intrusion nodes (PINs) and their probability as being the true source, and priority nodes within the PINs which warrant particular attention due to the potential higher consequence if impacted (e.g., hospitals), and (iv) an emergency response plan, acting to facilitate planning for closing valves and flushing impacted areas of the water distribution systems (WDSs).

As an essential part of a CWS, a sensor placement design methodology is described herein. Specifically, hydraulic and water quality models are applied to simulate the movement of a contaminant intrusion event. Hydraulic models require accurate nodal demand data; however, nodal demands are never known exactly and possess (possibly huge) uncertainty. This means models with fixed nodal demands at each hydraulic step will not describe the performance of a network, (e.g., the time delay to detect the presence of the contaminant). A sensor may or may not detect a specific event under nodal demand uncertainty; there may be uncertain flow directions, and velocities and thus contaminant transport. The first detection time (subsequently, referred to as time delay) of the event by a sensor network may differ appreciably as indicated in Shen and McBean (2011a), e.g., the time delay may be different for each of 100 demand realizations, and the 95 % quantile of the range of 100 values may be hours, indicating uncertain sensor performance for an intrusion event. As a consequence, it is crucial to incorporate nodal demand uncertainty in sensor placement design.

In this chapter, the impact of nodal demand uncertainty on sensor placement is assessed. The research, and ultimately the methodology to be useful are made possible with the leverage of parallel computing power available via a supercomputer, in this case, the Shared Hierarchical Academic Research Computing NETwork (SHARCNET).

5.2 Literature Review

It is essential to characterize a set of events which are to be detected as the first step in conducting sensor placement research. Shen and McBean (2011b) described the pros and cons regarding deterministic and random events sampling, and recommend a deterministic methodology. The random sampling approach cannot characterize representative events to rely upon (e.g., some trial events may be selected while those having large consequences are missed), making it hard to finalize a sensor network design.

A number of objectives have been proposed to select sensor locations, these include: (i) Minimization of expected time of detection (z1, i.e., minimize the average time delay from intrusion to first detection), (ii) Minimization of expected population affected prior to detection (z2, i.e., minimize the number of people exposed to contamination), (iii) Minimization of expected demand of contaminant water prior to detection (z3, i.e., minimize the consumption of contaminated water), (iv) Maximization of detection likelihood (z4, i.e., maximize the detection possibility of events) (Ostfeld et al. 2008), sensor detection redundancy (i.e., minimize sensor detection false positive rate), and time delay (Preis and Ostfeld 2008; Shen and McBean 2011b). The objective for time delay is different from objective z1 in defining the time elapsed from intrusion to first detection. The objectives z1, z2, and z3 are correlated with each other, and not competitive (Ostfeld et al. 2008), thus do not formulate a rigorous multiple objective optimization problem. As suggested in Shen and McBean (2011b), the two objectives, time delay and sensor detection redundancy, are competitive with each other and are potential candidates for multiple objective optimization.

NSGA-II is an algorithm proposed by Deb (2000) and Deb et al. (2002). The algorithm is applied by Preis and Ostfeld (2008), Austin et al. (2009), Preis et al. (2009), Weickgenannt et al. (2010), and Shen and McBean (2011b) in solving the sensor network placement problem. The algorithm e-NSGA-II is another multiple objective optimization solver, a variant of the NSGA-II. The details of the e-NSGA-II can be found in Kollat and Reed (2006), Reed et al. (2007). The solver is, in essence, continuous NSGA-II runs, and is demonstrated to outperform NSGA-II in identifying the true Pareto front, and therefore has been selected as the solver used herein.

Little research has been conducted on the impact of uncertainty on sensor placement design. Shen and McBean (2011a) did propose a way to undertake parallel simulation of multiple scenarios under nodal demand uncertainty using a super-computer. Furthermore, parallel computing is utilized in the multiple objective optimizations for sensor placement design herein, providing a way to study the design in reasonably short time.

5.3 Procedure

The sensor placement procedure is depicted in Fig. 5.1. The deterministic sampling method is used, where a set of events that happen at all WDS nodes, and every hour of the simulation duration, are sampled. Specifically, as an approach to investigate nodal demand uncertainty, nodal demands are assumed to follow the normal distribution. For each node, the demands at each hydraulic step are the multiplication of its base demand and a pattern factor (PF) which is normally distributed. By modeling in EPANET, for each node and hydraulic step, the mean of PF is the value from EPANET input file and the standard deviation is 10 % of the PF. To define a scenario or demand realization, random PFs at each step are generated. Multiple demand realizations are simulated with the supercomputer SHARCNET, and are stored in a SQLite database (each realization is stored in a table) for objective quantification by structural query language (SQL) sentence.

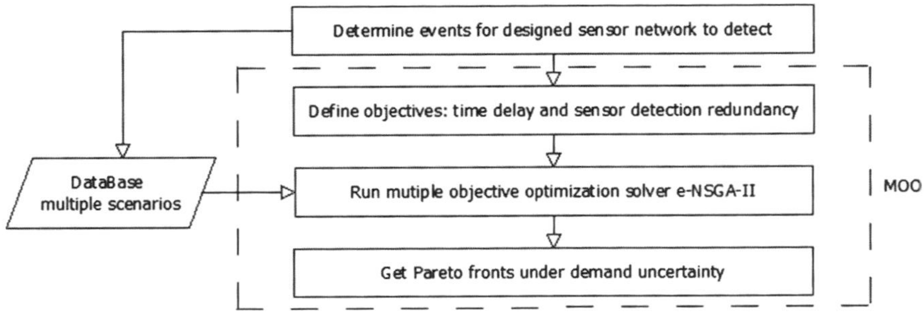

Fig. 5.1 Optimal sensor placement under nodal demand uncertainty

The two objectives in the Pareto optimality are defined as time delay ($F1$) and sensor detection redundancy ($F2$). Time delay is the summation of time elapsed for the set of events sampled. Sensor detection redundancy is the total event number that can be detected by at least two sensors. The mathematical equations are presented in Eqs. (5.1) and (5.2) below, following Shen and McBean (2011b):

$$\text{Minimize } F1 = \sum_{i=1}^{N_\text{d}} t_\text{d}(i) + N_\text{nd} t_\text{nd} \tag{5.1}$$

where $F1$ is the time delay, $t_\text{d}(i)$ is the time delay for the ith detected event, N_d is the number of detected events, t_nd is the time delay defined for each non-detected event, which is a large number to avoid interaction with $\sum_{i=1}^{N_\text{d}} t_\text{d}(i)$; the purpose is to identify the number of undetected events given the $F1$ value, and N_nd is the number of non-detected events. $F1$ is normalized to (0, 1) to feed into e-NSGA-II algorithm.

The second objective is to maximize $F2$, the monitoring system redundancy.

$$\text{Maximize } F2 = \frac{1}{N_\text{d}} \sum_{i=1}^{N_\text{d}} r(i) \tag{5.2}$$

where $F2$ is sensor detection redundancy, $r(i) = 1$, if the event is detected by at least two sensors, and $r(i) = 0$, otherwise.

From Eq. (5.2), a higher $F2$ value indicates more events are detected by at least two sensors, which reduces the possibility of a false alarm in event detection. If a single sensor identifies an event, it is possible (perhaps highly), this event is a false positive (Murray et al. 2012). With a second sensor (i.e., increased detection redundancy) identifying the same event, the false positive rate is greatly reduced. If frequent false positive alarms result, costly emergency responses would be triggered, and if false positives continue, people will begin to ignore the CWS. As well, with even more sensors detecting the same event, the false positive rate can be reduced further; however, large sensor numbers may be needed to acquire certain coverage, which increases sensor costs very quickly and this dimension goes beyond the scope of this chapter.

In the e-NSGA-II algorithm, the population (consisting of individuals) is evolved based on non-dominated sorts to compare goodness of individuals (i.e., set rank to each individual), and a set of online archived individuals (consisting of elites) to be 'injected' into current population to speed-up convergence; for details of the algorithm, refer to Reed et al. (2007). The method to deal with nodal demand uncertainty in the process of optimization is best illustrated with a simple example, as provided in Tables 5.1 and 5.2. Table 5.1 lists the $F1$ and $F2$ values in three different demand realizations of the same e-NSGA-II population individual (i.e., a trial sensor network design), within different demand realizations, the same sensor network design has different $F1$ and $F2$ values. However, to evolve a population in e-NSGA-II, an individual should have a single objective value, herein single $F1$ and single $F2$ value. Table 5.2 lists the objective values for the

Table 5.1 Objective values of $F1$ and $F2$ in three different scenarios (demand realizations) under nodal demand uncertainty

Objective	Scenario 1	Scenario 2	Scenario 3
$F1$	0.25	0.35	0.30
$F2$	0.65	0.70	0.75

Table 5.2 Objective values of $F1$ and $F2$ for optimization under best, worst, and average case

Objective	Best case Min ($F1$) Max ($F2$)	Worst case Max ($F1$) Min ($F2$)	Average case Avg. ($F1$) Avg. ($F2$)
$F1$	0.25	0.35	0.30
$F2$	0.75	0.65	0.70

population evolution. Three cases are designed: best, worst, and average, indicating the best, worst, and average performance for a sensor network under nodal demand uncertainty. In Table 5.2, for the best case, the minimum value is 0.25 among 0.25, 0.35, and 0.30 (the three $F1$ values in Table 5.1) and is selected to represent the best performance of actual $F1$ (since $F1$ is to be minimized) in a real application. The maximum value 0.75 of 0.65, 0.70, and 0.75 (the three $F2$ values in Table 5.1) is selected as the value of $F2$ (since $F2$ is to be maximized). Thus, in the best case, the $F1$ and $F2$ values, 0.25 and 0.75, are fed into the optimization; similarly, $F1$ and $F2$ values are 0.35 and 0.65 in the worst case, and 0.30 and 0.70 in the average case.

To evaluate the objectives $F1$ and $F2$, SQL sentences are constructed to query the database (including the sensor detection information under nodal demand realizations). SQLite is selected due to its capacity to execute SQL query sentences parallel in SHARCNET, which is critical since it dramatically reduces e-NSGA-II runtimes. The reduction in runtime is demonstrated in the Case Study section.

The parallel computing for e-NSGA-II is presented in Fig. 5.2. The basic idea is to parallel the quantification of objectives $F1$ and $F2$ for every individual of each population of e-NSGA-II, which is the most time-consuming unit in e-NSGA-II. The individuals are allocated evenly to the number of 'm' cores for evaluation. To schedule individuals among objectives evaluation cores, another core is dedicated to be solely responsible for scattering and gathering individuals to, and from, the m cores, and performing the other computational works of e-NSGA-II, with the exception of the evaluation of the objectives.

The best and worst cases will formulate range estimates for the actual Pareto front. To describe the 'distance' between two Pareto fronts, a metric is defined, which is a slight modification from Shen and McBean (2011b); to facilitate illustration, two Pareto fronts are denoted as A and B. The metric (distance between A and B) can be calculated in two steps: (i) compute points to Pareto front distance: calculate the Euclidean distance from a point in A to every point in B, pick up the minimum one as the distance to B; likewise to get the distance for each

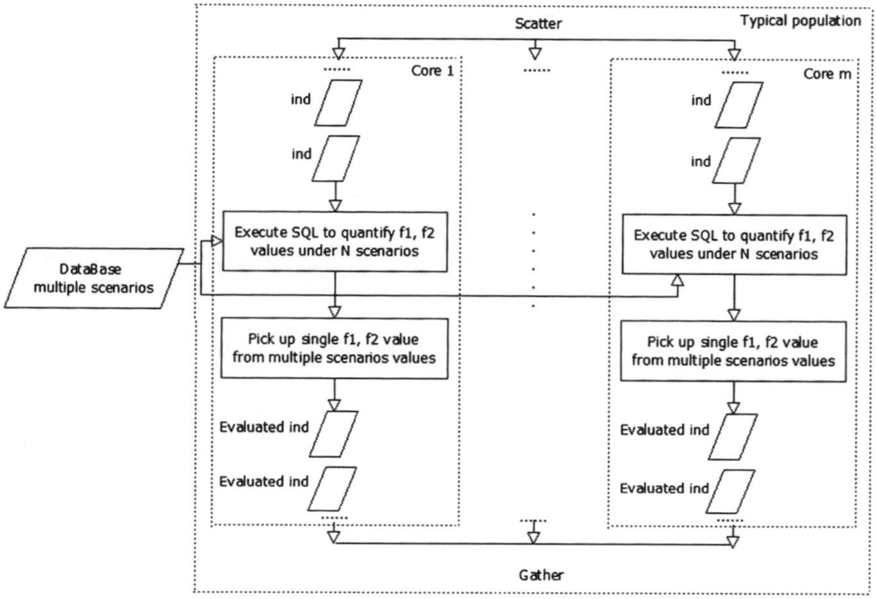

Fig. 5.2 Parallel computing of e-NSGA-II in SHARCNET

point in A to B, and (ii) compute Pareto front to Pareto front distance: set it as the median instead of the average value (used in Shen and McBean 2011b) of the distances from each point in A to B; this median can help to eliminate the impact of extreme values of points to Pareto front distances.

5.4 Case Study

The City of Goderich WDS is used herein, consisting of 285 nodes, 432 pipes, 1 reservoir, 2 tanks, and 1 pump. The wall clock time (optimization elapsed time) comparisons under serial and parallel computing are presented in Table 5.3. Specifically, to obtain a Pareto front, when 10 nodal demand realizations are

Table 5.3 Wall clock time comparisons of 10 and 20 scenarios under serial and parallel computing

Computational methods	Wall clock time	
	10 scenarios	20 scenarios
Serial	179.5 h	315.1 h
Parallel (51 cores)	3.6 h	6.4 h
Speed-up	49.9	49.9

applied for sensor network design, the serial computing would require 179.5 h to run e-NSGA-II; parallel computing can reduce the wall clock time significantly to 3.6 h by applying 51 cores. Note that one core is designated to scatter and gather individuals, and evolve (select, mutate, and crossover) current population; the other 50 cores are involved in evaluating the two objectives $F1$ and $F2$ with respective SQL sentences. Thus, the speed-up of parallel computing is 49.9 times (almost 50, the number of objectives evaluation cores) based on serial computing. With 20 demand realizations for sensor network design, the wall clock time is increased to 315.1 h from 179.5 h for serial computing due to 10 more scenarios are required to be evaluated; by applying 51 cores, the wall clock time is reduced to 6.4 h, or 49.9 times speed-up relative to the serial one.

To illustrate the trade-off between the two objectives $F1$ and $F2$ without any nodal demand uncertainty, only the 1st scenario is described. The Pareto front is presented in Fig. 5.3. Clearly, with increasing time delay, the sensor detection redundancy is increased; on the other hand, to obtain a shorter time delay, the redundancy has to be decreased, unavoidably causing larger false alarms of event detection.

With 10 demand realizations, the Pareto front under the best, average, and worst cases are developed and shown in Fig. 5.4. The points toward the upper left corner

Fig. 5.3 Pareto front with only the 1st scenario

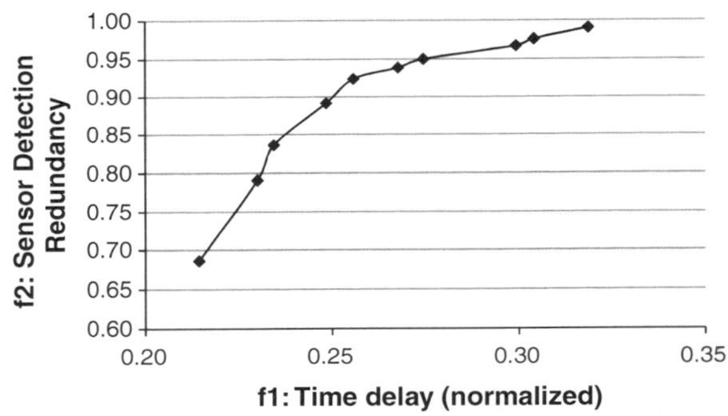

Fig. 5.4 Pareto fronts with 10 scenarios under three different cases

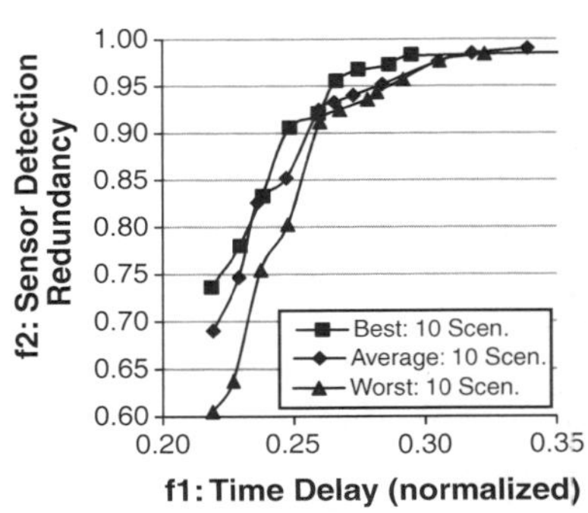

in Fig. 5.4 represent better performance since objective $F1$ is to be minimized, and $F2$ is to be maximized. The "Best 10 Scen" curve represents the best performance the sensor network can obtain under nodal demand uncertainty, and the "Worst 10 Scen" curve is the worst performance. These two curves represent a range of estimates where the actual Pareto front may lie, and the distance between the two curves is 0.0234. For an average case, the Pareto front resides within the range formulated by the best and worst performance curves.

By increasing to 20 scenarios, the Pareto fronts under the three cases are depicted in Fig. 5.5. Interestingly, the range is larger in comparison with the one in Fig. 5.4 (with distance 0.0306), due to 20 demand realizations expressing more nodal demand uncertainty than the ten scenarios. It suggests the e-NSGA-II must be run with more scenarios numbers to determine the number at which the range no longer expands, or the distance stops increasing. Comparisons between the Pareto fronts under 10 and 20 demand realizations are illustrated in Fig. 5.6, which demonstrates that minimal difference is found, indicating the Pareto front with other scenario numbers under the average case can serve as useful inferences.

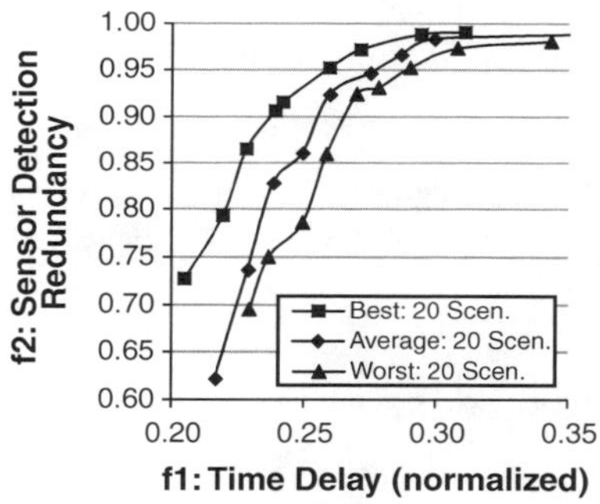

Fig. 5.5 Pareto fronts with 20 scenarios under three different cases

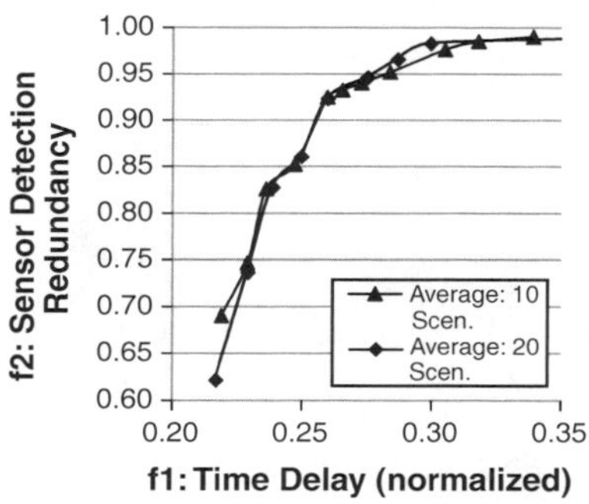

Fig. 5.6 Pareto fronts with different scenario numbers in average case

5.5 Conclusion

Optimal sensor placement is demonstrated with two objectives, time delay and sensor detection redundancy. The following points are concluded:

- Parallel computing provides a time-efficient way for database construction, for otherwise, huge runtimes will be involved, especially for large WDSs,
- Database management software SQLite is effective at supporting parallel execution of SQL sentences, and thus represents an efficient way to execute parallel SQL sentences,
- e-NSGA-II is paralleled by evaluating objectives in scattered processors, which is an efficient way to solve multiple optimization questions, particularly when the evaluation of objectives and constraints is very time-consuming,
- The constructed best and worst cases provide highly useful estimates of the range for the Pareto front in multiple objective optimizations under uncertainty, and thus better informs decision-makers than simply a single Pareto front curve with no reflection of uncertainty, and
- Nodal demand uncertainty represents a potentially important issue related to sensor placement. With more scenarios, the estimated range of actual Pareto front is enlarged, and may be relevant for rule characterization but the average case Pareto front is relatively constant.

Acknowledgments The financial support provided by NSERC Strategic Grand STPGP 336126-06 and the Canada Research Chair Program are gratefully acknowledged.

References

Austin RG, Choi CY, Preis A, Ostfeld A, Lansey K (2009) Multi-objective sensor placements with improved water quality models in a network with multiple junctions. World Environmental and Water Resources Congress, Great Rivers

Deb K (2000) An efficient constraint handling method for genetic algorithms. Comput Methods Appl Mech Eng 186:311–338

Deb K, Pratao A, Agarwal S, Meyarivan T (2002) A fast and elitist multiobjective genetic algorithm: NSGA-II. IEEE Trans Evol Comput 6(2):182–197

Kollat J, Reed P (2006) Comparing state-of-the-art evolutionary multi-objective algorithms for long-term groundwater monitoring design. Adv Water Resour 29(6):792–807

Murray S, Ghazali M, McBean E (2012) Real-time water quality monitoring: assessment of multi-sensor data using Bayesian belief networks. J Water Resour Plann Manage 138(1):63–70

Ostfeld A et al. (2008) The battle of water sensor networks (BWSN): a design challenge for engineers and algorithms. J Water Resour Plann Manage 134(6):556–568

Preis A, Whittle A, Ostfeld A (2009) Multi-objective sensor network model for integrated monitoring of hydraulic and water quality parameters. World City Water Forum, Incheon, Korea, 18–21 Aug 2009

Preis A, Ostfeld A (2008) Multiobjective contaminant sensor network design for water distribution systems. J Water Resour Plann Manage 134(4):366–377
Reed P, Kollat JB, Devireddy VK (2007) Using interactive archives in evolutionary multiobjective optimization: a case study for long-term groundwater monitoring design. Environ Model Softw 22(5):683–692
Shen H, McBean E (2011a) False negative/positive issues in contaminant source identification for water distribution systems. J Water Resour Plann Manage 137(3):243–248
Shen H, McBean E (2011b) Pareto optimality for sensor placements in a water distribution system. J Water Resour Plann Manage 137(3):243–248
Weickgenannt M, Kapelan Z, Blokker M, Savic D (2010) Risk-based sensor placement for contaminant detection in water distribution systems. J Water Resour Plann Manage 136(6):629–636

Chapter 6
Waterborne Transport Modeling of Radioactivity from the Fukushima Nuclear Power Plant Incident

William B. Samuels, Rakesh Bahadur and Christopher Ziemniak

Technical Terms and Definitions

ASW	Anti-Submarine Warfare
CMORPH	Climate Prediction Center Morphing technique
CMS	cubic meters per second
DOD	Department of Defense
DTRA	Defense Threat Reduction Agency
EROS	Earth Resources Observation and Science
GeoSFM	GeoSpatial Stream Flow Model
GIS	Geographic Information System
GRDC	Global Runoff Data Center
ICWater	Incident Command Tool for Drinking Water Protection
MEXT	Ministry of Education, Culture, Sports, Science and Technology
NAVOCEANO	Naval Oceanographic Office
NOAA	National Oceanic and Atmospheric Administration
NOAC	Naval Oceanography Antisubmarine Warfare Center
PET	Potential Evapotranspiration
SHARC	System for the Hazard Assessment of Released Chemicals
SRTM	Shuttle Radar Topography Mission
TEPCO	Tokyo Electric Power Company
USF	United States Forces Japan
USGS	United States Geological Survey

W. B. Samuels (✉) · R. Bahadur · C. Ziemniak
Science Applications International Corporation Center for Water Science and Engineering, 1710 SAIC Drive McLean, Virginia 22102, USA
e-mail: william.b.samuels@saic.com

R. Bahadur
e-mail: bahadurr@saic.com

C. Ziemniak
e-mail: ziemniakc@saic.com

R. M. Clark and S. Hakim (eds.), *Securing Water and Wastewater Systems*, Protecting Critical Infrastructure 2, DOI: 10.1007/978-3-319-01092-2_6,

6.1 Introduction

Japan has approximately 17,000 public water systems, of which 97.5 % serve less than 50,000 people. Of the systems that serve less than 50,000 people, 93.8 % serve less than 5,000 people. Therefore, most of the water utilities in Japan could be classified as small (JWWA 2009). Once water supplies and systems are established they may be vulnerable to drought, earthquakes and terrorism. Unlike droughts which are described as a creeping phenomenon, the damage associated with earthquakes is concentrated in time and space. In 1906 an earthquake in San Francisco caused numerous pipes to rupture and caused drowning of dozens of residents when broken water pipes flooded the Valencia hotel (Clark, Deininger, 2001). It was impossible to control the firestorms that spread through the area and entire buildings exploded in a huge firestorm during which temperature were reported to reach 2,000 °F (1093.2 °C).

In 1995, a major earthquake directly hit the city of Kobe, Japan. The quake lasted seconds and 4,069 people died, 14,679 were injured, and 222,127 people were moved into evacuation shelters. There were 67,421 fully collapsed structures of which 6,985 were burned to the ground and there was a city wide power failure and a nearly city wide water supply failure (Clark and Deininger 2001). On Monday, April 11, 2011 in the Hamadōri region of Fukushima, Japan an earthquake of 9.0 triggered tsunami waves that reached heights of up to 40.5 m (133 ft) in Miyako and which, in the Sendai area, travelled up to 10 km (6 mi) inland. At least 1.5 million households were reported to have lost access to water supplies after the earthquake but by March 2011, this number had fallen to 1.04 million (http://wikipedia.org/wiki/2011). In addition to water systems many other types of critical infrastructure were affected by the earthquake.

This paper describes the development of a model that was used to simulate the surface water transport of radioactivity released from the Fukushima Dai-ichi nuclear power plant after the tsunami and earthquake. It provides an example of an approach that might have direct application to the failure of other critical infrastructure such as water supply systems during natural or manmade disasters.

6.1.1 Earthquake and Tsunami

On March 11, 2011, the strongest earthquake in almost a century occurred off the northeast coast of Japan (see Fig. 6.1). More than 21,000 people were declared dead or missing in the devastation from this 9.0 earthquake and the ensuing tsunami. Numerous nuclear plants in Japan automatically shut down. The closest nuclear plant impacted by the tsunami, Fukushima Dai-ichi, was crippled, subsequently leading to releases of radioactive steam and waste water (Pullen et al. 2012). The events at reactors 1, 2, and 3 have been rated at Level 7 on the United Nations Atomic Energy Agency scale, which indicates major release of radioactive

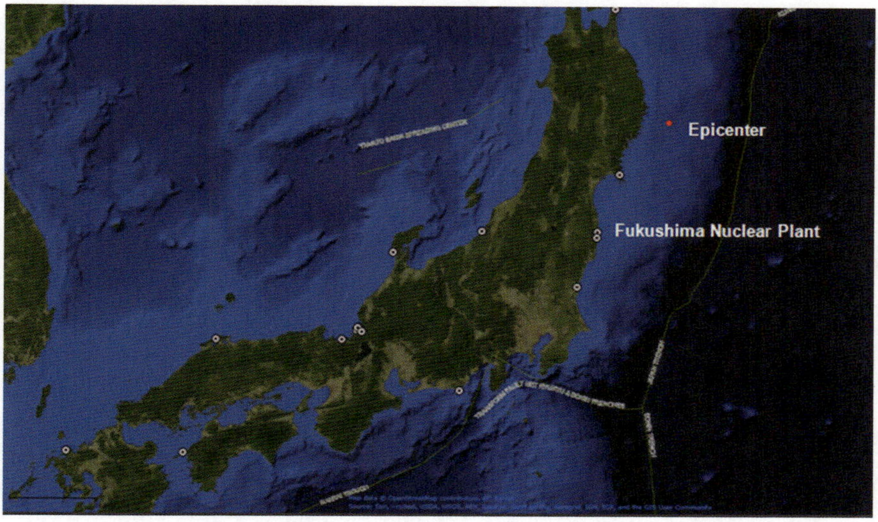

Fig. 6.1 Map showing the earthquake epicenter and the location of the Fukushima nuclear power plant

material with widespread health and environmental effects requiring implementation of planned and extended countermeasures. Operation Tomodachi was initiated by the United States Armed Forces to assist and support Japan in disaster relief. The Defense Threat Reduction Agency's (DTRA) Reachback Team supported this operation by providing radioactive waterborne hazard modeling. DTRA's Technical Reachback capability provides 24/7 operational support and planning for consequence assessment collateral effects as well as force protection. Reachback is a team of hazard modeling subject matter experts with the ability to leverage expertise across the Department of Defense (DOD).

6.1.2 Application of Modeling Tools

DTRA's initial response was an intensive 2 week multi-disciplined, multi-organizational effort to adapt two existing waterborne transport models (Samuels et al. 2012; Ward 2012): Incident Command Tool for Drinking Water Protection (IC-Water) and the System for the Hazard Assessment of Released Chemicals (SHARC), for use during Operation Tomodachi (Meris and Trigg 2012). This effort represents the use of science and technology tools and subject matter expertise to provide urgent decision-support information to the United States Forces Japan (USFJ) Commander. Ocean environmental data were provided by Navy Oceanography Office (NAVOCEANO), Naval Oceanographer Anti-Submarine Warfare (ASW) Center (NOAC), at Stennis, MS. On-shore and off-shore monitoring data from the Tokyo Electric Power Company (TEPCO) and the Ministry of Education, Culture, Sports, Science and Technology (MEXT) were used to characterize the source term.

6.2 Methodology

6.2.1 Hydrologic Model

The first step in modeling the surface water transport of radionuclides from the Fukushima Nuclear Power Plant was to build a hydrologic model of the watersheds surrounding the plant. This was accomplished using the GeoSpatial Stream Flow Model (GeoSFM) developed by the U.S. Geological Survey (Asante et al. 2007). GeoSFM consists of a GIS-based module used for model input, data preparation, and the rainfall-runoff simulation model (see Figs. 6.2 and 6.3). The inputs to GeoSFM include hydrologic and terrain data sets from remotely sensed data. This would include: digital elevation model, land use, soils, catchment boundaries, stream network, precipitation, and evapotranspiration. Parameterization of the basins' hydrologic properties is accomplished through the use of three data sets describing the surface topography, land cover, and soils. In addition, the literature searches are performed to gather additional hydrologic data to fill data gaps (this may yield additional local data that can be used to enhance the digital

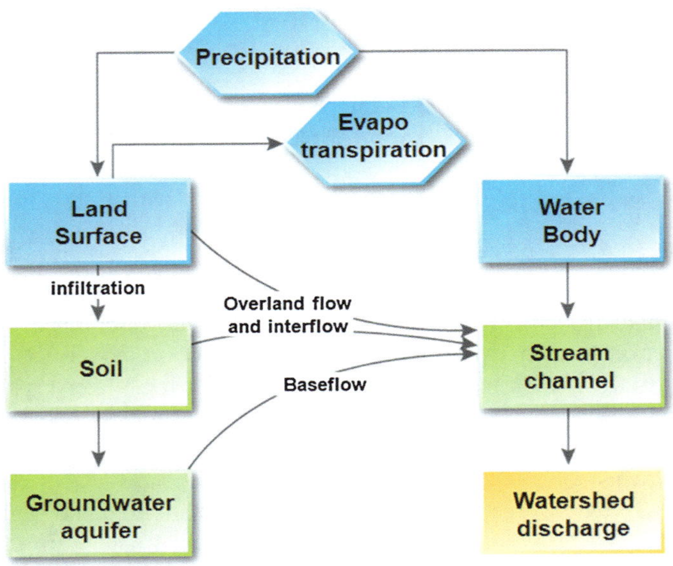

Fig. 6.2 GeoSFM schematic diagram (Asante et al. 2007)

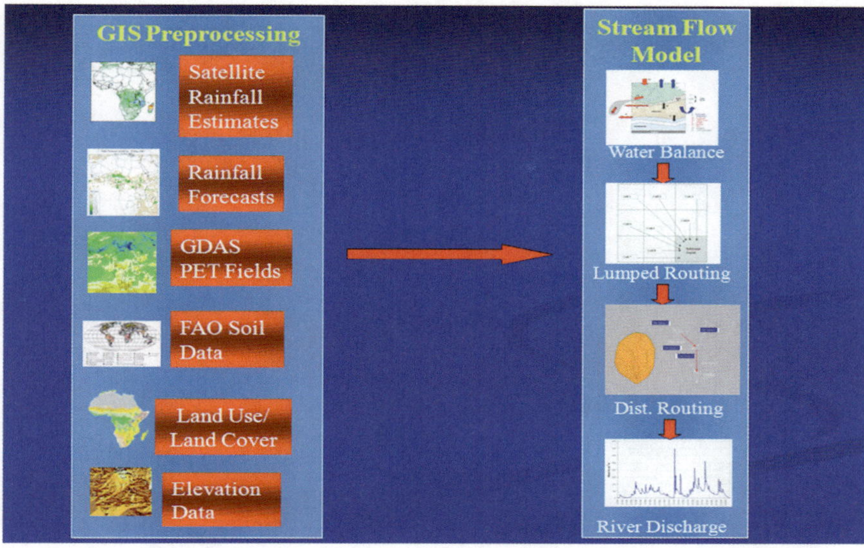

Fig. 6.3 GeoSFM data inputs and model processes (Asante et al. 2007)

data sets and to perform model calibration). Once this process is complete, river discharge is calculated using hydrologic modeling techniques and a geospatial model.

The USGS Earth Resources Observation and Science (EROS) Data Center has developed a hydrologic modeling system (Asante et al. 2007) which consists of an operational data processing system and a geospatial stream flow model. GeoSFM's data processing system generates evapotranspiration and precipitation data from a variety of remotely sensed sources on a daily basis. To allow for rapid implementation in data scarce environments, widely available terrain (WWF 2007), soil (FAO 1997) and land cover (USGS 2006) data sets are used for model setup and initial parameter estimation. GeoSFM performs geospatial preprocessing and postprocessing tasks as well as hydrologic modeling tasks within an ArcView GIS environment. Figures 6.2 and 6.3 show, respectively, a schematic diagram of the model and the data inputs/model processes. Precipitation is an input to both the land surface and the river network. Evapotranspiration accounts for loss from vegetation and the soil layer. Overland flow, interflow, and baseflow provide inputs to the stream reaches where flow is routed downstream.

6.2.2 Data Inputs

Much of the input data for GeoSFM are derived from satellite observations. For example, the terrain data used to build the catchments and river network,

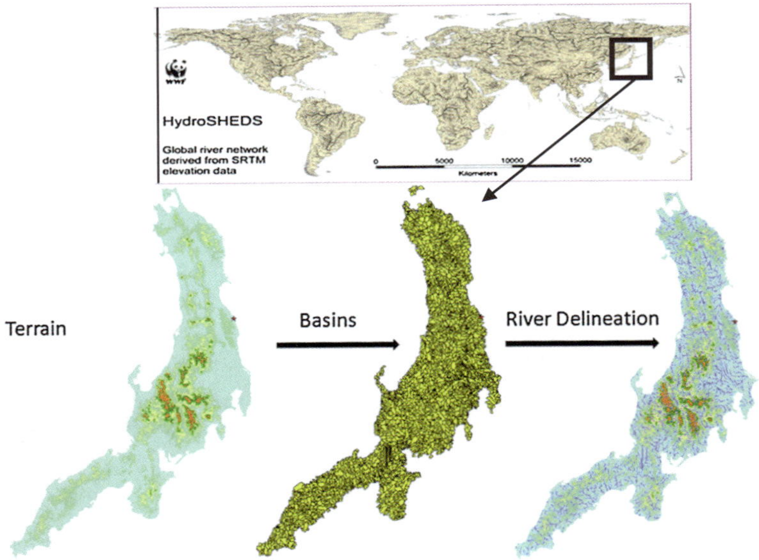

Fig. 6.4 HydroSHEDS data used to derive catchments and the river network for Japan

HydroSHEDS, are derived from the Shuttle Radar Topography Mission (SRTM) at 3 arc-sec resolution. The original SRTM data have been hydrologically conditioned using a sequence of automated procedures. Existing methods of data enhancement and newly developed algorithms have been applied, including void-filling, filtering, stream burning, and upscaling techniques. Manual corrections were made where necessary. Preliminary quality assessments indicate that the resolution of HydroSHEDS significantly exceeds that of existing global watershed and river maps (WWF 2007). HydroSHEDS vector and raster data sets include: stream networks, watershed boundaries, drainage directions, and ancillary data layers such as flow accumulations, distances, and river topology information (see Fig. 6.4).

Rainfall-runoff calculations are dependent on both land use/land cover and soils data. The Digital Soil Map of the World (FAO 1997) was derived from an original compilation at 1:5,000,000 scale and contains attributes for the soil associations that are used to set hydraulic parameters that govern interflow, soil moisture content, and deep percolation to the ground water table. Rates at which subsurface layers release water to the stream network also depend on these physical soil attributes. Global Land Use/Land Cover which is a factor in determining roughness coefficients and overland flow was provided by the USGS (USGS 2006). Figures 6.5 and 6.6, respectively, show the maps of these databases for Japan.

Daily precipitation was obtained from the National Atmospheric and Administration (NOAA) Climate Prediction Center Morphing technique (CMORPH) (Joyce et al. 2004). CMORPH produces global precipitation analyses at very high

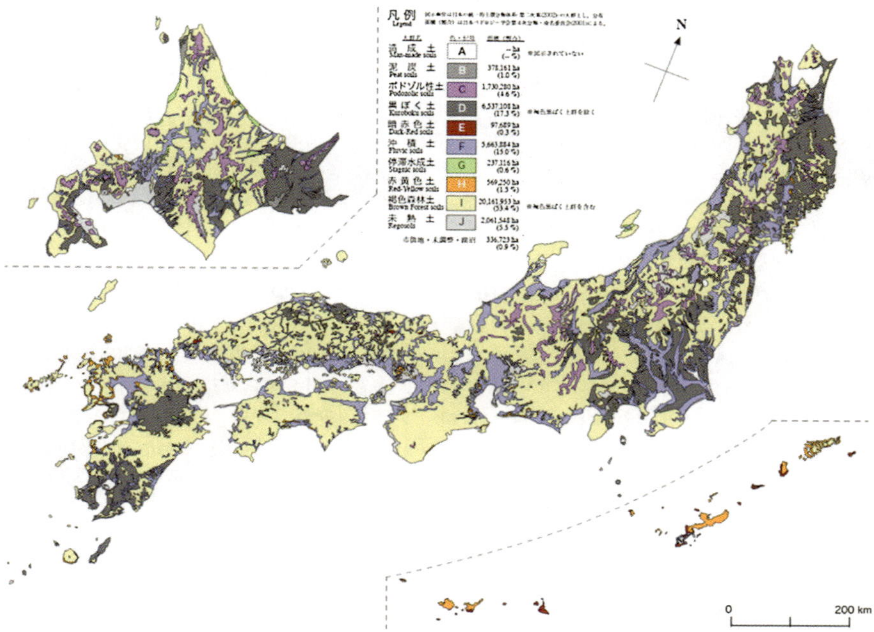

Fig. 6.5 Digital soils map for Japan

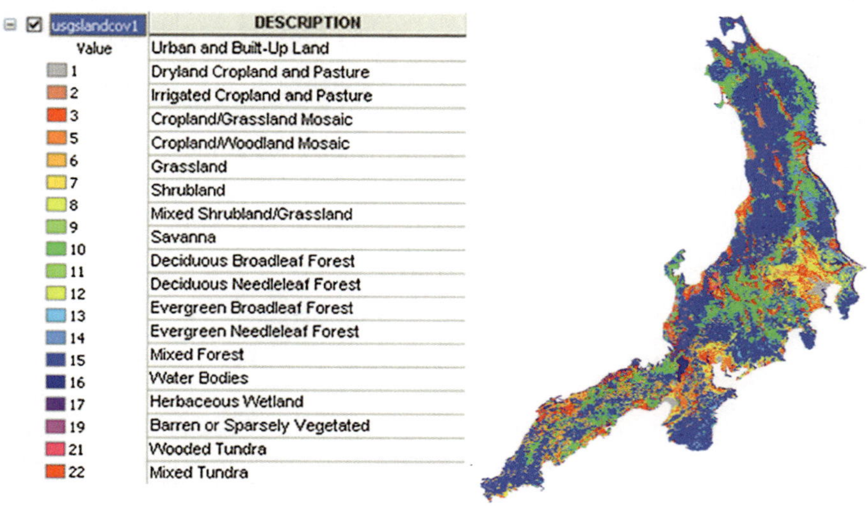

Fig. 6.6 Land use/land cover for Japan

spatial and temporal resolutions. This technique uses precipitation estimates that have been derived from low orbiter satellite microwave observations exclusively, and whose features are transported via spatial propagation information that is obtained entirely from geostationary satellite IR data. CMORPH is not a precipitation estimation algorithm but a means by which estimates from existing microwave rainfall algorithms can be combined. Therefore, this method is extremely flexible such that any precipitation estimates from any microwave satellite source can be incorporated. CMORPH data are available in GIS format on a ¼ × ¼ degree grid.

Daily PET was obtained from the USGS Global Data Assimilation System. PET is the maximum extraction rate from soil and is based on air temperature, atmospheric pressure, wind speed, relative humidity, and solar radiation (long wave, short wave, outgoing, and incoming). The daily PET is calculated on a spatial basis using the Penman–Monteith equation using the formulation of (Shuttleworth 1992). GeoSFM converts PET to actual daily evapotranspiration based on antecedent soil moisture conditions. PET is available on a 1 × 1 degree grid (USGS 2007).

6.3 Flow Modeling

The GeoSFM model was used to generate hydrologic parameters for the modeled area. Benchmarking of the hydrology (flow) model was done using data from the Global Runoff Data Center (GRDC) (GTN-R 2007). Figure 6.7 shows the location of the flow observation stations near the plant.

The GeoSFM model was run with daily precipitation and evapotranspiration data over a 1 year period of record to establish base flows to compare with the observed data downloaded from the GRDC. Average daily flow was used for the comparison between the observations and model output as shown in the Table 6.1. Visual inspection of the results show close agreement in the average flow values.

6.4 Calculation of Travel Time and Dilution

The ICWater model (Samuels et al. 2006) was used for the radionuclide transport. The hydrologic inputs for the model were provided by the GeoSFM. At the time of the incident observed radionuclide data were available only for air, dust, soils, leafy vegetables, and at a few places for drinking water in a 50 km (31.1 mi) radius around the plant. Because of the lack of surface water quality monitoring data, a two step approach using a conservative tracer as the source term was used for the initial radionuclide transport modeling to determine the following parameters:

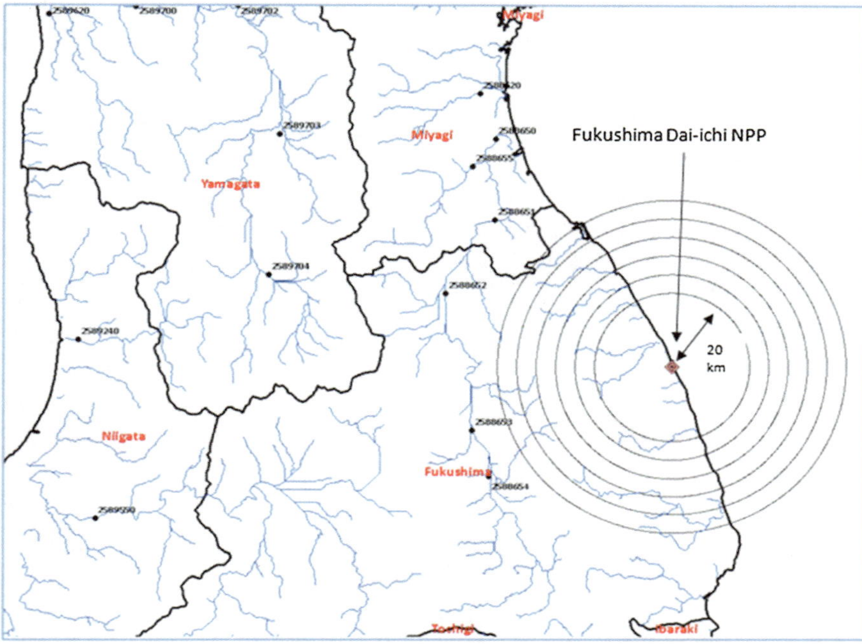

Fig. 6.7 Map showing flow observation stations on rivers near the plant

Table 6.1 Comparison between the observed and simulated flows in several rivers near the plant

Observation Station	Observed Flow (CMS)	Modeled Flow 2006 (CMS)
2588650	140.87	118.6
2588651	124.5	92.3
2588652	107.46	71.8
2588653	64.85	46.3
2588654	25.51	31.6
2588655	27	17.7

- Time of travel along major river reaches in the 50 km (31.1 mi) radius zone
- Dilution of contaminants in the 50 km (31.1) radius zone

ICWater was run for different river reaches and the results of the simulation runs are summarized in Fig. 6.8. Travel times ranged from 6.6 to 22.5 h and dilution ratios ranged from none (plug flow) to 14.

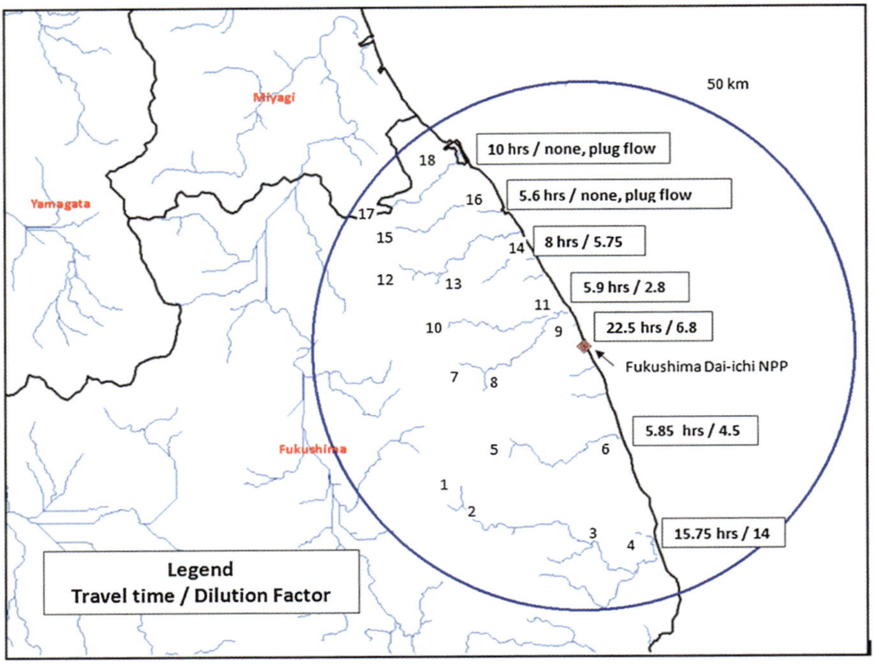

Fig. 6.8 Map showing the river IDs, travel time and dilution factor at the coastal discharge of each river

6.5 Radionuclide Transport

To perform radionuclide transport in rivers from nuclear fallout on the landscape, the following parameters were input to ICWater:

- An area source term was developed for two watersheds northwest of the plant-based on soil monitoring data
- Daily rainfall over the area was used to calculate surface runoff to mobilize the radioactivity into the rivers

For this analysis, the data for radionuclide monitoring in soil samples were downloaded on April 24, 2011 from Japanese Ministry of Education, Culture, Sports, Science and Technology (MEXT 2011). Figure 6.9 shows concentration (in soil, 5 cm (2.0 inch) (surface layer) of the radionuclide Cesium-137 at sampling locations near the plant. These soil concentrations were used to establish the initial conditions for the runoff of radioactivity from two watersheds northwest of the plant. These two watersheds were selected since the Cesium-137 levels (soil)

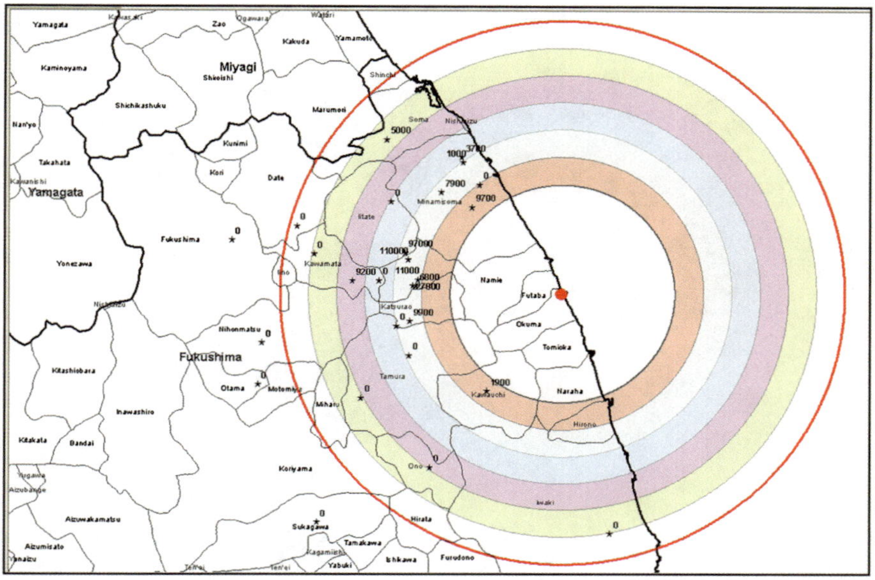

Fig. 6.9 Map showing Cs^{137} concentration (Bq/kg) at the soil monitoring locations on April 24, 2011. Prefecture boundaries are also shown

were highest, respectively, 100,000 Bq/kg and 9,900 Bq/kg. This simulation was performed using actual rain event on April 24, 2011. The rain data were downloaded from NOAA's CMORPH data repository (CMORPH 2011). The rain file contains a global estimate of 24 h accumulated precipitation in GIS ESRI GRID format with a spatial resolution of 0.25 degrees. The ICWater model was run for 24 h and the downstream trace from the simulation is shown in Fig. 6.10. Table 6.2 shows the results of the two simulations from location 1 (north river) and location 2 (south river). The breakthrough curves at the coastal discharge for each river are shown in Fig. 6.11.

These breakthrough curves were used as source term input to the SHARC coastal transport model to calculate the seaward transport and dispersion of radionuclides from river discharge. Example results from these SHARC runs are shown in Fig. 6.12.

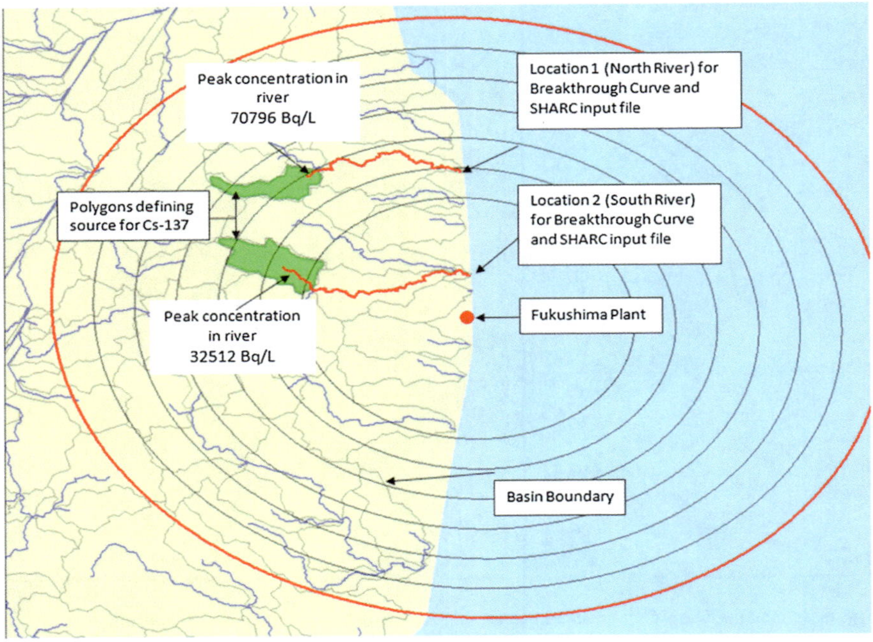

Fig. 6.10 Map showing the watersheds used for source term characterization based on the soil monitoring data

Table 6.2 Results from model simulations of radionuclide runoff and transport downstream

Location	North River	South River
Soil monitoring in watershed	100,000 Bq/kg	9900 Bq/kg
Peak conc. (headwaters)	70,786 Bq/L	32,512 Bq/L
Peak conc. (Ocean discharge)	12,377 Bq/L	2,228 Bq/L
River dilution factor	5.7	14.6

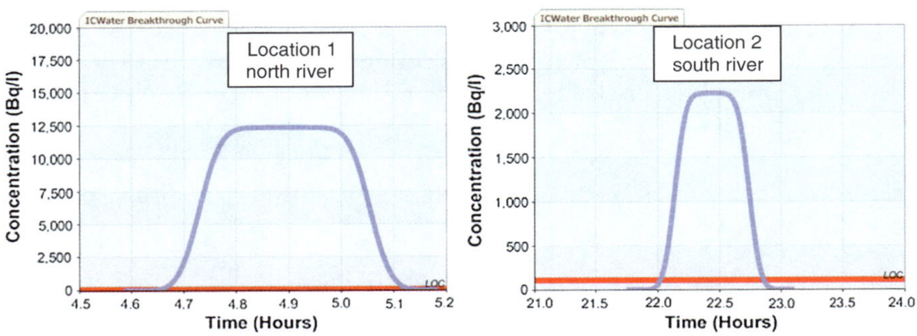

Fig. 6.11 Breakthrough curves for locations 1 (north river) and 2 (south river)

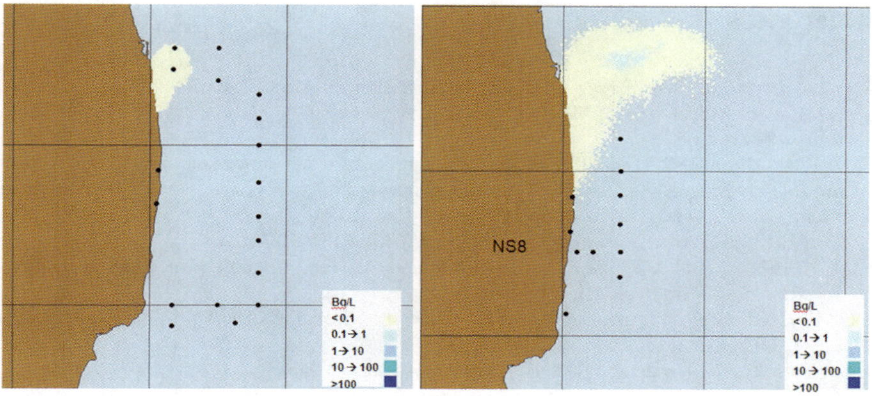

Fig. 6.12 Transport and dispersion of Cesium-137 using ICWater breakthrough curves as input to the SHARC coastal model. Coastal monitoring stations are also shown (Ward 2012)

6.6 Summary and Conclusions

The damage to the Fukushima Dai-ichi Nuclear Power Plant resulted in a significant release of radiological material to the atmosphere, surrounding land and local marine waters. Ground deposition (especially from the NW corridor) may have contributed to significant hazards from water runoff in the vicinity of the plant. A proof-of-concept integration of the ICWater riverine model and SHARC coastal model demonstrated an effort to account for runoff, transport and coastal river discharge from radioactive "hotspots" near the plant. ICWater output was used as a source term for input to the SHARC coastal ocean transport model. Shortly after the impacts of the earthquake and resulting tsunami on the power plants were known, these models were providing daily operational forecasts to the DTRA Reachback Team. An effort is underway by DTRA to transition this Fukushima ad-hoc response to a stand-by capability so it would be ready in advance if and when another event occurs. In summary, the main tasks achieved by this project were:

- A river model was built for Japan
- Modeled river flows were benchmarked with historical gage observations
- Travel times and dilution factors were calculated for rivers near the plant
- The radionuclide source term was estimated from monitoring data collected on land and in the sea near the plant
- Downstream tracing and breakthrough curves were calculated for radionuclide runoff from watersheds
- Time series concentration data were exported from ICWater for input to the SHARC coastal ocean model

References

Asante KO, Artan GA, Pervez S et al (2007) Technical Documentation of the Geospatial Streamflow Model (GeoSFM), U.S. Geological Survey U.S. Geological Survey Open-File Report 1441, p 65

Clark RM, Deininger RA, Minimizing the vulnerability of water supplies to natural and terrorist threats. In: Proceedings of the American water works association's IM Tech conference Atlanta, GA, 8–11 April 2001, pp 1–20

CMORPH (2011) ftp://ftp.cpc.ncep.noaa.gov/fews/CMORPH/GIS/

FAO, (1997) Digital soil map of the world and derived properties on CD-ROM, http://www.fao.org/AG/agl/agll/dsmw.htm

GTN-R, 2007 Global Terrestrial Network for River Discharge, http://gtn-r.bafg.de/servlet/is/Entry.2492.Display/

Japan water works association (2009) Water supply data in Japan

Joyce RJ, Janowiak JE, Arkin PA et al (2004) CMORPH: A method that produces global precipitation estimates from passive microwave and infrared data at high spatial and temporal resolution, J Hydromet 5: 487–503 http://en.wikipedia.org/wiki/2011

Meris RG, Trigg J (2012) Operation Tomodachi radiological waterborne hazards: Littoral and open ocean dispersion. In: Proceedings in the American Meteorological Society, New Orleans, LA, 22–26 January 2012

(2011) MEXT http://radioactivity.mext.go.jp/en/monitoring_around_FukushimaNPP_monitoring_out_of_20km/

Pullen J, Chang J, Hanna S, (2012) Air/sea transport, Dispersion and fate modeling for the Fukushima crisis. In: Bulletin of the American Meteorological Society. (in press)

Samuels WB, Bahadur R Surface water transport of radioactivity from the Fukushima nuclear power plant incident. In: Proceedings at the American Meteorological Society, New Orleans, LA, 22–26 January 2012

Samuels WB, Bahadur R, Monteith M, Amstutz D, Pickus J, Parker K, Ryan D (2006) NHD, Riverspill and the development of the incident command tool for drinking water protection (ICWater). Water Resour Impact 8(2):15–18

Shuttleworth, WJ, (1992) Evaporation. In D. Maidment (ed.), Handbook of Hydrology. McGraw-Hill

USGS, 2006. Global Land Cover Characterization (GLCC), http://eros.usgs.gov/products/landcover/glcc.html

USGS, 2007. Global Potential Evapotranspiration (PET), http://igskmncnwb015.cr.usgs.gov/Global/product.php?image=pt

Ward M (2012) Rapid development and implementation of a waterborne radiological transport and fate model in support of the recent events at the Fukushima nuclear power plant. In: Proceedings at the American Meteorological Society, New Orleans, LA, 22–26 January 2012

WWF (2007) HydroSHEDS Overview, http://www.worldwildlife.org/hydrosheds

Chapter 7
Quantitatively Assessing Water Asset Reliability in the Netherlands: 15 Years of Experience

R. H. S. Beuken, H. de Kater and J. H. G. Vreeburg

Technical Terms and Definitions

ALEID	Network model developed by KWR
BTO	Joint research program of the Dutch water companies
FMECA	Failure mode effects, and criticality analysis techniques InfoWorks-proprietary network model
KWR	Research organization owned by the drinking water companies in the Netherlands
NRW	Non revenue water
PWN	Dutch water company
SynerGEE	Proprietary network model
Vewin	Association of Drinking Water Companies in the Netherlands
WML	Watermaatschappij Limburg (Dutch Water Company)

Some of the material in this chapter was presented at LESAM 2009—the Leading Edge conference on Strategic Asset Management—held 11–13 November 2009, Miami, USA.

R. H. S. Beuken (✉) · J. H. G. Vreeburg
KWR Watercycle Research Institute N.V, Nieuwegein, The Netherlands
e-mail: Ralph.Beuken@kwrwater.nl

J. H. G. Vreeburg
e-mail: JanVreeburg@kwrwater.nl

H. de Kater
Evides N.V, Rotterdam, The Netherlands
e-mail: H.deKater@evides.nl

R. M. Clark and S. Hakim (eds.), *Securing Water and Wastewater Systems*, Protecting Critical Infrastructure 2, DOI: 10.1007/978-3-319-01092-2_7,

7.1 Introduction

Drinking water in the Netherlands is supplied by ten water companies. One of these companies is municipality based, and the other nine are limited liability companies with the associated municipalities and provinces as shareholders. The total annual production of drinking water is 1,141 million m^3 (40,294 million ft^3), which is distributed to 7.66 million connections of which 96 % are metered. The water quality is of a high standard and distributed without chlorine. The total network length is 117,000 km (72,704 mi) and the non-revenue water (NRW) is estimated at 5 %. The average price of drinking water is 1.14 €/m^3 (0.03 €/ft^3) (excl. taxes). For all figures presented in this paragraph, see www.vewin.nl.

Drinking water companies have the responsibility for providing their customers with sufficient quantities of potable drinking water at adequate pressures. The quality of drinking water, in the Netherlands, is based on a large number of parameters which are specified in legal standards. The legal standards governing quantity and pressure are much less detailed then those for quality. In the previous Dutch Drinking Water Act, only general descriptions were given, stipulating that water companies are required to supply water with sufficient quantity and pressure.

Toward the end of the 1980s, a number of problems highlighted the need to establish a quantitative regulatory framework for providing drinking water. These included:

- Several disasters which received a great deal of public attention, such as the Tsjernobyl nuclear melt-down and the fire at Sandoz, which resulted in large amounts of toxic chemicals polluting the River Rhine.
- The expansion of drinking water networks due to a merging of water companies and closing of smaller water production plants.
- A more critical attitude on the part of customers toward water supply interruptions.

In order to deal with these concerns the Association of Drinking Water Companies in the Netherlands, the Vewin, established a committee in 1989, with the goal of developing a method for evaluating drinking water supply reliability. In 1994, a regulatory framework proposed by the committee was adopted, here referred to as the Reliability Framework. This framework requires every water company to develop a so-called Supply Plan that describes how drinking water is to be supplied under normal conditions and how water would be supplied in case of a disturbance. The plan also prescribes a quantitative approach for assessing the reliability of the supply system.

The Dutch drinking water sector, which is characterized as publicly owned, but privately managed, embraces the idea of self-regulation. Regulation initiated by the sector, has the advantage that it is realistic and well suited to the sectors needs. During the past 15 years, water companies in the Netherlands have applied the Reliability Framework and have made, where necessary, improvements. The challenge is, however, to convince government authorities that regulations initiated by the drinking water sector do indeed encourage an improved supply to the public. In the recently adopted Drinking Water Act, the basic principles of the framework have been included, which indicates that the Government has confidence in this self-regulating principle.

In preparation for the new Drinking Water Act, KWR, the research organization owned by the drinking water companies in the Netherlands, evaluated the use of this quantitative guideline by the water companies. Afterwards, a workshop was organized to facilitate sharing of knowledge and practices on this subject.

This paper briefly describes the framework and the quantitative guidelines. A more comprehensive description is given in Vreeburg et al. (1994, 1998). Later in the manuscript we provide an overview of the experiences of the Dutch water companies with the guideline and provide examples to illustrate its application.

7.2 Framework Supply Plan

The Supply Plan is the document in which a water company describes how the drinking water supply is guaranteed under three circumstances:

1. Supply under normal conditions: In this section, water companies provide a description of the water system and indicate how standards for quantity and pressure are met. This section also contains measures and procedures on how to react to failures. Failures are defined as the breakdown of a water system element which can be repaired within 24 h or which have only local effects.
2. Supply in case of a breakdown of one major element for more than 24 h: This section deals with the breakdown of major elements of the water supply system that cannot be repaired within 24 h. These events are referred to as calamities. Due to their low chance of occurrence and the high impact of their effects, these types of breakdowns cannot be considered as normal events. Examples include a burst of a major water main or the breakdown of a pumping station. The quantitative guidelines are intended to assess the reliability of supply, when a calamity occurs.
3. Supply in case of a disaster: This type of circumstance refers to events in which the public water supply is interrupted on a large scale and for more than 24 h. Examples include the breakdown of a complete water treatment plant or the breakdown of multiple elements in the water distribution network.

Other aspects to be described in the Supply Plan include a communications plan and procedures to be followed in order to provide customers with emergency water. If the supply of drinking water through the normal distribution network is not possible, water companies are obligated to provide customers, with at least 3 L of emergency water per person per day within 24 h. The Supply Plan is updated regularly, typically every 4 years, and must be approved by the Inspectorate of the Ministry for Housing, Spatial Planning, and the Environment. The framework is not meant to be a rigid instrument. Therefore, it is possible to discuss deviations with the Inspectorate if the outcome of the reliability assessment would result in investments which are beyond the capability of the water company.

7.3 Reliability Assessment

A basic assumption underlying a reliability assessment is that every part of a water system may fail, even though the probability may be low. Therefore, the assessment focuses only on the effects of a failure and cannot be regarded as a full risk analysis. When the assessment concept was established in the 1990s, little information was available on the probability of failure.

In the assessment process every element is systematically taken out of service and the effect on water supply is evaluated. The criterion for evaluation is defined as follows: *In case of failure of one element of the drinking water supply system, the remaining capacity of supply to centers of demand should be at least 75 % of the maximum daily demand.* It has to be noted that when this criterion was defined, network models generally consisted of the major mains, therefore the reliability assessment of the water distribution network is restricted to the most important mains. However, it is advisable to use a detailed model, since smaller mains can provide a significant contribution to transporting the required amount of water to centers of demand.

The main terms in this definition include:

- *Element*: a part of the drinking water supply system that can be isolated. For example, a pipe segment between valves or an element of a treatment plant that can be bypassed or completely isolated. As the effects of calamities are assessed, the assessment focuses only on the parts of the system that cannot be repaired within 24 h.
- *Centers of demand*: a cluster of connections with a demand equivalent to 2000 house connections.
- *Remaining supply capacity*: the amount of water that can be supplied to the most severely affected centre of demand.
- *75 % of the maximum daily demand*: the assessment is based on a 24 h demand cycle, meaning that during peak hours the supply can be less than 75 % of the hourly demand if this is compensated during hours with a lower demand. The value of 75 % is based on the assumption that during a calamity, due to effective communication, daily water use can be reduced to 75 % during days with peak demand.

A hydraulic network model is indispensable when assessing the capability of a system for complying with the reliability criterion. The hydraulic model used in this assessment must be:

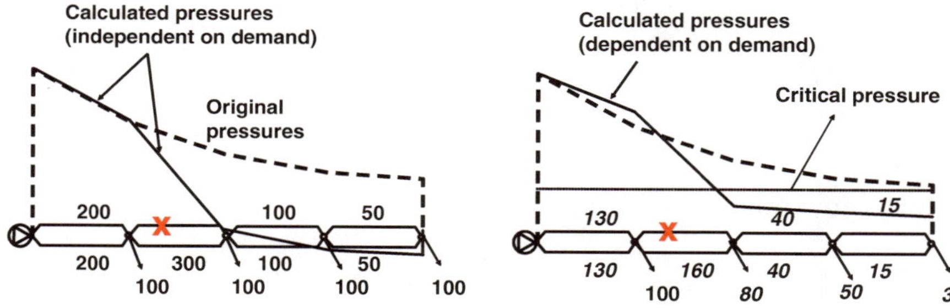

Fig. 7.1 A simple network consisting of 4 identical loops, with a total demand of 400 m^3/h. At the *left*, as a result of a pipe break, a model with pressure independent demand calculates a high hydraulic resistance in the neighboring pipe, resulting in low (and even negative) pressures at the more remote nodes. At the *right*, due to pressure dependent demand, all demands are decreased when the pressure is below a critical value. This results in a more realistic model, where the available water is spread over more connections

- capable of making a 24-hour calculation;
- capable of calculating the remaining supply capacity;
- organized so that all major elements are modeled separately;
- organized so that each separately modeled demand is equivalent to that of about 2000 households.

An important feature in the reliability assessment is the principle of pressure-dependent demand, meaning that if a low pressure occurs the supply decreases and the original demand is not fully satisfied. It is assumed that below a certain threshold (mostly set at 200 kPa (29.0 psi)) supply has a linear relation with the pressure. This implies that in case of a failure the amount of available water can be distributed over more connections. The dependency of demand on local pressure requires an extraiteration step in the calculation. The principle is illustrated in Fig. 7.1.

The assessment is performed by applying a n-1 approach to the major elements of the drinking water system. For reasons of efficiency, a two step approach is chosen. At the first step, for the maximum demand situation each relevant element is shut-off successively and a static calculation is made to evaluate the consequences for supply at centers of demand. If in this static calculation the supply criterion is not met, a dynamic (24 h) calculation is performed. If the total daily demand in a cluster is below 75 %, the element is labeled as critical and measures are to be taken.

Typical measures are the introduction of redundancy for specific elements, connections to neighboring distribution networks, splitting up of production plants into separate lines, power back-up for pumps and production plants, installation or manipulation of valves or assuring repair within 24 h.

In the 1990s, nearly all water companies in the Netherlands used the network model ALEID. This model was developed by KWR and especially equipped for making reliability assessments. Currently, most water companies in the

Netherlands use more sophisticated and commercial packages (most popular in the Netherlands are InfoWorks and SynerGEE). These packages perform comparable calculations to the ALEID model.

7.4 Evaluation of 15 Years of Reliability Assessment

In 2008 within the Joined Research Programme of the Dutch water companies (BTO), KWR evaluated the use of the framework for reliability assessment by the water companies (van den Boomen 2008). All of the water companies reported that they regard the framework as a useful instrument to identify those parts of the supply system where failures can have relative high consequences. Although, water companies apply the framework, some have their own company specific interpretation.

Most water companies do not see the need to extend the assessment within a national legal framework to include the probability of failure, in order to transform it into a full risk analysis. Although there is currently more information available concerning the probability of failure of system components, this information relates mostly to distribution main installation. However, little information is available on the probability of more complex incidents which could result in an emergency situation.

In the period from 1994 to 2000, all Dutch water companies conducted a reliability assessment of their supply systems. The Supply Plan is updated every 4–6 years. Since water supply systems do not change frequently, this schedule for updating the Supply Plan is found acceptable.

After conducting a reliability assessment, the supply system of one company was found to be fully reliable. Other companies made modifications to their systems, of which the most common were:

- the construction of redundant trunk mains;
- the construction of connections to other supply zones;
- increasing the capacity of treatment steps, pumps, or reservoirs;
- the separation of production plants (including energy supply) into parallel treatment trains.

In addition to structural modifications, water companies also took non-structural measures in order to be prepared for incidents and to reduce the period of interruption, such as:

- development of policies and procedures for dealing with incidents;
- strategic stocks of the most important and vulnerable parts;
- contracts for rapid intervention with specialized contractors;
- training on the preparedness for calamities.

7.5 Examples of Reliability Assessment

During a workshop held in November 2008, water companies presented applications of the reliability assessment (Beuken and Vloerbergh 2009). Four examples are presented in this paper.

Evides supplies drinking water to the South-West of the Netherlands, including the city of Rotterdam. The production plant Berenplaat has an annual production of approximately 100 mil m^3 (3,530 mil ft^3) and is the largest treatment plant in the Netherlands. Because Berenplaat is so large, if it fails, then other production plants will not be able to backup the required demand. In 2006, a major renovation of the plant was completed, which included a major effort to ensure the reliability of supply from this plant.

The original configuration and lay-out of the Berenplaat plant was not optimal. Although treatment trains in the plant were separated, an incident could result in a failure of both treatment trains. For example, Berenplaat is located next to one of the tributaries of the River Rhine and in case of a severe flood major failures could occur since important equipment was located below ground. An assessment of the Berenplaat plant resulted in the following major improvements :

- the total production capacity of 18,000 m^3/h (63,5580 ft^3/hr) was split into two fully independent treatment trains;
- all electrical installations were relocated aboveground;
- fire detection and extinguishing capability was installed;
- an additional clear water reservoir was constructed.

Much attention was given to fully separate the treatment trains. Both treatment trains have separate power supplies (including back-ups), dosing facilities and control installations. Each treatment train feeds a separate clear water reservoir and a pumping station and these pumping stations are located in separate buildings.

The reliability assessment prescribes that in all centers of demand, 75 % of the daily maximum demand must be satisfied. Since this is required in the nearby town of Rotterdam as well in the more remote centers of demand, the absolute minimum capacity of Berenplaat must be more than 75 %. The assessment showed that Berenplaat must be able to provide a minimal demand of 13,000 m^3/h (459,160 ft^3/hr), or 87 % of the average hour of the maximum day.

The Berenplaat plant is designed so that in case of a breakdown of a treatment train, the capacity of the remaining treatment train can be increased from 9,000 m^3/h (317,880 ft^3/hr) to 13,000 m^3/h (459,160 ft^3/hr). The treatment process consists of the steps: micro-sieves, sludge blanket clarifiers, double layer filtration, UV-disinfection, and activated carbon filtration. It is possible to split this process into two parallel treatment processes. In that case, one process consists of micro-sieves, sludge blanket clarifiers and double layer filtration (Treatment 1), and the other of UV-disinfection and activated carbon filtration (Treatment 2). In this emergency situation additional chlorination will be applied. Obviously, in this situation the drinking water quality will not meet the companies' standards.

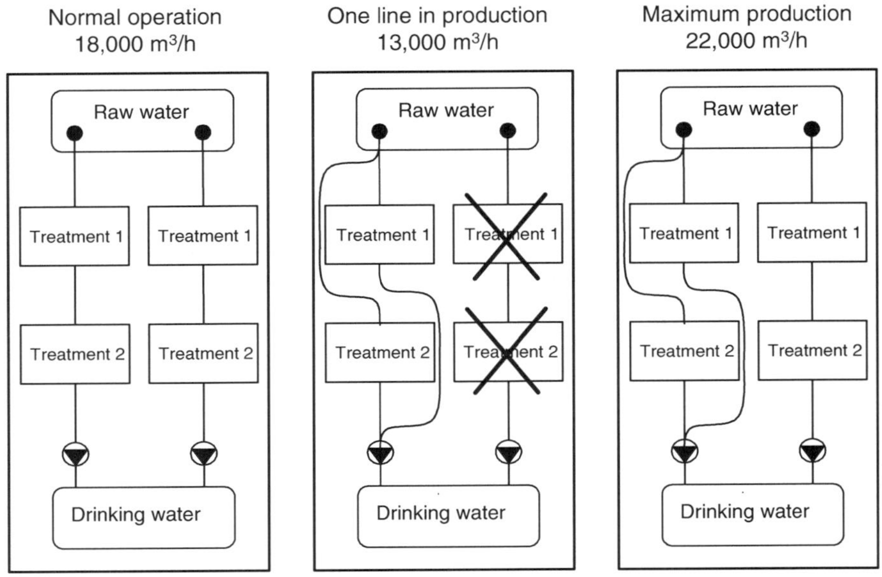

Fig. 7.2 Three configurations of Berenplaat, during normal operation, with only one line in production and in case of maximum production. Treatment 1 consists of micro-sieves, sludge blanket clarifiers and double layer filtration, and Treatment 2 of UV-disinfection and activated carbon filtration

However, it has been determined that in this exceptional situation a temporary supply of substandard quality is acceptable. This policy is discussed with and approved by the Inspectorate.

If extra capacity is required in an emergency situation, e.g., in the case of complete failure of one of the smaller pumping stations, the configuration can be altered so that the capacity increases to 22,000 m^3/h (777,040 ft^3/hr)) (Fig. 7.2).

In order to achieve a reliable supply, it was found that modifications to the distribution network were needed as well. Two culverts connecting Berenplaat to Rotterdam were found to be weak points in the system. These culverts, made of 1400 mm (55.1 in) steel and laid in 1963, cross an intensively used tributary of the Rhine. The distance between these culverts is less than one meter and both culverts can be isolated by valves. Nevertheless, it was found that both culverts could fail due to one accident, e.g., a ship with a loose anchor. In order to protect the Rotterdam supply against such a failure it was decided to construct a new crossing using horizontal drilling.

PWN supplies drinking water to the Province of North-Holland, with the exception of the Amsterdam area. PWN made a reliability assessment of the Bergen production plant and the required capacity to guarantee minimal sufficient supply to the centers of demand. The distribution network was analyzed using the network calculation model SynerGEE. SynerGEE has a so-called reliability module, which is comparable with the calculation for pressure dependent demand as presented at this paper.

It was concluded that the minimum required capacity of the pumping station at Bergen is 3,750 m^3/h (132,412.5 ft^3/hr) at 450 kPa (65.3 psi), corresponding to a plant production capacity of 2,200 m^3/h (77,682 ft^3/h) or 50 % of the capacity required during normal operation. Further analysis of the plant identified the sand filters as playing a crucial role in the reliability of the plant's production.

The next step was to identify which failures of elements in the sand filter building have a negative impact on the reliability of supply. This analysis resulted in a number of measures. Afterward, risk analysis applying the FMECA (Failure Mode, Effects, and Criticality Analysis) was performed to identify which corrective actions are to be executed first.

Vitens is the largest water company in the Netherlands, supplying the Central, East, and North of the country. Vitens supplies 330 mil m^3 (11,652.3 mil ft^3) yearly through 47,500 km (29,516.5 mi) of mains. Vitens uses the network calculation model InfoWorks and at their request a modification has been made to perform a pressure dependent calculation over 24 h. The analysis is based on clusters of 2000 connections. By applying a so-called critical linked analysis, every main that can not recover within 24 h in case of a calamity is taken out successively and the effect on reliability is calculated. The calculations showed no problems with the reliability of the network. Much effort was required to obtain the appropriate level of data availability and data quality

WML supplies the Province of Limburg in the South-East of the Netherlands. Part of the supply area includes the hilly zones in the South of Limburg. Maintaining reliability is complex in these hilly zones. The distribution network serving this area is complex and consists of 33 different pressure zones. Many reservoirs are located on the tops of hills, and therefore the pressure in many zones is constant and at a high level. Calculation of pressure-dependent demand has been modelled and has shown that a main break and the subsequent closure of a section results in limited areas with pressures lower than 200 kPa (29 psi) where a decrease in demand is to be expected. This implies that distribution networks in hilly areas need to be more robust in order to preserve a reliable supply. Another important aspect of reliable water networks in hilly areas, is the increased probability of negative pressures and consequently the intrusion of pollutants into the network.

7.6 Strategic Asset Management and Reliability Assessment

Alegre (2008) defines strategic asset management as: 'a multidimensional approach that may be defined as the corporate strategy and the corresponding planning and systematic and coordinated activities and practices through which an organization optimally manages its assets, and their associated performance, risks and expenditure over their lifecycle'. This definition defines the three focus areas of strategic asset management, notably: performance, risks, and expenditure. The reliability assessment as described in this paper is an instrument to evaluate the robustness of the water supply system as a whole against major failures. It can therefore be regarded as a valuable instrument for risk assessment, however, it does not take the probability of failure into account. Water supply systems are in general large and complex and nowadays, more information has become available concerning the reliability of individual elements. A quantitative risk analysis of the complete system is still a difficult task for most water companies. A major obstacle is the major investment required in order to develop information systems for acquiring sufficient and reliable data. Another obstacle is that information on complex incidents which could result in a calamity is scarce, since calamities such as the breakdown of a pumping station occur very infrequently and when they occur, are often the result of simultaneous incidents or a chain of events.

Water companies in the Netherlands assess the reliability of their supply systems based on the Reliability Framework, originating in the 1990s. This framework has been implemented by the companies and has been the basis for achieving more reliable supply systems, during the last 15 years. Initially, the ALEID network model incorporated a module that was utilized to assess the reliability of distribution networks and the required capacity for production plants, pumping stations, and reservoirs. More recent and commercial network models currently being used by many water companies, have the potential for supporting these calculations.

This experience indicates that a proactive approach by water companies can actively address problems and can have a positive impact on the water sector. In the example described in this paper, a joint initiative resulted in the development of framework for reliability assessment, which have been the basis for further regulations. If water companies collectively define their needs, companies, and research institutes will be better equipped to acquire useful knowledge to develop new and very useful tools.

7.7 Conclusions

This paper describes 15 years of experience with the quantitative assessment of water supply system reliability in the Netherlands. The following conclusions are drawn:

1. The Reliability Framework and the corresponding quantitative assessment as applied by water companies in the Netherlands provides useful insights and provides guidance for improving the reliability of water supply systems. Typical measures resulting from this assessment could be infrastructure modifications or improved operating procedures.
2. During the last 15 years in the Netherlands some major problems have occurred, however, none have resulted in a interruption of supply for more than

24 h. These results justify confidence in the robustness of supply systems in the Netherlands.

3. The framework for reliability is a product of self-regulation by the drinking water sector in the Netherlands. The framework is an initiative by the sector and has been recently integrated into the new Drinking Water Act.
4. The quantitative assessment of reliability should be regarded as an instrument to evaluate the robustness of the total supply system against failure. The reliability assessment focuses purely on the effects of a failure and can therefore not be regarded as a risk assessment. Based on experiences in the Netherlands, a focus on effects seems to be effective. Parts of the system identified as having lower reliability, can at later stage be subject to more comprehensive risk analysis.
5. The framework is not meant to be rigidly applied. Though reliability is a major aspect of decision making for improving water supply systems, it is not the only one. Therefore, in certain cases it should be possible to deviate from the guideline but these deviations must be approved by an independent body.
6. A specific module for reliability assessment has been developed within the network calculation model ALEID, which during the 1990s was used by most of the water companies in the Netherlands. Nowadays, more commercial and sophisticated models are used. In these packages, comparable calculations can be made.
7. The reliability assessment applies to incidents that effect larger parts of the system, and have a duration of more than 24 h. As the level of service increases, water companies tend to give more attention to incidents of a lower magnitude. The Reliability Framework can be regarded as a basic asset management instrument to evaluate the robustness of the supply system. Further improvements for increasing the level of service are to be expected with the development of asset management techniques, which is a major topic within the drinking water sector in the Netherlands.

References

Alegre H (2008) Strategic infrastructure asset management: concepts, "schools" and industry needs, Water Asset Management International, March 2008, IWA Publishing

Beuken RHS, Vloerbergh IN (2009) Proceedings of the Workshop Reliability, November 18th, 2008, KWR, Nieuwegein, BTO2009.004(s) (in Dutch)

Boomen M. van den (2008) Reliability of drinking water systems, an evaluation of the current practice at water companies in the Netherlands. KWR, Nieuwegein, BTO2008.020(s) (in Dutch)

Vreeburg JHG, van den Hoven TJJ, Hoogsteen KJ (1994) A quantitative method to determine reliability of water supply systems. Water Supply 12(1–2):7.9–7.13

Vreeburg J, Rosenthal L, Poortema K, Hoogsteen K (1998) Risk management: integral approach in the Netherlands. Water Supply 16(1–2):472–474

Chapter 8
Environmental Management of the Puyango–Tumbes River Basin in Ecuador and Peru

Napoleon Puño

8.1 Introduction

The problem of polluted and mismanaged watersheds is becoming increasingly acute throughout the world. One such watershed is the Puyango–Tumbes river basin, which is located within Ecuador and Peru. This bi-national watershed, which drains into the Pacific Ocean, is one of the most important regions along the Pacific coast in South America. It encompasses a large diverse land area, parts of which are devoted to agriculture. The Puyango–Tumbes River is an important source of water for irrigating crops that are grown in the agricultural regions. This is especially important for crops grown in the downstream valleys located on the north and south sides of the river.

Clearly the Puyango–Tumbes river basin is an important source of fresh water; however, it is no stranger to problems of environmental pollution. A diverse collection of pollutants discharged from mining and agricultural activities, and from domestic sources, appear in the form of wastewater and solid waste. Evidence of pollution is especially apparent in the lower portions of the basin, particularly in areas close to the city of Tumbes. Rice fields on the north bank of the Tumbes River (Tumbes refers to the 90 km (60 mi) downstream portion of the Puyango–Tumbes River) and the mangrove ecosystem downstream from the city of Tumbes are heavily polluted. Bioaccumulative contaminants and biomagnified pollutants, such as heavy metals, have been identified in rice kernels and in shell fish. These pollutants pose a significant risk to the health of persons consuming these food products. Another fundamental human health risk is due to the fact that people living near the river drink water from this source. Residents in the city of Tumbes drink water from the Tumbes River that has been treated with a conventional filtration system. Since a number of pollutants adhere to water colloidal elements, they are difficult to remove with such conventional treatment.

N. Puño (✉)
Universidad Nacional de Tumbes, Av. Universitaria s/n, Pampa Grande, Tumbes, Peru
e-mail: mrsjoule@hotmail.com

R. M. Clark and S. Hakim (eds.), *Securing Water and Wastewater Systems*, Protecting Critical Infrastructure 2, DOI: 10.1007/978-3-319-01092-2_8,

In an effort to identify beneficial uses of the resources provided by the Puyango–Tumbes River and to address pollution problems such as those described above, a report—An Environmental Management Plan for the Puyango–Tumbes River—has been prepared. The main objective has been to prepare a plan to achieve a sustainable society based upon an optimal and rational use of natural resources, with a special attention to water. Success in achieving this objective would provide a multigenerational security through the wise use of land, water, and natural resources in the basin.

A bi-national activity involving Peru and Ecuador to improve water quality in the Puyango–Tumbes watershed may come as a surprise to those remembering the long history of border disputes between these two countries. Over a period of more than 150 years, a number of battles were fought between the two countries over disputed territory, particularly in the Amazon region. The last such military engagement, which occurred in the mid-1990s, was followed by the signing of a comprehensive peace accord on October 26, 1998. The agreement was ratified by the congresses in both nations, finally bringing an end to the dispute. Since then, trade between the two countries has increased substantially and other kinds of cooperative endeavors have occurred, including the development of plans for productive, social, and environmental development. This recent period of cooperation between the two countries has paved the way for them to join in efforts to improve the water quality in the Puyango–Tumbes basin.

8.2 Description of The Puyango–Tumbes River Basin

Peru and Ecuador legally share use of the Puyango–Tumbes River. The annual average volume of water entering the Pacific Ocean from the river is approximately 3,400 × 10^6 m^3 (120,054 × 10^6 ft^3). Less than 20 % of this resource is currently used for economic development, such as agricultural production and power generation; therefore, significant opportunities exist for beneficial uses.

The Puyango–Tumbes River drains an area of approximately 4,850 km^2 (1872.58 mi^2); about 60 % lies in Ecuador and 40 % in Peru (Fig. 8.1). The river drops 532 m (1,745 ft) from its origin in the Andes to its mouth at the Pacific Ocean. Some of the river's tributaries rise at a much higher elevation in the Andes Mountains. The river's length is approximately 230 km (142 mi). Currently, a portion of the river flowage is used to irrigate land in both nations, but much greater possibilities exist. There is a potential for irrigating up to 70,000 ha (172,900 acres) of land in Ecuador and about 90,000 ha (222,300 acres) in Peru. The natural flow variation from year-to-year and season-to-season requires regulation, but it would be necessary to create large reservoirs for this purpose. Several desirable sites for such structures have been identified along the river. In the process of regulating river flows by dams and diverting water to irrigate agricultural land, large amounts of hydropower could also be generated. Control of the river could eliminate flood damage that regularly occurs in the lower reaches

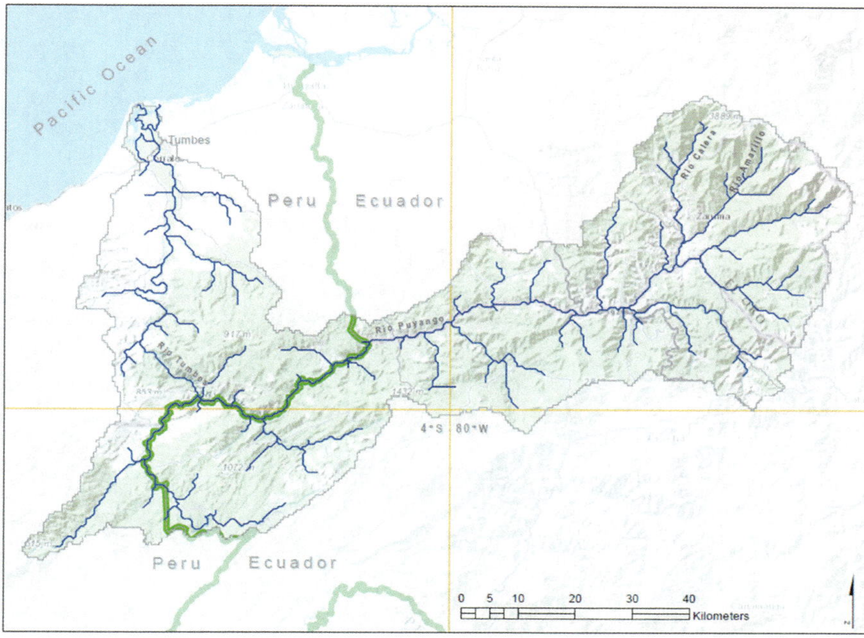

Fig. 8.1 The Puyango–Tumbes watershed

where important agricultural areas and Peruvian population centers are located. Geographically, the basin is centered near the intersection of Latitude 4°S and Longitude 80°W.

8.2.1 Rainfall and Hydrography

The Puyango–Tumbes basin has two clearly defined rainfall periods: a wet period that starts in December of a given year and goes to April of the following year. The dry period covers the months from May to November. More than 85 % of the rainfall occurs during the months from December to April; March is typically the wettest month and August is the driest. Temporal rainfall data are shown in Table 8.1.

There is a significant variation in temperature over the basin. The annual average temperature in the lower part of the basin is 24 °C (75.2 °F); in the higher elevations, 2,700 m (8,856 ft) above sea level, it is 15 °C (59 °F).

The Puyango–Tumbes River belongs to the Pacific Hydrographic system which has its sources in the Republic of Ecuador. The river originates in the Cordilleras and Cerro Chilla ranges of the Andes. Seasonal rainfall in these high basin areas feeds the river's tributaries and its main stem. Beginning at its source, the first

Table 8.1 Monthly and yearly mean precipitation amounts in the Puyango–Tumbe Basin (millimeters, mm)

	Jan	Feb	Mar	Apr	May	Jun	Jul	Aug	Sep	Oct	Nov	Dec	Annual
Peru													
Lower basin (0–250 m)[a,b]	48.3[c]	103.0	116.9	77.3	24.8	5.3	2.2	0.6	1.4	2.8	6.0	13.7	402.3
Middle basin (250–1,000 m)	74.0	185.3	192.9	135.8	42.5	7.4	3.0	0.9	2.4	5.2	5.1	35.8	690.3
Weighted average[d]	59.7	139.6	150.7	103.3	32.7	6.3	2.5	0.7	1.8	3.9	5.6	23.5	530.3
Ecuador													
Lower basin (0–250 m)	149.5	228.4	196.7	141.6	66.3	29.7	15.4	9.7	13.9	18.2	26.2	64.0	767.9
Middle basin (250–1,000 m)	185.1	285.0	300.4	235.8	84.1	23.0	6.5	3.3	12.2	18.3	30.0	114.6	1054.9
Upper basin (>1,000 m)	289.1	368.0	374.5	285.0	135.2	43.9	13.7	7.0	29.3	52.3	74.6	177.7	1619.1
Weighted average[d]	212.9	301.8	304.6	233.6	97.7	31.4	10.7	5.8	18.3	29.7	44.2	125.7	1187.1
Basin													
Middle of the basin	149.8	234.8	241.2	179.9	70.9	21.0	7.3	3.7	11.5	19.0	28.3	83.6	916.4

[a] Meters above sea level
[b] To convert from meters to feet multiply by 3.28
[c] To convert from mm to inches multiply by 0.039
[d] Average precipitation times the area of influence divided by the total basin area. Area of influence estimated by Thiessen Polygon Method

100 km (62 mi) of the river lies entirely in Ecuador; this section is called Puyango. After changing its direction, the course of the river marks the border between Ecuador and Peru. The common border between the two countries, defined by the river, is 40 km (24.8 mi) in length and is referred to as Puyango–Tumbes. The final 90 km (55.8 mi) section, which lies entirely in Peru, carries the name Tumbes and enters the Pacific Ocean at Punta Mal Pelo. The river has monthly average flows ranging from 1244.2 m^3/s (43,933 ft^3/s) to 7.7 m^3/s (271.9 ft^3/s). This information is based on records at the El Tigre Station, located 25 km (15.5 mi) upstream from the city of Tumbes, during the period from 1965 to 2004. The El Tigre Station measures the drainage from approximately 4,380 km^2 (1,691.12 mi^2) of the basin, a portion which is 90 % of the total basin area of 4,850 km^2 (1872.60 mi^2).

8.2.2 *Flora*

Ecuador and Peru, the two countries that share the Puyango–Tumbes basin, are rich in biodiversity. Ecuador's land area comprises approximately 0.2 % of the world's land area, yet it is estimated that its territory is home to 10 % of all plant species on the planet. Meanwhile, Peru has a large diversity of plants in its forest frontier and is among the 15 countries in the world with the most diverse terrestrial and aquatic ecosystems, species, and genetic resources. In the Puyango–Tumbes basin itself, several plant communities can be distinguished: Mangroves, Chaparral Forest, Dry Forest, Montane Rain Forest, Montane wet forest, riparian, and sandy and rocky.

Mangroves are forests that exist in a transition zone from sea to land. They develop in marshy areas with extreme soil conditions that form at the junction between salty and fresh water in river estuaries. Mangroves protect living larvae and juveniles of various species of fish and shellfish. They provide protection from floods and tidal wave impacts, provide habitat for many bird species, and help to control erosion. Mangroves exist in northwestern Peru in the state of Tumbes and in southern Ecuador.

Chaparral forests occur at intermediate altitudes between the coast and the hills. Since this zone is very dry, with annual rainfall amounts between 125 and 250 mm (4.88 and 9.75 in), the plants that grow are able to survive with little water. The vegetation consists of forest shrubs and small trees with an herbaceous understory. It is likely that this vegetation type has arisen following a degradation of a dry forest as a result of overgrazing by goats. During the period from the early part of the last century to the 1940s, carob forests in this area were harvested to produce charcoal for urban consumption in the city of Lima.

Dry forests, existing at elevations between 50 and 300 m (164 and 984 ft) in mountainous terrain with steep gradients, are populated by plants and trees that can survive even though they receive no rain for several months. These plants and trees develop in conditions that are similar to those in deserts, but they are subtly altered by a short rainy season that makes its appearance in the early summer. Ceiba trees, a dominant species in the dry forest, have learned to live with water scarcity and survive long periods of drought by storing water in their thick outer bark.

In higher elevations, rainfall is much greater. The montane rain forest, located at elevations between 300 and 2,000 m (984 and 6,560 ft), experiences annual rainfall amounts in the range of 1,000–2,000 mm (39–78 in.). Among the dominating species in this forest are walnut and cedar; some of these trees grow to a height of 30 m (98.4 ft). The trees tend to be top heavy because they are usually coated with epiphytes, nonparasitic plants that derive their nutrients and moisture from the air and rain. This type of forest occurs in about 38 % of Ecuador.

The montane wet forest exists at an elevation between 2,000 and 3,700 m (6,560 and 12,136 ft) and the annual rainfall amounts there exceed the amounts that occur in the montane rain forest. A large part of the precipitation, which ranges from 1,000 to 2,500 mm (39 to 97.5 in.), comes from supersaturated air moisture and fog. This forest is very important for water retention. The tree species are similar to those in the montane rain forest. In some areas, the forest land has been almost totally destroyed and converted into pasture for livestock.

The riparian zone is the area along the banks of the rivers. The vegetation that occurs there is extremely important for curbing bank erosion and for providing habitat for aquatic species. In the lower basin, vegetation in the riparian zone is scarce due to agricultural practices. Willow is the most important tree species in the middle basin. Wastewater from mining operations affects the riparian vegetation along the entire river. Excessive rainfall that occurred during the 1997–1998 "El Niño" caused excessive flooding and drastically changed the morphology of the river and riparian ecosystems. The floods, persisting for nearly a year, killed fruit trees and stimulated the growth of a new type of shrub. At certain places where the Puyango–Tumbes River meanders, sandy and rocky ledges are formed. In these areas, water lilies often grow and flourish.

8.2.3 Fauna

A very large number of animal species exists in the Puyango–Tumbes river basin. In studies of animals in the basin, 32 mammal species, 181 bird species, 12 reptile species, 10 river crustacean species, and 46 species of river fish have been identified. In addition, 104 species of marine fish and 91 aquatic invertebrate species are known to exist in the basin. Four species, endemic to the basin and in danger of extinction, are the Tumbes crocodile, Tumbes howler monkey—so-called because of the powerful loud roars it emits in defense of its territory—northwest otter, and wildcat. All four species are protected in both Peru and Ecuador. Although protected by law, the Tumbes crocodile and northwest otter are still being hunted illegally. Other interesting animals found in the basin are ocelots, margays, anteaters, gray neck squirrels, vultures, peccaries, bearded turkeys, large woodpeckers, and partridges.

Two reserves have been established to conserve and protect flora and fauna. The Amotape Hills National Park, located on the north bank of the Tumbes River, was established in 1975. Its purpose is to promote wise management of forest resources and to preserve animal species. The Tumbes Reserved Zone, located in the Department of Tumbes, was established in 1994 with the aim of conserving and protecting representative samples of Pacific rain forest.

8.2.4 Economic Activities in Basin

In the upper basin of the Puyango River in Ecuador, small-scale gold mining has occurred since the colonial period. During the first half of the twentieth century, the South American Development Company (SADCO) was formed to mechanize the gold mining process. Before going out of business in 1950, SADCO produced an estimated 108 metric tons (119 short tons) of gold. The Ecuadorean government purchased SADCO's assets and formed a new company called the Associated Mining Industrial Company (CIMA). From 1950 to 1965, CIMA produced 12 metric tons (13.23 short tons) of gold. After 1965, due to a lack of investment in exploration and development, production was low and the company filed for bankruptcy in 1978 (Fig. 8.2).

During the period from 1979 to 1994, various organizations in Ecuador attempted to develop more efficient mining operations. The focus of this effort was in an area called "Project Portovelo". However, these projects were able to employ only a small number of workers from the CIMA labor force. As a result, a virtual state of anarchy existed and an uncontrolled proliferation of small-scale mining operations emerged. At the present time as many as 400 of these small underground mines are in existence. Many of them are family operations; some are located at the sites of old industrial mines. The technical and administrative management of these mining operations is very primitive and, as a result, contaminated sediments arising from these operations enter the Puyango River and are flushed downstream as far as the metropolitan area of Tumbes.

Some environmental improvements began to occur in the mid-1990s when a few modern and well-organized mining companies were established. These small- to medium-scale operations have had a production level of slightly less than one metric ton per year and, currently, have a few more years of proven reserves.

Agriculture is another important part of the economy in the Puyango–Tumbes basin. The dominant crops in the Peruvian part of the basin are bananas, rice, beans (cowpea), lemons, soy, and mangos. The land used to produce these crops comprises 61 % of the total number of hectares in the Department of Tumbes. In the period from 1990 to 2000, the production of rice more than doubled with 43,000 metric tons (47,399 short tons) harvested at the beginning of this period and 92,000 metric tons (101,412 short tons) at the end. However, during the same period banana and lemon production declined. Banana production was more than halved, declining from 110,000 metric tons (121,253 short tons) to 47,000 metric

Fig. 8.2 Gold mining sites are located along the highlighted Amarillo (*Yellow*) and Calera Rivers

tons (51,808 short tons). Lemon production fell from 5,900 metric tons (6,504 short tons) to 1,030 metric tons (1135.4 short tons); the latter representing about one-fifth of the former amount. The full record of production during this 11-year period is presented in Table 8.2.

Table 8.2 Annual crop production in the department of Tumbes (metric tons)

Years	Crop[a]							
	Bananas	Lemons	Mangos	Rice	Corn	Soy	Beans	Total
1990	110,385	5,909	560	42,877	1,086	2,372	23	163,212
1991	134,666	6,806	568	45,012	496	434	13	187,994
1992	21,424	1,300	703	7,544	1,444	55	10	32,480
1993	39,958	2,623	612	54,744	1,045	2,704	140	99,149
1994	76,053	1,257	518	35,437	674	133	1,701	115,773
1995	76,268	1,232	322	48,430	645	522	851	128,270
1996	71,887	591	494	51,413	466	93	38	124,982
1997	69,236	1,680	116	69,247	535	549	78	141,441
1998	3,750	669	428	12,106	2,092	92	86	19,223
1999	18,065	1,355	430	86,042	2,310	1,635	401	110,238
2000	47,446	1,033	171	92,588	2,471	388	188	144,285

[a] To convert from metric tons to short tons multiply by 1.1023

The productivity rates for rice and bananas grown in Tumbes are significantly above the national averages for Peru. Soy and bean production rates are slightly above the national average, while lemon, mango, and corn production rates are well below the national average. In Table 8.3 productivity rates for crops grown in the Department of Tumbes are shown in comparison with Peru's national productivity rates for the same crops. On an international level, it is noteworthy that Peru has a higher yield per hectare of rice than each of the following countries: Chile, Columbia, Bolivia, Ecuador, Brazil, Thailand, and Viet Nam.

The livestock sector in the Department of Tumbes consists of goats, sheep, pigs, and cattle. Poultry production is also important; there are more chickens than goats, pigs, sheep, or cattle. During the period from 1990 to 2000, goat and sheep numbers increased, while cattle and pig numbers remained about the same. The poultry numbers also declined. Full details are presented in Table 8.4.

An interesting new agricultural activity is beekeeping. The Peruvian Ministry of Agriculture has introduced a beekeeping program in the upper valley of Tumbes that includes the production of honey and its by-products. There are more than 25 beekeepers, each of whom manages 20 hives, for a total of 500 hives.

Table 8.3 Productivity rates of the principal crops in the department of Tumbes in comparison with the productivity rates for Peru (kg/ha)

Product	Tumbes[a] Output	Peru[a] Output
Bananas	19,882	11,507
Lemons	5,021	10,329
Mangos	3,696	11,163
Rice	7,722	6,590
Corn	3,067	3,555
Soy	1,844	1,455
Beans	1,355	919

[a] To convert from kg/ha to lbs/acre multiply by 0.893

Table 8.4 Animal production in the department of Tumbes, 1990–2000

Years	Production (thousands of animal units)					
	Goats	Sheep	Hogs	Cattle	Milk[a] production (thousands of metric tons)	Chickens
1990	44.3	3.5	13.8	16.7	0.5	128
1991	35.0	2.0	15.0	15.0	0.4	200
1992	34.0	2.5	14.5	12.0	0.2	300
1993	35.8	2.5	15.0	14.5	0.3	290
1994	36.0	2.5	14.8	14.0	0.2	280
1995	60.0	4.5	14.0	14.9	0.2	285
1996	70.4	4.9	12.4	15.1	0.3	184
1997	69.9	4.9	12.4	15.0	0.7	154
1998	83.8	5.9	14.8	16.4	0.2	100
1999	88.5	4.7	16.2	17.0	0.3	110
2000	71.2	7.2	13.8	15.3	0.3	95

[a] To convert from thousands of metric tons to thousands of short tons multiply by 1.1023

Water from the Tumbes River is needed in the Department of Tumbes to support agricultural crops. The south bank of the river is fully supported with pumping resources. The north bank (Hearer, Higueróen, Hualtacal, and Playa Rica) is provided with water through a system that pumps water into a canal that starts at La Peña. From there water flows by gravity along the canal's 25 km (15.5 mi) length. The canal has a capacity to provide the water needed to irrigate 6,000 ha (14,820 acres) of agricultural land. Irrigation in Tumbes is supported by the Tumbes Users Board which brings together 21 Irrigation Commissions formed by 6,337 users.

The agricultural crops—mainly bananas, rice, soy, beans, and lemons—and meat from livestock animals—goats, pigs, and sheep—produced in the Department of Tumbes provides for the needs of the local population. Three markets are located in the Department of Tumbes, one in each of the provinces; the most important of these is the market in the city of Tumbes. Food products that are produced in excess of local needs are sold in neighboring Ecuador and in parts of Peru to the south of Tumbes.

8.3 Description of Watershed Problems and Issues

The main sources of pollutants in the Puyango–Tumbes River Basin are artisanal mining activities in its upper reaches, discharges from agricultural activity, raw sewage discharges from the domestic sector, and solid wastes from the urban and industrial sectors.

8.3.1 Mining

Cyanide and heavy metals used in mining are the major pollutants entering the upper reaches of the river in the higher elevations. The most serious problem is that the metal refining often takes place very close to the bank of the river, mainly in the presence of water. A tailings dam, built very close to the river, places the residue practically in the river without protection. The main mine pollution sources and their locations are listed in Table 8.5 and rivers along which mining operations occur are highlighted in Fig. 8.2.

The photograph shown in Fig. 8.3 displays clear visual evidence of pollution problems arising from mining activities. A typical mining operation is shown in the photograph in Fig. 8.4.

Table 8.5 Inventory of sources of mining contaminants

Contaminant source	Receiving body	Location
Industrial mining company, (Birapampa)	Calera River	Piñas, Ecuador
Gold extraction at mineral processing plants, (Pache locality)	Calera River	Zaruma, Ecuador
Gold mining, (Plant Minanka)	Amarillo (Yellow) River	Portovelo, Ecuador
San Antonio and Espibala III Mines	Amarillo River	Portovelo, Ecuador
Gold mining plant, Gavior (Salty Sector)	Amarillo River	Portovelo, Ecuador
Processing plant, (Mafre II)	Amarillo River	Portovelo, Ecuador
Beneficial mining plants, (George Vaquez, Virgin of the Cloud 1, 2, and 3)	Amarillo River	Zaruma, Ecuador
Jorupe mill, (Chilca)	Arcapampa Ravine, Calera River	Zaruma, Ecuador
Beneficial mining plants, (Queen of the Cisne)	Calera River	Zaruma, Ecuador
Beneficial mining plant, (Alexander Group)	Calera River	Zaruma, Ecuador
Beneficial mining plant, (Queen of Fatima II)	Calera River	Zaruma, Ecuador
Mine and beneficial mining plant, (Ecuador Mary Jane Gold Mine (Molluncay Area))	Calera River	Zaruma, Ecuador
Beneficial mine separation plant, Efren Pasto (Molluncay Area)	Calera River	Zaruma, Ecuador
Separation mill and plant, Cianuración Cosme Apolo	Calera River	Zaruma, Ecuador
The Consorcio Orquidea Plant, Amlatminas S.A.	Calera River	Zaruma, Ecuador
Betsabeth mining area	Remache Ravine, Calera River	Atahualpa, Ecuador
Final settling processor, George Plazzio (by Huertas Pacchas)	Palto River	Atahualpa, Ecuador
Beneficial mining and milling plant, Velázquez (by Huerta Pacchas)	Calera River	Atahualpa, Ecuador
Beneficial plant, Queen of the Cisne (Mill and Cianuración)	Calera River	Zaruma, Ecuador

Fig. 8.3 Confluence of the Pindo and Amarillo (*Yellow*) rivers

Fig. 8.4 Active mining area along the Calera River

8.3.2 Domestic Untreated Sewage

Many people live near or on the bank of the Puyango–Tumbes River or one of its tributaries. They contribute raw (untreated) sewage and other types of domestic wastes to the river. The primary contaminant sources are listed in Table 8.6. One of the major dischargers is the Coloma pumping station, a source of raw domestic sewage from the city of Tumbes. The effluent from the Coloma pumping

Table 8.6 Principal releases of domestic wastewater to the Puyango–Tumbes river

Contaminant source	Receiving body	Location
Domestic wastewater releases from the city of Piñas	Piñas River	Piñas, Ecuador
Domestic wastewater pump discharge by the San Roque customer	Piñas River	Piñas, Ecuador
Discharges by the san Jacinto population, hospital pampas and shoreline populations, right and left	Tumbes River	Tumbes, Peru
Service wastewater releases from the Pontonero district (250 m from a potable water source)	Tumbes River	Tumbes, Peru
Domestic wastewater discharge from the Coloma pump provided by the city of Tumbes	Tumbes River	Tumbes, Peru

Fig. 8.5 Effluent from the Coloma pumping station

station (a photograph is shown in Fig. 8.5) contains high concentrations of nutrients, heat resistant coliform, and parasites. Nearby is another source of contaminated effluent. This source is near a drinking water intake, so it poses a clear risk to human health. Some treatment occurs at this source, but there is no assurance that 100 % of the pathogens are removed.

8.3.3 Agriculture

Runoff from agricultural areas is another source of contaminated water. Nitrates, nitrites, phosphates, heavy metals, and pesticides are contained in water which drains from the fields into the river. Table 8.7 presents information about

Table 8.7 Primary releases of wastewater from active agricultural areas in the Puyango–Tumbes river basin

Contaminant source	Receiving body	Location
Agricultural drainage, Rugged Urcos	Tumbes River	Corrales, Tumbes, Peru
Rough Las Peñas, right shore (low and high Brujas)	Tumbes River	Tumbes, Peru

pollutants from agricultural sources. Figures 8.6 and 8.7 provide visual evidence of pollutants arising from agricultural sources.

Intensive shrimp farming exists in the lower reach of the Puyango–Tumbes River. Large volumes of water are needed to fill shrimp ponds and after a period of use this water is discharged into the river. Many products, including disinfectants and antibiotics, are used when raising shrimp. These products are included in the wastewater from the shrimp ponds, but additional products are often added by unscrupulous managers and employers. Examples of such products are vehicle crankcase oil, domestic sewage and other types of domestic wastes, and fuel. Table 8.8 provides information about sites where contamination from shrimp farming occurs.

Fig. 8.6 Panoramic view of an agricultural drain

Fig. 8.7 Disposal of a pesticide container in an agricultural drainage

Table 8.8 Primary releases of wastewater from active shrimp ponds to the Puyango-Tumbes river basin

Contaminant source	Receiving body	Location
Shrimp pond discharge, Domingo Rodas SAC	Tumbes River	Tumbes, Peru
Shrimp pond discharge, Negusa	Tumbes River	Tumbes, Peru
Shrimp pond discharge, Santa Elena	Tumbes River	Tumbes, Peru

8.3.4 Domestic and Industrial Solid Waste

People living near the banks of the main stem of the Puyango–Tumbes River or one of its tributaries often dump solid waste into the stream. This problem exists because of population growth, lack of environmental education, and failure by municipal authorities who by law have responsibility for solving this problem. Table 8.9 identifies the primary sites where riverbed littering occurs, but there are many other places where people engage in these irresponsible practices.

8.3.5 Monitoring Contaminants

A wide range of parameters that describe specific features of the contaminated river have been identified and measured. This information was collected between 2004 and 2008 at eight monitoring stations in the Peru downstream region. The locations of these monitoring stations are shown in Fig. 8.8. Readings were taken

Table 8.9 Primary releases of solid waste to the Puyango–Tumbes river basin

Contaminant source	Receiving body	Location
Waste of domestic solids (Portovelo and Zaruma Cantons)	Amarillo River	Portovelo, Ecuador
Informal wasting of solids on the right and left shores by the San Jacinto and Pampas hospital populations	Tumbes River	Tumbes, Ecuador
Waste of domestic solids, Corrales Ravine	Corrales Ravine, Tumbes River	Corrales, Tumbes, Peru

and samples were collected at monitoring stations twice in a given year, once during the dry period and once during the wet period, but it was not the case that this was done at all stations each time. Some information was obtained at the monitoring station sites; other information was obtained from laboratory analysis of samples collected in the field. Twenty-one parameters were monitored during the collection period: temperature, pH, electrical conductivity, total dissolved solids, chlorides, phosphates, sulfates, nitrates, biochemical oxygen demand, dissolved oxygen, cyanide, coliforms, arsenic, cadmium, copper, chromium, iron, manganese, mercury, lead, and zinc.

Five parameters, arsenic, cadmium, iron, manganese, and lead, were identified as dominant contaminants. Information about these contaminants is displayed in Figs. 8.9, 8.10, 8.11, 8.12, 8.13, 8.14, 8.15, 8.16, 8.17, 8.18. Instead of showing graphs for the remaining parameters, summaries are provided in the following paragraphs.

In most years, the temperatures during the dry season ranged from 25 to 28 °C (77 to 82.4 °F). However, in 1 of the 5 years temperatures were higher, falling in the range of 29–31 °C (84.2–87.8 °F). During the rainy season, the temperatures were consistently between 24 and 27 °C (75.2 to 80.6 °F) in the mid to downstream reaches, but in the upstream a high of 29 °C (84.2 °F) was recorded in 1 year and 31 °C (87.8 °F) in another. As to pH, the values during the dry season ranged from 7.2 to 8.7, with higher values upstream. During the rainy season, the readings in all years except one ranged from 7 to 8.5. In the other year, there was a low reading of 6 at one station.

The electrical conductivity during the dry season was fairly steady around 200 μS/cm in the middle portions of the stream, but spikes occurred in both the upstream and downstream readings, with values as high as 1,200 μS/cm. During the rainy season, the values ranged between 100 and 500 μS/cm, except for 1 year when there was a high reading of 1,000 at a downstream station. Information is available on total dissolved solids during 2 years of the rainy season. In 1 year, the values were steady in a range between 100 and 200 mg/l. In the other year, there was a considerable variation with a high downstream value of 450 mg/l and a high upstream value of 900 mg/l. No dissolved solids information was acquired during the rainy season.

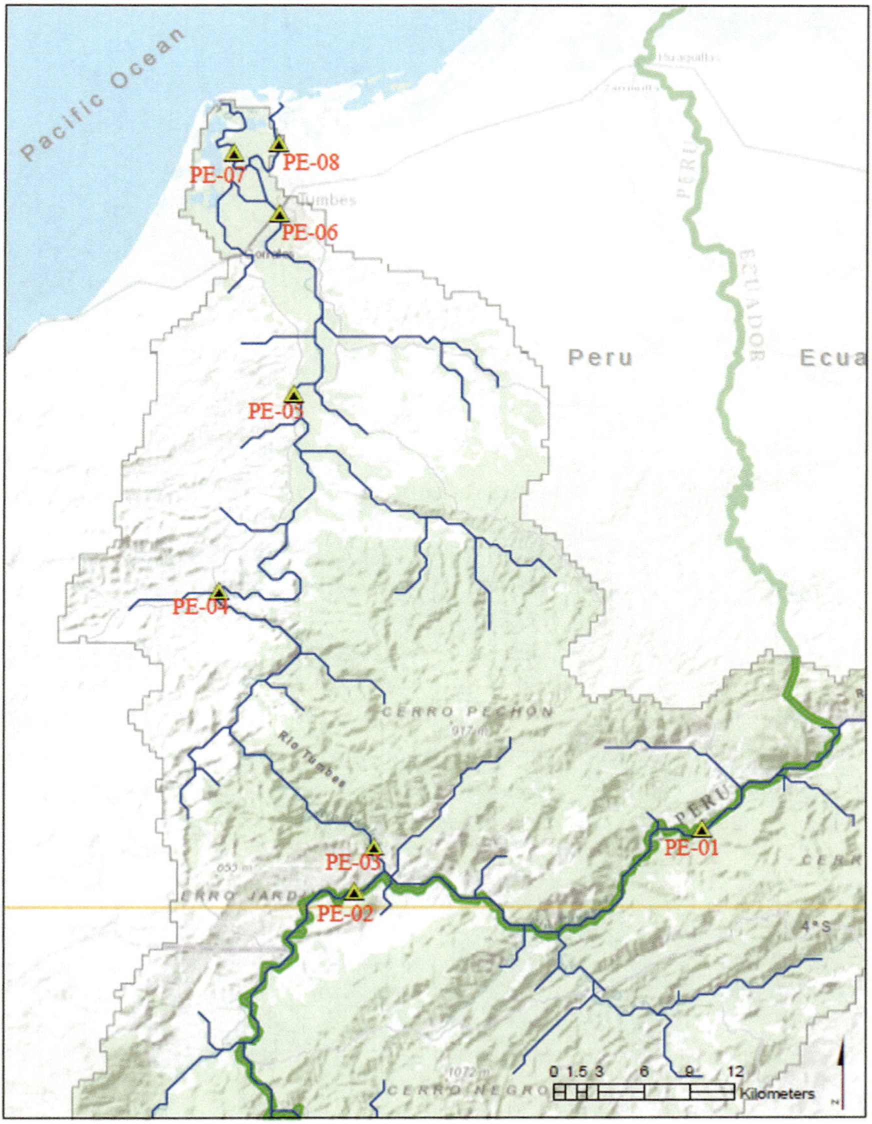

Fig. 8.8 Monitoring stations on the Tumbes river

During the dry season, chlorides were steady within a range of 10 and 20 mg/l in the middle portions of the stream, but spikes as high as 80 and 100 mg/l occurred in the upstream and downstream areas. The readings during the rainy season were similar. For phosphates, in 1 year during the dry season, readings

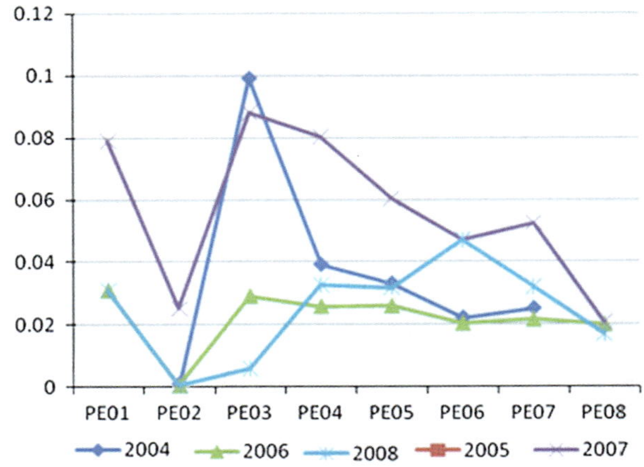

Fig. 8.9 Monitoring data for Arsenic during the dry season (mg/l)

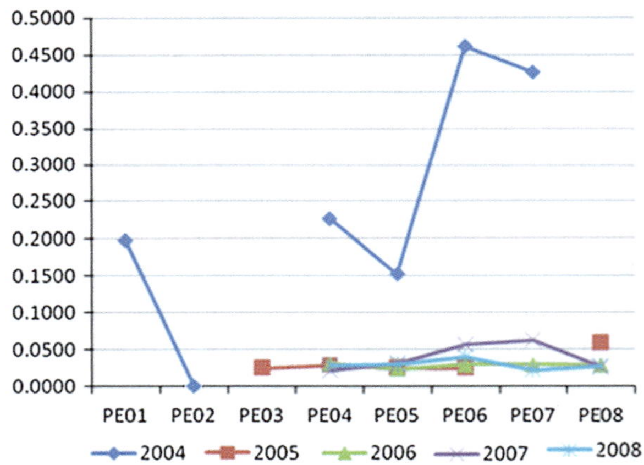

Fig. 8.10 Monitoring data for arsenic during the rainy season (mg/l)

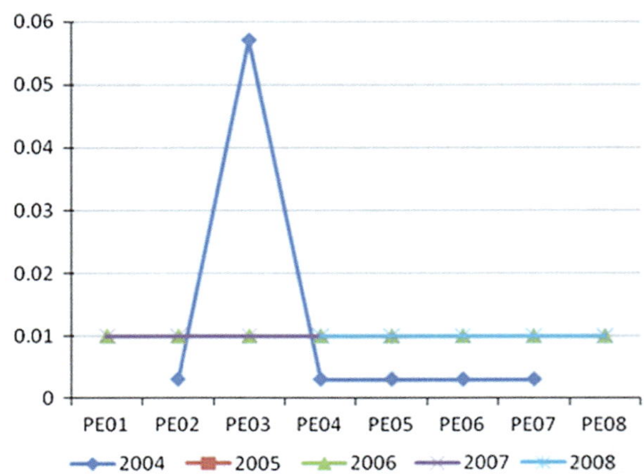

Fig. 8.11 Monitoring data for cadmium during the dry season (mg/l)

Fig. 8.12 Monitoring data for cadmium during the rainy season (mg/l)

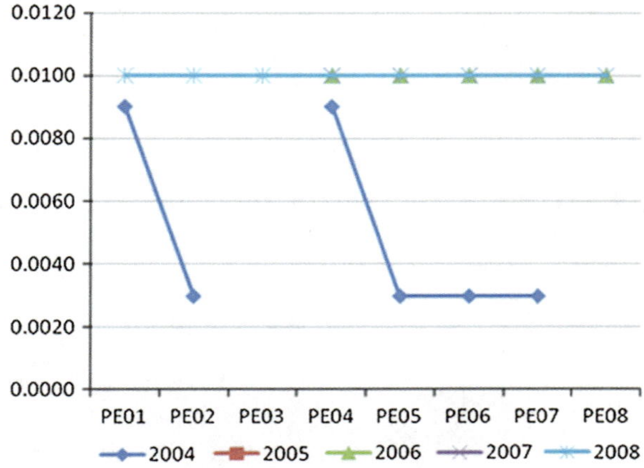

Fig. 8.13 Monitoring data for Iron during the dry season (mg/l)

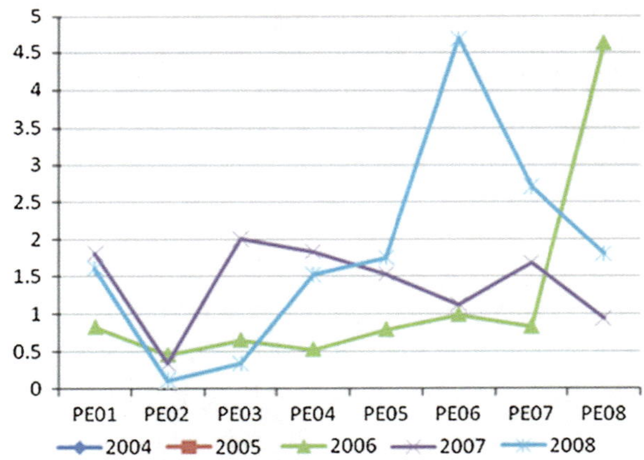

Fig. 8.14 Monitoring data for iron during the rainy season (mg/l)

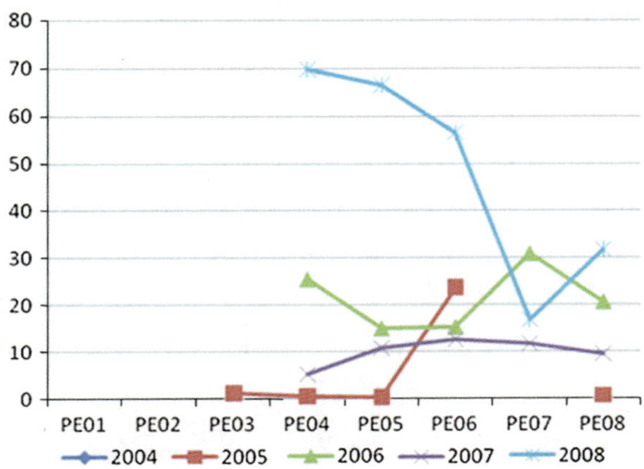

Fig. 8.15 Monitoring data for manganese during the dry season (mg/l)

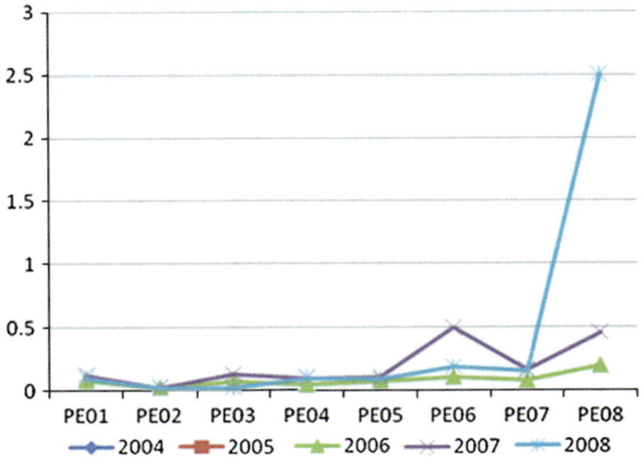

Fig. 8.16 Monitoring data for manganese during the rainy season (mg/l)

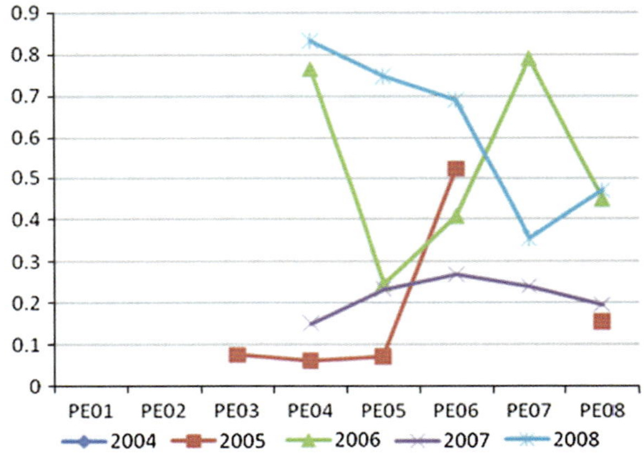

Fig. 8.17 Monitoring data for lead during the dry season (mg/l)

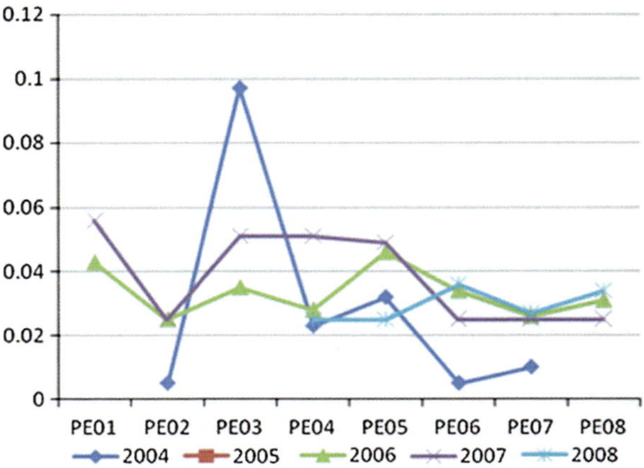

Fig. 8.18 Monitoring data for lead during the rainy season (mg/l)

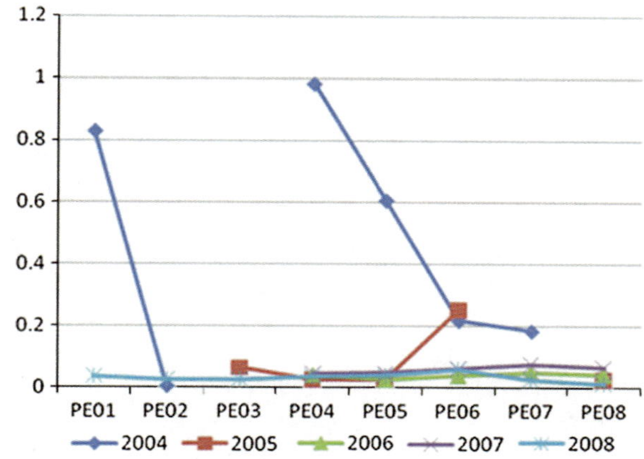

were steady over the course of the stream at approximately 0.1 mg/l. In other years, a great deal of variation occurred with one reading as high as 1.0 mg/l. During the rainy season, the readings were fairly steady with values in the range of 0.2–0.4 mg/l. An exception occurred in 1 year when a value of 0.9 was recorded at an upstream station.

In the dry season, sulfate values were mostly in a range between 20 and 100 mg/l, but in 2 of the 5 years values as high as 300 mg/l were recorded at downstream stations. During the rainy season, values for one of the 5 years were steady in a range between 10 and 20 mg/l, but in other years there was considerable variation with values ranging from 25 to 150 mg/l. During the dry season, nitrates varied considerably over the course of the stream and from 1 year to the next; values ranged from 0 to 4.0 mg/l. The values also jumped around during the rainy season with values ranging from 0.3 to 4.0 mg/l.

Biochemical oxygen demand values were recorded during 1 year of the dry season. The values ranged from 0.4 to 2.0 mg/l. In the dry season, dissolved oxygen values varied over the course of the stream and over the 5-year period; the range was 4.0–7.2 mg/l. During the rainy season, the values were steady in some of the years ranging from 4.5 to 7 mg/l, but in two of the years a greater variation occurred with values as high as 8.5 and 10 mg/l.

Data on cyanide were collected in 2 of the 5 years during the dry season. In 1 year, the values were steady over the course of the stream ranging between 0 and 0.01 mg/l. In the other year, there was a large variation over the course of the stream with values between 0 and 0.09 mg/l. During the rainy season, readings were taken in only 1 year. The values were considerably lower than those observed during the dry season, ranging between 0 and 0.006. During the dry season, coliforms varied from 1 to 1,000,000 MPN/100 ml, with the larger values recorded at downstream stations. During the rainy season, the coliforms tended to be steady over the course of the stream, but there was a considerable variation from 1 year to another with values ranging between 10 and 1,000,000 MPN/100 ml.

During the dry season, copper readings varied over the course of the stream and from 1 year to another. The values ranged between 0 and 0.08 mg/l. The readings from two of the years during the rainy season were steady from upstream to downstream with values between 0 and 0.1 mg/l. In the other 3 years, the values jumped around, ranging from 0 to 1 mg/l. Data for chromium were collected in two of the years during both the dry and rainy seasons. In each case, the values held steady at 0.05 mg/l.

Mercury readings were taken in the dry season in only 1 of the 5 years. The values were steady at 0.0001 mg/l. During the rainy season data were collected in 3 of the 5 years; again the values were steady at 0.0001 mg/l. As to zinc, in each of the 5 years during the dry period the values were highly variable over the course of the stream. The values ranged from 0 to 0.16 mg/l. The readings in three of the 5 years during the rainy season were quite steady with values ranging from 0 to 0.3 mg/l. The values for the other 2 years were more varied with spikes in one of the years reaching 1 mg/l and 2 mg/l in the other.

8.4 Restoration and Management Plans: Moving Toward A Future with Increased Water Security

As has already been discussed, the pollution of the Puyango–Tumbes river basin is extensive. The set of contaminants contributing to the basin's pollution is diverse and there has been a long history of environmental neglect. A management plan has been prepared with the goal of reducing the contaminants and, thereby, moving toward a future with an improved and more secure source of fresh water. The management plan contains the following six specific activities:

(1) Relocate the mining industries that are now located along the banks of the Amarillo and Calera Rivers—both tributaries of the Puyango–Tumbes River—in the Ecuadorean Cantons of Portovelo and Zaruma to a metallurgical mining industrial park, which would be built to contain and eliminate shoreline contamination. Dredge river shorelines to remove contaminated sediments.
(2) Reforest areas in the Ecuadorean uplands which have been inappropriately cleared and used for other purposes.
(3) Build river shoreline defenses where needed to protect against erosion. Planting grass and shrubs and building retaining walls and rock jetties are alternative methods for stabilizing the soil on the river banks. Alternative remedies must be chosen with care based upon the nature of the erosion problem at a given location.
(4) Design and build wastewater treatment plants for urban locations that currently release raw sewage from residential and commercial sources into the river. The need for such a treatment plant is particularly urgent in the metropolitan area of Tumbes.

(5) Design and build curbside solid waste collection and recycling systems in urban areas; again the metropolitan area of Tumbes is an important place to begin. Such solid waste collection systems are important in their own right, but they would also provide an alternative to the present practice of dumping commercial and household waste into the river.
(6) Develop and implement an environmental education program for citizens of all age groups living in the basin.

In the following paragraphs, additional details are provided for each activity contained in the management plan. The specific goals are also stated.

The mining relocation activity would focus on the relocation of artisanal miners presently operating on the banks of the Amarillo and Calera Rivers. A metallurgical industrial park, equipped with basic services and a system for treating tailings and controlling solid waste, would be constructed. The specific goals are:

(a) Within 3 months, establish the legal foundation (enact an ordinance or law) to ban gold mining companies from the banks of the Amarillo and Calera Rivers.
(b) Remove approximately 500,000 m^3 (655,000 yd^3) of sediment from 12 km (7.44 mi) of shoreline on these two tributaries within 6 months.
(c) Complete studies and designs for the industrial park and build it within 2 years.
(d) Within 1 year after the construction of the industrial park, relocate the miners by moving them from the banks of the two rivers to the industrial park.

The reforestation of the upper watershed would begin with a study of the deforested areas that would lead to the development and implementation of a plan for the restoration of native species in these land tracts. The specific goals are:

(a) Prohibit logging near the banks of the rivers that form the upper and middle reaches of the Puyango and Tumbes Rivers. The prohibition order would be put into effect within 6 months.
(b) Quantify the areas of greatest deforestation in the basin within 6 months.
(c) Build three nurseries within 1 year that would provide stocks of native plant species for the restoration of deforested tracts of land.
(d) Over the next 10 years reforest tracts in the most vulnerable areas at the rate of 1,000 ha (2,470 acres) per year.

In the shoreline defense program techniques designed to enhance the protection of river banks will be employed. In some cases, grass and shrub vegetation will be planted on river banks. These plants would support livestock and slow soil erosion in times of floods. In other cases, the construction of earth retaining walls or rock jetties may be needed. These improvements could be made continuously. The specific goals are:

(a) Within 6 months, identify areas most vulnerable to water and wind erosion along the Puyango–Tumbes River.

(b) Increase the issuance of permits for the extraction of aggregates (construction materials) in riverbeds and streams.
(c) Apply sustainable techniques to protect river banks.

Design, build, and operate conventional municipal wastewater treatment plants to provide sewage treatment in populated areas where untreated effluent is currently discharged into the receiving water body. The sites for the treatment plants are shown in Fig. 8.19. The specific goals are:

(a) Within 4 years, design, build, and operate conventional wastewater treatment—primary and secondary (stabilization ponds)—for the following cities along the river: Zaruma, Portovelo, Pampas de Hospital, San Jacinto, Corrales, and Tumbes.
(b) Reuse the treated wastewater in reforestation programs and municipal green areas. This would be accomplished in within 5 years.

Curbside solid waste pickup and recycling programs would be established with the aid of a compensation plan. Landfills would also be designed and constructed to receive and properly dispose of nonrecyclable waste. These landfill facilities would be located in the vicinity of generator cities within the basin. The specific goals are:

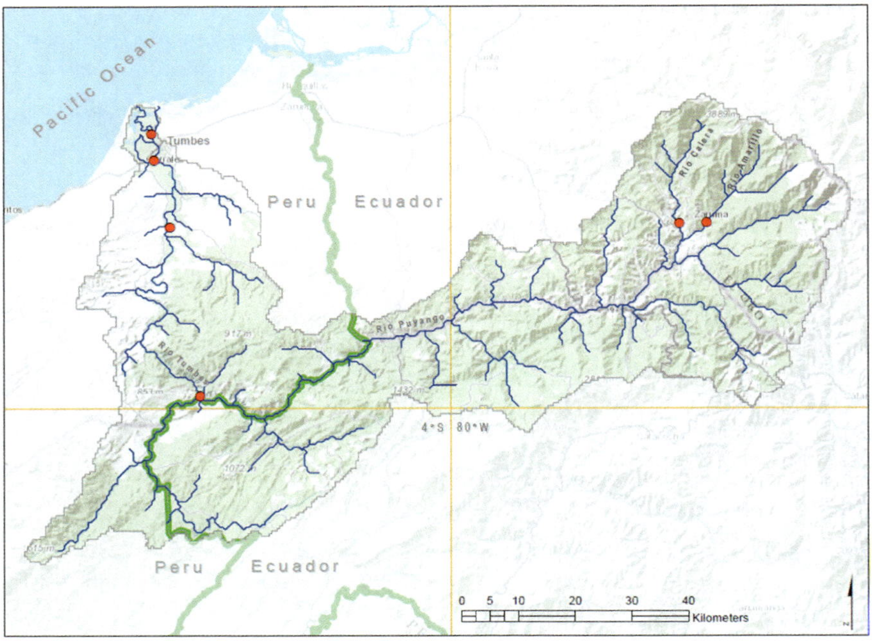

Fig. 8.19 Locations for the new wastewater treatment plants

(a) Within 2 years initiatives are taken to develop curbside solid waste pickup programs that include recycling of appropriate materials. These initiatives, based upon a household compensation plan, would be implemented within 2 years in the following cities: Zaruma, Portovelo, Pampas de Hospital, San Jacinto, Corrales, and Tumbes.
(b) Design and build six sanitary landfills during a 5-year period.

Community-wide environmental education programs would be established to initiate and strengthen environmental awareness and concern on the part of all citizens living in the basin. These programs would address land, air, and water issues related to the basin. The specific goals are:

(a) Initiate and strengthen the environmental conscience of the population in the Puyango–Tumbes River Basin.
(b) Implement an environmental education program throughout the basin.

8.5 Compliance Plan

The proposed schedule for the construction activities and the investment required to implement the Environmental Management Plan for the Puyango–Tumbes River System is shown in Table 8.10.

8.6 Conclusions and Recommendations

Four specific conclusions have been derived from this study of the Puyango–Tumbes river basin. They are:

(1) The main pollution sources in the Puyango–Tumbes river system are the effluents from gold mining and refining operations in the upper basin (Amarillo and Calera Rivers), the runoff from agricultural activities in the lower basin, the generation of domestic and commercial wastewater and solid waste in the cities of Portovelo, Zaruma, and Tumbes, and the effluent from shrimp ponds downstream from the city of Tumbes.
(2) The levels of the main pollutants affecting water quality in the Puyango–Tumbes basin exceed the A1 water type environmental quality standards. The list of pollutants and their pollution levels are: coliforms (above 0 MPN/100 ml), arsenic (above 0.01 mg/l), cadmium (above 0.0003 mg/l), iron (above 0.3 mg/l), manganese (above 0.1 mg/l), and lead (above 0.01 mg/l).
(3) The proposed environmental management plan for the Puyango–Tumbes river basin consists of six distinct programs. The compliance period for this proposed plan is 6 years. The total financial investment required to implement the recommended actions is 100.93×10^6 new soles (approximately 35×10^6 $US).

Table 8.10 Implementation activities for the environmental management plan: schedule and costs

Description of activities and means of mitigation in the environmental management plan		Annual costs (thousands of new soles[a])						Total
		2013	2014	2015	2016	2017	2018	
Mining relocation activity	Prohibition of new mining construction in the Amarillo and Calera Rivers	10	0	0	0	0	0	17,510
	Removal of sediments	1,500	0	0	0	0	0	
	Design and construction of the Metallurgical mining park	1,000	5,000	5,000	0	0	0	
	Mine planning	0	0	0	5 000	0	0	
Reforestation of upper watershed	Prohibition of tailings	10	0	0	0	0	0	5,710
	Analysis of the deforested area	100	0	0	0	0	0	
	Construction of tree nurseries	0	600	0	0	0	0	
	Reforestation	0	1,000	1,000	1,000	1,000	1,000	
Shoreline defense program	Analysis of vulnerable areas	100	0	0	0	0	0	15,110
	Increased use of permits	10	0	0	0	0	0	
	Application of riverine defense techniques	0	3,000	3,000	3,000	3,000	3,000	
Municipal wastewater treatment plants	Design and construction of treatment plants	5,000	5,000	5,000	5,000	5,000	0	33,000
	Reuse of treated water	0	2,000	2,000	2,000	2,000	0	
Solid Waste pick-up and recycling programs	Implementation of recycling compensation program	2,000	2,000	0	0	0	0	29,000
	Design and construction of sanitary landfills	5,000	5,000	5,000	5,000	5,000	0	
Environmental education		100	100	100	100	100	100	600
Total project costs		14,830	23,700	21,100	21,100	16,100	4,100	100,930

[a] To convert new soles to $US multiply by 0.347

Based on this study, five recommendations are also presented. They are:

(1) It is important to establish an additional monitoring station upstream from the creek hunting grounds. The added monitoring station should be located between stations PE02 and PE03 because there is some evidence that readings at these monitoring stations may not be accurate. A nonuniform pollution plume that often exists near a tributary that enters the river in the vicinity of the stations gives rise to the accuracy concerns.
(2) The monitoring plan should be institutionalized and the necessary monitoring facilities should be put in place so that strict compliance to the planned schedule can be enforced.
(3) The management plan should begin when feasibility studies for the Puyango–Tumbes irrigation project begin, since a dam that is to be a part of this project will be located in the Cabo Inga transfer to the river from the Zarumilla River.
(4) The above recommendations are announced to help plan for the expected generation in a few regions of heavy metals. These heavy metals will be found in both the water column and in the sediments.
(5) This study should be extended to determine if there is evidence of a direct relationship between metal contaminants in the Puyango–Tumbes River and a cause of deaths that have been reported by the Direcciöen Regional de Salud Ambiental, DIRESA (Regional Directorate of Environmental Health).

8.7 Map References and Information

Maps throughout this chapter were created using ArcGIS® software by ESRI. ArcGIS® and ArcMap™ are the intellectual property of ESRI and are used herein under license. Copyright © ESRI. All rights reserved. For more information about ESRI® software, please visit www.esri.com

Topographic Map: ESRI—Available at http://server.arcgisonline.com/ArcGIS/rest/services/World_Topo_Map/MapServer

River Network: HydroSHEDS with edits by Hans Bremer – Available at http://hydrosheds.cr.usgs.gov

Drainage Basins: HydoSHEDS—Available at http://hydrosheds.cr.usgs.gov

World Country Polygons: ESRI—Available at ArcGIS Online

World Latitude and Longitude: ESRI—Available at ArcGIS Online

Acknowledgments The author thanks H. Jack Day and Robert Wenger from the University of Wisconsin-Green Bay for translating and editing his manuscript that was written by him in the Spanish language. Their effort in putting the material in a form appropriate for this book is very much appreciated. I also thank Hans Bremer, a staff member at the Biodiversity Center at the University of Wisconsin-Green Bay, for preparing the maps contained in the chapter based on originals from my doctoral dissertation.

Chapter 9
Water Scarcity in Asia and its Long-Term Water and Border Security Implications for Australia

Gurudeo Anand Tularam and Patrick Marchisella

9.1 Introduction

A review of the current literature on the sustainability of natural resources provided several findings that have critical relevance to the motivation of this study. One was that none of the studies reviewed were related to the topic of this paper. Another was that although water scarcity statistics exist for the regions considered, few studies have considered longer term implications for proximate countries such as Australia in terms of the data. However, it was noted that the sustainability of water is rapidly gaining interest within the environmental, ecological and economical sciences (Gleick 1993; Tularam 2010; Tularam and Ilahee 2007). Indeed, many governmental bodies worldwide are now starting to place priority on long-term water sustainability in environmental research (Tularam and Properjohn 2011; World bank 2012). This is particularly evident in Asia, where in some regions the ratio of percentage of water used annually compared to the annually renewable supply of water has decreased in value significantly within the last 20 years. Hydrologists are now forecasting that if the negative rate of change of this ratio continues similarly to that of the last 20 years, by 2025 some moderately industrialised and developed regions of East Asia will be unable to satisfy the minimum level of fresh water required for annual human consumption per capita (Alcamo et al. 2000; Gleick 1993). This forecast assumes no increase in current population levels within the region, and thus it becomes immediately obvious that if increasing population levels were considered in the forecast, the results would be even more alarming

G. A. Tularam (✉) · P. Marchisella
Environmental Futures Centre, Science Environment Engineering and Technology [ENV], Griffith University, Brisbane, Australia
e-mail: a.tularam@Griffith.edu.au

P. Marchisella
e-mail: patrick.marchisella@griffithuni.edu.au

R. M. Clark and S. Hakim (eds.), *Securing Water and Wastewater Systems*, Protecting Critical Infrastructure 2, DOI: 10.1007/978-3-319-01092-2_9,

As part of a larger project on investigating water, food and financial sustainability in the Asian region and long-term implications for border security in Australia, preliminary research concerning water security in Asia has recently been conducted by the authors in two areas; the first being an investigative review of current and long-term water sustainability in Asia, and the second being developing mathematical models and techniques to model human dispersal in such conditions. This manuscript only concerns the investigation into the water security aspect in Asia and it implications for Australian boarder security in the longer term, while modelling is considered in another paper. The main aim here is to review and critically analyse the current water scarcity situation in Asia in terms of population growth patterns as well as forecasts or predictions made by major authorities. The method used is quantitative based on water data with a critical comparative analysis of the literature, in terms of the longer term impacts on Australia for any future water crises in Asia, based on current conditions in proximate Asian countries.

Australia is regarded to be the driest inhabited continent on earth, with respect to naturally occurring renewable water sources and total land mass. However, coastal Australia is abundant in naturally occurring renewable resources (Tularam and Ilahee 2010). It has been forecasted that coastal regions will be able to satisfy the minimum level of fresh water required for annual human consumption per capita for a considerably longer time interval, assuming no population increase (Alcamo et al. 2000). If a moderately increasing population is assumed, the forecasted length of time until the same ratio observes a trend towards a negative rate of change within Australia is still less of a concern than that of some Asian regions. Alternatively, if a significant population increase is assumed, as would be observed if the population of a neighbouring nation that has become water-short desires to migrate to the closest region that is abundant in water resources, the forecasted length of time until the ratio observes a trend towards a negative rate of change would be dramatically decreased (Alcamo et al. 2000; Gleick 1993).

The current literature on reasons for large-scale human migration showed that a motivating factor for migration to a new region is often referred to as either a "push" or a "pull" factor, depending on the origin and nature of the factor (Kainth 2009; Higgins 2008; Dzvimbo 2003; Parkins 2010). Push and pull factors are evident in nearly all recorded incidents of large-scale human migration. The motivation for large-scale human migration is most commonly motivated by beliefs amongst inhabitants that a region different from their own and geographically accessible (from their current region) is more abundant in some qualitative and/or quantitative lifestyle elements than that of their current region. Such beliefs thereby underpin the concept of push and pull factors as motivators for large scale-human migration.

This manuscript is presented in the following manner: First, the sample of the Asian countries selected is presented and the selection motivation explained. This is followed by a section on pull and push factors that determine human migration from home countries. Next, a discussion is presented on Australia and its current situation in terms of water statistics as well as the impacts on population increases on water and other resources, as well as security. A summary of the findings and conclusion is then presented with a final section on implications and limitations conclude this paper.

9.2 Regions Considered

The regions of continental Asia considered in this research were India, Pakistan, Nepal, Bangladesh, China and Indonesia. These regions were selected for several reasons incorporating geography, economy, population, water stress, and model design. Spatial and temporal conditions that will determine the geographic and time intervals on which the model will be defined also contributed to the reasoning of the selection of these regions.

Economic considerations and current statistics are provided here per region considered. As the number of categories of statistics related to the focus of this project is immense, the statistics chosen to be investigated for each region were population (United Nations 2011), renewable freshwater availability per capita per year (m^3), total annual renewable internal freshwater resources (billion cubic metres), and total annual percentage of water withdrawn from internal resources (comprising agricultural, domestic and industrial withdrawals) (World Bank 2012). It is generally agreed that the minimum amount of water needed per annum for human survival is 1,000 m^3 (35,318.3 ft^3) per capita, with a region being designated as 'water scarce' when the observed availability of water fit for human consumption in that region falls to this amount or less (Reddy et al. 2004). Thus, each region here is also noted as being either likely or unlikely to be designated as a water scarce region by 2025.

9.2.1 India

The level of freshwater in India has been low for a while now and Table 9.1 shows that in 2025 it will be designated as a water scarce country. India's primary source of water is rainfall, which varies spatially and temporally each year mainly due to the widely varied climatic regions of the country. Climates ranging from arid to wet tropical combined with seasonal extremes with respect to prolonged dryness in

Table 9.1 India's annual freshwater availability, withdrawal and scarcity designation

Population	Annual freshwater availability per capita (m^3)	Annual renewable internal freshwater resources (bcm^3)	Annual water withdrawals (percentage of internal resources)	Designation as 'water scarce' by 2025
$1,224,614 \times 10^3$	1,252	1446	40	Yes

To convert from m^3 to ft^3 multiply by 35.318

cooler months, and sustained rainfall in monsoon periods directly affect the consistency of renewal of freshwater resources (Reddy et al. 2004). Groundwater and surface water constitute the majority of India's water resources, however surface water is present in far greater volumes than ground water in terms of resources accessed for freshwater withdrawal (Tularam and Murali 2009; Tularam 2012). The majority of rivers in the region are not perennial, with the exception of the Ganges and Brahmaputra, which together provide 58.5 % of the country's total surface water resources and support withdrawals of freshwater for more than a third of the country's population (Raj 2010). Correspondingly, renewal and storage of freshwater sourced from the remaining river networks is dependent on seasonal climate behaviour, which is often erratic and unpredictable, particular in the warmer (monsoon) period.

Contributing also to the stress on water availability in India is the rate of growth of the population. Currently, India's population is estimated to be growing at a rate of 1.3 % (World Bank 2012). Although this rate has seen consistently decreasing values in recent years, the uneven spatial distribution of India's population with respect to stress placed on internal water resources implies that if a growth rate larger than or the same as that currently observed is maintained, India will be designated as water scarce by 2025 (Tularam and Murali 2009; Tularam 2012). Increasing population levels not only contribute to total annual domestic withdrawals on internal renewable freshwater resources, but also agricultural and industrial withdrawals on the same resources over longer time intervals. Groundwater is the primary water source for irrigation of India's agricultural areas, with more than 50 % of the total agricultural land area relying on groundwater to sustain crop production. Withdrawals for irrigation purposes are increasing as a result of rapid urbanisation and population growth, and consequently, groundwater levels are rapidly becoming inadequate to continue supplying the land area currently sustained (Raj 2010).

Also of particular concern in the balance of water usage and renewal in India is the issue of water storage. The freshwater delivery and extraction network in India is still experiencing large inefficiencies with respect to the volume of freshwater from runoff and rainfall that can be adequately stored for future consumption. These inefficiencies are exacerbated by increasing stress placed on the current water delivery network by urbanisation and population growth (Brooks 2007).

9.2.2 Pakistan

Current water stress in Pakistan is dire as shown in Table 9.2. Pakistan's primary source of water is groundwater, due to drastic seasonal contrast in climates. Rainfall is observed during summer months, however, the vast majority of this fall occurs mainly on coastal areas, with very little being recorded throughout middle and northern regions, and the duration of each season—and subsequently the duration of rainfall periods—varies dramatically with location (Blood 1994). The

Table 9.2 Pakistan's annual freshwater availability, withdrawal and scarcity designation

Population	Renewable freshwater availability per capita (m^3)	Renewable internal freshwater resources (bcm^3)	Annual water withdrawals (percentage of internal resources)	Designation as 'water scarce' by 2025
173,593,380	323	55	80	Yes

To convert from m^3 to ft^3 multiply by 35.318

Indus River, originating in western China and flowing primarily south-westerly in direction, provides the majority of surface water for domestic, agricultural and industrial regions situated along and around the rivers course through the Indus basin and southwest Pakistan. The majority of this consumption is due to irrigation, needed by the agricultural regions lying to the northwest and southeast of the river basin. The infrastructure of this irrigation network has suffered severe deterioration since its implementation, resulting in only approximately 36 % of the withdrawn water reaching agricultural land (Tularam and Murali 2009; Tularam 2012). As such, the system is massively inefficient, requiring large amounts of water to be withdrawn so that agriculture in all regions relying on the river basin receives adequate amounts of water for crop production.

The upper (northern) regions of the Indus basin have observed significant population growth in recent years, placing increasing consumption stress on groundwater in the area. Combined with recent deforestation within the region, this has contributed to increased levels of sediment build up in downstream water reservoirs, storage facilities and irrigation networks, thereby decreasing the quality of water available for domestic, agricultural and industrial purposes (All and Benjaminsen 2004). As such, renewable freshwater for consumption is approaching a critically low level of availability, as the quality and quantity of water supplied by Pakistan's primary river source is quickly diminishing under population- and agricultural-related stress, and rainfall is inadequate in providing a significant source of renewable water due to the seasonal climatic extremes of the region.

9.2.3 Nepal

Nepal is one of the few Asian nations with currently adequate water resources (see Table 9.3). It is unlikely to be designated a water scarce region by 2025, and currently records one of the highest annual freshwater availability per capita statistics of all the Asian nations. Nepal's currently adequate level of water resources is largely due to its geographic location—it's northern and eastern borders are situated along the meeting of the Ganges and Brahmaputra river basins. A monsoonal climate from June to September provides more than 75 % of the annual rainfall, and total annual surface runoff due to this rainfall is currently estimated to be 225 bcm (billion cubic metres) (7.94662×10^{12} ft^3).

Table 9.3 Nepal's annual freshwater availability, withdrawal and scarcity designation

Population	Renewable freshwater availability per capita (m^3)	Renewable internal freshwater resources (bcm^3)	Annual water withdrawals (percentage of internal resources)	Designation as 'water scarce' by 2025
29,959,364	6734	198	5	No

To convert from m^3 to ft^3 multiply by 35.318

Adequate amounts of extractable groundwater in the southern plains and northern mountainous areas also contribute to total water resource levels (Nepal National Committee of ICID 2000). However, Nepal's situation is worthy of investigation. Annual freshwater availability per capita has been steadily decreasing since 1986, with the current statistic being almost half of the value of the same statistic recorded in 1986. Interestingly, whilst Nepal's population growth rate is high in comparison to other Asian nations, it remained relatively stable (maximum increase of 0.01 %) from 1986 to 1999, and since then has been observed to decrease steadily (World Bank 2012). This places Nepal as somewhat of an exception amongst the other Asian countries with respect to freshwater availability per capita compared to population growth rate. Figures 9.1 and 9.2 illustrate the decline in values of annual renewable internal freshwater resources and population growth rate of Nepal over the last 22 years, respectively (data sourced from World Bank 2012). Basic linear regression plots are also included as interpolative references.

The decline in levels of annual renewable internal freshwater resource per capita in recent years have been largely attributed to a combination of factors relating to irrigation and geography. The southwestern planar areas constitute the majority of Nepal's agricultural land, and irrigation to the agricultural areas is

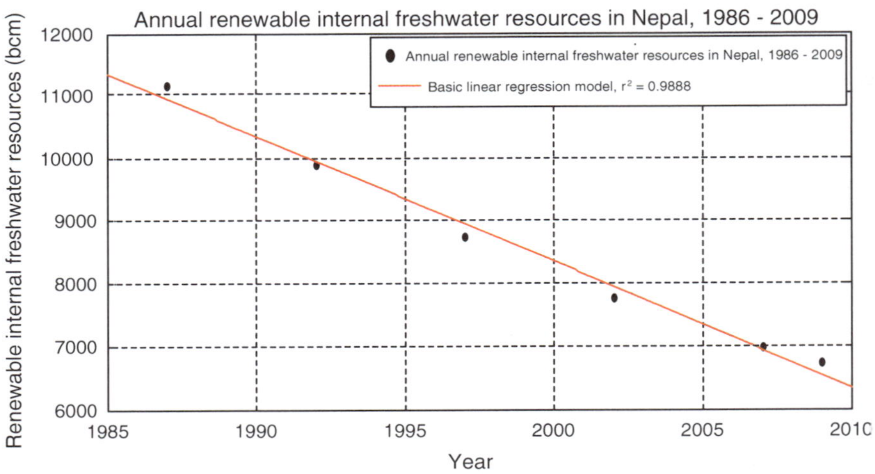

Fig. 9.1 Decline in annual renewable internal freshwater resources in Nepal, 1986–2009

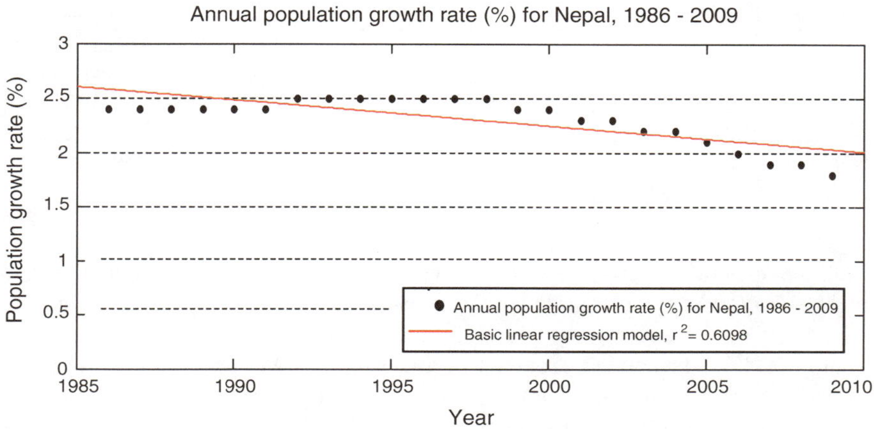

Fig. 9.2 Annual population growth rate in Nepal, 1986–2009

provided primarily from valleys in the mountainous Brahmaputra basin region that spans the northeastern border. However, deterioration of irrigation infrastructure due to minimal maintenance and increased water usage for domestic sources primarily in the south has resulted in less than half of the total agricultural land area receiving adequate levels of water via irrigation. The mountainous region of the northeast also poses geographic difficulties for water availability with respect to rural populations of the area being unable to access adequate quantities of freshwater provided by domestic delivery networks situated mainly in the southwest of the country (Tularam and Murali 2009; Tularam 2012). Waterborne disease in rural areas—due to poor hygiene habits amongst inhabitants and lack of sanitisation facilities—is also a contributing factor to the decline of freshwater resources per capita, as has been reported for some time (Pradhan et al. 2005).

The declining growth rate is mainly attributed to birth control and regulation policies implemented by governments over the last 15–20 years. An increase in the recorded use of contraception over this time interval also contributes to the falling growth rate—a direct result of the establishment of a National Health Education, Information and Communication Centre (NHEICC) administered by the Department of Health in 1993 (World Health Organisation—South East Asia Region 2005).

9.2.4 Bangladesh

Table 9.4 shows that water is also scarce in Bangladesh. Bangladesh's water scarcity is largely due to the geography, climate and socio-domestic status of the region. The majority of Bangladesh's land area lies at low elevation above sea

Table 9.4 Bangladesh's annual freshwater availability, withdrawal and scarcity designation

Population	Renewable freshwater availability per capita (m^3)	Renewable internal freshwater resources (bcm^3)	Annual water withdrawals (percentage of internal resources)	Designation as 'water scarce' by 2025
148,692,131	714	105	3	Yes

To convert from m^3 to ft^3 multiply by 35.318

level, and spans the junction and lowest reaches of the Ganges, Brahmaputra and Meghna rivers. As such, Bangladesh is a vastly planar region affected by similar climate-related water scarcity issues found in regions with similar geography throughout Asia. Prolonged droughts during summer (March through May), dry winters (November through February) and a seasonally long monsoon period (June through October) constitute a climate whereby the annual amount by which naturally occurring water sources are renewed is unstable and unpredictable. In particular, intense storms occur over coastal areas during early summer and late in the monsoon period, which frequently cause infrastructural damage to populated coast line regions in the form of flooding and high-velocity winds (Embassy of Bangladesh–Bahrain 2011). Such annual climatic events immediately affect the infrastructure of water networks and storage facilities, thereby affecting the availability and supply of freshwater for domestic consumption. Salinity levels in drinking water sourced from natural resources along the coast of Bangladesh have also risen in recent years due to storm- and flood-related damage to water network infrastructure, rising sea levels as a result of climate change, and freshwater withdrawals from upstream locations (Kahn et al. 2011).

In relation to its high population density (with respect to spatial boundaries), Bangladesh currently records 31.5 % of its population as living below the international poverty line (in 2008 World Bank recalibrated the international poverty line as being an income of <US$1.25 per day) (Kabir 2011; Bauer et al. 2008). Although this has dropped from 40 % in 2005 (Kabir 2011), and Bangladesh's growth rate seems to have plateaued at 1.1 % in recent years (World Bank 2012), the rapid urbanisation of densely populated areas has slowed the negative change in percentage of population living below the poverty line. As a result, the percentage of the population that is able to access freshwater resources in regions of high population density is also decreasing rapidly, as urbanisation is restricting the accessibility of freshwater to poverty ridden inhabitants (Ahmad 2005).

9.2.5 China

China is facing the prospect of being water scarce by 2025 (Table 9.5) if current population-, industrial-, agricultural- and climate-related impacts on freshwater supply and renewal are maintained. China's landscape is geographically and

Table 9.5 China's annual freshwater availability, withdrawal and scarcity designation

Population	Renewable freshwater availability per capita (m^3)	Renewable internal freshwater resources (bcm^3)	Annual water withdrawals (percentage of internal resources)	Designation as 'water scarce' by 2025
1,338,299,512	2,113	2,813	20	Yes

To convert from m^3 to ft^3 multiply by 35.318

climatically diverse. Arid, temperate to sub-arctic planar regions constitutes the majority of land area in the north, whilst mountainous, water-rich sub-tropical and temperate climatic zones span the remaining western, southern and eastern borders, respectively. China's population is densest around the coastal southern and eastern regions, with the majority of arable, cultivated land also being situated along these borders (Hays 2008).

Water sources in southern China are renewed primarily by glacial melt from the Himalayas. Such sources include the Yangtze and Yellow rivers. However, pollution, consistent over-withdrawal, and climate change-related effects are contributing to the lessening of the quality and quantity of these freshwater sources. In particular, pollution from agricultural and industrial sources is placing increasing stress on freshwater sourced from the nations' river networks. From 2002 to 2006, a relatively short-time interval, it was reported that approximately 63 billion tonnes of wastewater, comprised of 62 % pollutants from industrial sources, and 38 % poorly treated or raw sewage from municipalities and rural areas, flowed into China's river network each year (Turner 2006). In 2011, the Chinese government reported that 43 % of state-monitored rivers contain levels of toxic pollution high enough so that human contact is ill-advised (Hays 2008). In particular, freshwater availability from the Huai, Hai and Huang river basins is decreasing as a result of large-scale dumping of industrial and agricultural waste. These basins provide freshwater for domestic consumption to the densely populated regions along the east, southeast and south coasts of China, and as such, freshwater supply for more than half of the countries' population is being affected by pollutants (Tularam and Murali 2009; Tularam 2012).

Northern China is currently experiencing rural population growth, which is in turn placing stress on the small percentage of freshwater resources (7.6 % of the country's total freshwater resources) located in the region (Xia et al. 2005). Approximately 40 % of both China's population and cultivated land area are situated in the north, and with freshwater usage by industry and agriculture in the area increasing as a result of population growth, freshwater availability in the region is forecasted to decrease dramatically if current usage trends persist or increase. In particular, groundwater levels in northern China are reported as depleting rapidly due to agricultural and industrial expansion, diminishing by 1–3 m (3.2–9.8 ft)/year (Upali et al. 2005).

9.2.6 Indonesia

Along with Nepal, Indonesia is currently one of only a few Asian nations showing adequate water resources (Table 9.6). As an archipelago situated between two continents, annual variation in Indonesia's climate is mainly due to seasonal and spatial rainfall patterns, which are a result of monsoonal periods from June to September and December to March. Whilst rainfall is seasonally and spatially frequent across the nation, the majority of this fall occurs during the December to March monsoon period, with less occurring during the June to September period. Annual average precipitation (in millimetres) within Indonesia is currently estimated to be approximately 2702 mm (106.4 in.) (World Bank 2012). With respect to total land area, the distribution of this rainfall ranges from more than 2,000 mm (78 in.) annually in western and northern regions such as Sumatra, Java, Bali and Kalimantan, to <1,000 mm (39.4 in.) annually in regions closer to Australia, such as Nusa Tengarra and eastern Java. Whilst current water availability in Indonesia in terms of usage and renewal of resources is adequate, stress on water availability is accumulating due to factors such as population growth, pollution and contamination, urbanisation and climate change.

As with many other developing Asian nations, urbanisation and economic advancement within Indonesia is having an increasing impact on the country's renewable water resources. In Indonesia, this impact is most apparent in the form of pollution. Of the total percentage of water withdrawals, latest reports show that 93 % is attributable to agriculture, 6 % to domestic and 1 % to industrial sources. In 2008, it was reported that of the total pollution of Indonesia's ground and surface water resources, 76 % was attributable to industry, 7 % to agriculture and 17 % to domestic sources (Napitupulu and Gunawan 2008). The domestic contribution to water pollution is relatively high when compared to other Asian nations, with 45 % of the population (approximately one hundred million people) not having access to adequate sanitation and waste disposal facilities, and 53 % withdrawing water for consumption and domestic purposes from resources contaminated by raw sewage (Napitupulu and Gunawan 2008). Levels of waste from industrial and domestic sources are increasing in both groundwater and surface water resources, with untreated wastewater being discharged into domestic waterways—a direct result of urbanisation and economic advancement in recent years (Tularam and Murali 2009; Tularam 2012).

Table 9.6 Indonesia's annual freshwater availability, withdrawal and scarcity designation

Population	Renewable freshwater availability per capita (m^3)	Renewable internal freshwater resources (bcm^3)	Annual water withdrawals (percentage of internal resources)	Designation as 'water scarce' by 2025
239,870,937	8,504	2,019	6	No

To convert from m^3 to ft^3 multiply by 35.318

Whilst Indonesia's climate is relatively stable in terms of average annual temperature, with variation being attributed primarily to monsoonal seasons, in 2009, Yusuf and Francisco (2009) reported that areas of Indonesia are critically vulnerable to environmental impacts related to climate change—in particular, the geographical locations of Jakarta and western Java place them at high risk of experiencing extreme climate change-related effects, such as flooding and landslide. These regions represent two of the highest spatial population densities of the Asian nations, and as such, current water delivery and storage networks are experiencing stress related to population growth and urbanisation. Extreme climatic events would place further stress on these networks, and thereby directly impact the accessibility and availability of freshwater in regions of high population density.

The regions considered in the above review are all very highly populated and apart from two countries most have serious water availability problems. In fact, many of them have been predicted to be labelled water scarce by 2025. If the conditions related to water are so dire now and is predicted to be highly critical within the next 10–15 years, then what could be the position in the next 50 or so years. More importantly, how the inhabitants of the regions cope with the even more critical water availability conditions at that time assuming the population increases as well is an interesting consideration.

9.3 Push and Pull Factors: Reasons for Migration and Security Implications for Australia

Large-scale human migration is a well-studied phenomenon (Adams et al. 1998; Gamble et al. 1996, 2000). Historically, motivation for large-scale human migration has come from varied sources, ranging from exploration and evolution in the earliest recorded cases, to political, economic, socio-economic, environmental, demographic, and oppression- and conflict-related sources in recent years. Here large-scale human migration motivated by sources relating to scarcity and abundance of natural resources is considered.

The push and pull factors may be categorised into various groups, depending on the context and qualitative or quantitative nature of the factor (Kuzmin and Tankersley 1996). Whilst lacking rigorous definitions, push factors may be interpreted in the general sense as 'conditions that bring about the desire to leave a current place or region of dwelling on a permanent basis', and similarly, pull factors may be interpreted in the general sense as 'conditions or beliefs that bring about the desire to relocate to a new, specific place or region of dwelling on a permanent basis'. Tables 9.7 and 9.8 illustrate how these factors may be categorised in terms of their nature, and include a qualitative or quantitative contextual description of that factor (Oglethorpe et al. 2007; Skellam 1951).

Whilst these Tables 9.7 and 9.8 display obvious correlations between corresponding push and pull factors, the correlations do not necessarily enforce dependence of one factor on the other in order for migration to take place (Schoorl

Table 9.7 Categorisation of push factors, with contextual descriptions

Factor	Contextual description/s
Socio-economic	Sustained quantitative degradation of lifestyle—lack of employment opportunities, poor or low income, little or no capital or financial security, poor levels of education or academic opportunity, little or no prospect of entrepreneurial opportunity, increase in inadequacy of land to sustain industry or agriculture, poor living conditions (unforced)
Oppressive	Sustained qualitative degradation of lifestyle due to religiously, politically or economically related hostility or conflict, poor living conditions (forced)
Environmental	Natural disasters, increase in the frequency of extreme climatic events, pollution or forced degradation of habitat, depletion of natural resources due to climate-related effects or over-usage with little or no renewal
Demographic	Sustained qualitative and quantitative degradation of lifestyle due to over-population, over-development and/or urbanisation
Personal	Desire to leave current habitat on a permanent basis for purposes relating to personal reasons and/or ambitions

Table 9.8 Categorisation of pull factors, with contextual descriptions

Factor	Contextual description/s
Socio-economic	Belief that a different, specific habitat will provide improved economic prospects: employment opportunities; higher achievable income; opportunities for capital gain; increased financial security; higher levels of accessible education; improved quantities of adequate land for industry or agriculture; improved living conditions
Safe harbour	Belief that a different, specific habitat harbours political, economic and/or religious views that are in better agreement with or better tolerate the individual's political, economic and/or religious views: little or no prospect of being subject to oppression in the new habitat; improved living conditions
Environmental	Belief that a different, specific habitat experiences less occurrences of extreme climatic events and natural disasters: safer environment for population growth and socio-economic development; improved living conditions
	Belief that a new, different habitat sustains improved levels of accessible natural resources: adequate resources for population growth and socio-economic advancement; improved ratio of usage versus renewal of natural resources; low levels of pollution and degradation of land; improved living conditions
Demographic	Belief that a different, specific habitat promotes improved living conditions due to lack of being over-populated, over-developed and/or urbanised
Personal	Belief that personal reasons and/or ambitions can be satisfied within a new, specific habitat

et al. 2000). Desire to leave a particular habitat and relocate to another may be a result of one factor influencing an individual, but with no contributing influence from a separate but related corresponding push/pull factor. An example of this would be an individual pursuing capital gain abroad despite having adequate and secure financial income and capital assets in their current place of dwelling. However, in the context of the depletion of natural resources as a push factor, two corresponding pull factors, each contributing equally to the desire to migrate, can be reasonably assumed:

- adequate levels of natural resources in the new dwelling; and
- any new dwelling must display a sustained absence of the causes of natural resource depletion compared to the current dwelling.

Whilst being reasonably intuitive, these assumptions also translate to a property of density dependent diffusion modelling—that being the observed and well documented property of diffusion of a defined quantity towards areas of little or no density of that quantity, within spatial-boundary and temporal conditions (Murray 2002; Steele and Hazelwood 2003; Steele 2009).

As such, some properties of large-scale human migration motivated by depletion of natural resources may be considered as 'density' dependent (Philibert 2006)—that is, as the migration continues within a closed region, the behaviour of these properties at each point is directly affected by the same effects as those causing water scarcity in the original habitat. This and other mathematical properties and considerations of migration due to depletion of natural resources will lead to mathematical modelling considerations for another paper. The countries considered in this preliminary research comprise a total population of more than 3.1 billion people. Countries within close proximity of these regions that display adequate availability and projective sustainability of water include Myanmar, Malaysia, Laos, Bhutan and Australia. Table 9.9 considers some basic demographic and economic statistics of these countries, including renewable freshwater availability per capita for each (World Bank 2012). Level of income, improved urban water sources and improved urban water facilities including sanitation are collectively intended as a means of determining the current general developmental status of each country. Arrows indicate an increasing or decreasing trend in the value of the quantity over the last 10 years.

Upon comparing and contrasting the values of these quantities with respect to depletion of natural resources as a push factor for migration, several assumptions

Table 9.9 Basic demographic and economic statistics—including renewable freshwater availability per capita—of countries within close proximity of the considered regions

Country	Population	Renewable freshwater availability per capita	Current population growth rate (%)	Improved urban water sources (percentage of population with access)	Improved urban water facilities including sanitation (percentage or urban population with access)	Level of Income
Myanmar	47,963,012	21,071 m^3 ↓	0.8 ↑	75	86	Low
Laos	6,200,894	31,151 m^3 ↓	1.4 ↓	72	86	Lower middle
Bhutan	725,940	109,295 m^3 ↓	1.7 ↓	99	87	Lower middle
Malaysia	28,401,017	20,752 m^3 ↓	1.6 ↓	100	96	Upper middle
Australia	22,328,800	22,413 m^3 ↓	1.7 ↑	100	100	High

To convert from m^3 to ft^3 multiply by 35.318

can be reasonably made about the likelihood of these countries as target habitats. Firstly, however, some known properties of migration motivated by the depletion of natural resources must be examined. Historically, in the case of large-scale human migration caused by extreme climatic, environmental, political or oppressive events, the following two properties have been observed and noted:

- a considerable majority of migrants travel only until a new and suitable habitat is found (Olsson 1965; Ravenstein 1885); and
- a pre-determined destination may be evident depending on the income status of the individual at the time of migration, and generally, migrants prefer a region that is of the same or higher developmental status as that of their origin country (Keeley 2009).

As such, the proximity of habitats that display an adequate shortage of the migration-motivating conditions, the income status of the individual, and the developmental stage of the target regions all play critical roles in the decision of a target habitat, and can be thought of as sub-pull factors of sorts.

In this analysis, in terms of the shortest distance between borders, Myanmar, Malaysia, Laos and Bhutan are relatively close to all regions considered in this study, whilst Australia is closest to Indonesia. It may then be reasonably assumed that with the exception of Indonesian migrants (and not including refugees, displaced citizens and/or illegal immigrants), a migrant's income status will affect their capability to consider Australia as an accessible target destination. With the exception of China, the considered regions consist of third world and low to middle income countries, and so it follows that of the total population of these regions, only a certain percentage would be of adequate enough income to consider Australia as a possible habitat. Therefore, stratification of migrants by income status must be considered when investigating push and pull factors in the case of large-scale migration due to depletion of natural resources.

When assessing the statistics presented, along with the water sustainability and accessibility status of each of these regions, it is clear that Australia's current status with respect to the usage and renewal of this natural resource is comparatively better. Whilst Table 9.9 indicates that Myanmar, Laos, Bhutan and Malaysia also display better statistics with respect to water consumption and renewal than that of the considered regions, consideration of the properties discussed above implies that Australia will also be a target destination in the event of migration caused by water scarcity in these regions.

9.4 Discussion: Australian Situation

From 2002 to 2011, the number of immigrants to Australia from Asian regions per year shows a generally increasing trend compared to previous years. Table 9.10 displays the net permanent addition to the Australian population over this time interval, along with the total number and percentage of the net permanent addition

Table 9.10 Net permanent addition to the Australian population from 2002 to 2011 along with the percentage of the net permanent addition attributed to migrants from southeastern, south and southwestern regions of Asia per year. Net permanent addition attributed to humanitarian reasons for immigration per year is also included

Years	Net permanent addition to Australian population	Percentage of net permanent addition attributed to migrants of southeastern and southern Asian origin	Net permanent addition attributed to humanitarian reasons for immigration	Percentage of humanitarian intake attributed to migrants of southeastern and southern Asian origin
2002–2003	125 860	26.96	10 185	4.07
2003–2004	148 884	26.56	10 938	3.72
2004–2005	167 319	25.92	17 528	3.47
2005–2006	179 807	26.12	16 964	6.80
2006–2007	191 907	28.49	14 158	18.14
2007–2008	205 940	29.65	11 729	30.68
2008–2009	224 619	29.76	14 854	31.39
2009–2010	208 921	31.65	14 553	29.75
2010–2011	213 409	30.91	13 976	28.56

attributed to migrants from south and southeastern regions of Asia per year. The percentage of the net permanent addition attributed to humanitarian reasons for immigration per year is also included as a means to determine the proportion of the permanent addition that would have migrated due to socio-economic, oppressive, and environmental push and pull factors (Australian Government: Department of Immigration and Citizenship 2011).

Amongst other observations, assessment of these quantities with respect to migration motivated by socioeconomic, oppressive, and environmental push and pull factors shows the following trends:

- over the interval 2002–2011, a generally increasing trend was observed not only in the value of the percentage of net permanent addition attributed to migrants of southern and southeastern Asian origin, but also in the value of the percentage of humanitarian intake attributed to migrants of southern and southeastern Asian origin between 2004 and 2009, the increase in the latter being particularly rapid over that period;
- the years between 2009 and 2011 is the only interval in the last 10 years where net permanent addition attributed to migrants of southern and southeastern Asian origin, percentage of net addition attributed to humanitarian reasons for migration, and percentage of humanitarian intake attributed to migrants of southern and southeastern Asian origin all decreased simultaneously.

By these observations alone, it would be reasonable to assume that over the last 10 years, Australia has become considerably more attractive and accessible as a target destination for permanent migration motivated by socioeconomic, oppressive and environmental push and pull factors.

Although Australia's current status with respect to water usage and renewal seems adequate in comparison to the nations considered in this study, investigation has provided somewhat less positive findings in terms of an adequate state of sustainability and usage in the long term. Water resources in Australia are comprised of groundwater and surface water resources, with surface resources providing the majority of domestically and industrially consumed water. However, due to frequent and varied climatic influences on surface water, groundwater continues to provide the most consistent source of water for a majority of agricultural purposes. Surface water is largely comprised of rivers, estuarine networks, wetlands and lakes. In terms of total usage sustained by groundwater and surface water collectively, and amongst total domestic, agricultural and industrial uses, agriculture continues to record the highest values of usage, followed by domestic and industrial usage values (respectively) per year in recent years (Australian Bureau of Statistics 2011).

In 2010, the Water Services Association of Australia published a report on the implications of population growth in Australia on urban water resources. Included in the report were three projections, each for urban water consumption in both 2026 and 2056, and each assuming population increases that were either minimal, in line with trends observed in 2009, or rapid (respectively) until the year of each projection. In these projections, *urban water consumption* incorporates residential, commercial, municipal and industrial water uses. The projection for 2026 is immediately worthy of consideration as this is the approximate time by which certain regions considered in this study may be deemed water scarce. Table 9.11 presents the projected urban water consumption for 2026 per capital city for each assumed level of population increase, compared with reported usage in 2009 (reproduced from *Implications of population growth in Australia on urban water resources* Water Services Association of Australia 2010).

It may be reasonably assumed that the increase in Australia's population over the period 2009–2026 is unlikely to be less than that observed in 2009, and so it would follow that the change in urban water usage over this interval is likely to be between 42 and 49 %. Whilst at first this seems somewhat acceptable when compared to the current and long-term implications on water availability and sustainability faced by certain regions considered in this study, such a comparatively acceptable projection can also be reasonably considered to act as a pull factor in the event of large-scale permanent migration due to water scarcity.

As discussed throughout the reviews of water usage and renewal for each considered region, population increase can dramatically affect the rate of urbanisation and development of a region, thereby placing increased stress on water delivery and sanitation networks. Along with the urban sector, agriculture in Australia could face implications in terms of water availability as a result of population increase. Agriculture usage already constitutes the largest percentage of total freshwater withdrawals in Australia. Surprisingly, yearly statistical records measuring total water consumption each for the domestic, industrial and agricultural sectors are not available, and as such, identifying any trend in usage changes for these sectors on a yearly basis is difficult due to inadequate amounts of data.

Table 9.11 Projected urban water consumption for 2026 per capital city (gL-gigalitres; 1 gL = 1.0×10^{12} cm^3)

City	2009 Usage (gL)	Projected usage in 2026 assuming rapid population increase (gL)	Percentage change from 2009 usage (%)	Projected usage in 2026 assuming population increase similar to that observed in 2009 (gL)	Percentage change from 2009 usage (%)	Projected usage in 2026 assuming minimal population increase (gL)	Percentage change from 2009 usage (%)
Sydney	492	620	26	613	25	605	23
Melbourne	360	535	48	511	42	494	37
South-East Queensland[a]	223	536	141	494	122	494[a]	122[a]
Adelaide	138	178	29	174	26	175	27
Perth	250	308	23	284	14	265	6
Canberra	42	66	57	59	41	53	27
Total	1505	2242	49	2136	42	2,086	39

[a] Southeast Queensland 2026 projections were sourced from the 2009 Southeast Queensland Water Strategy. No minimal growth population projection has been formulated within the 2009 Southeast Queensland Water Strategy. Thus, the projected values for population increase similar to that observed in 2009 have also been used for the case of minimal population growth (*Implications of population growth in Australia on urban water resources*, Water Services Association of Australia 2010)

Table 9.12 Percentage of total water consumption in Australia attributed to agricultural usage for certain years between 2002 and 2010 (%) (Australian Bureau of Statistics)

Years	Percentage of total water consumption attributed to agricultural usage (%)
2000–2001	67
2004–2005	65
2008–2009	54
2009–2010	52

However, total water consumption for each of these sectors has been recorded for certain 1 year interval since 2000, and Table 9.12 presents the percentage of total water consumption in Australia attributed to agricultural usage for certain years between 2002 and 2010 (Australian Bureau of Statistics 2011).

Firstly, it must be noted that if a negative trend were to be assumed and extrapolated from this data, then it has been reported and is generally agreed that such an assumptive negative trend in total agricultural water usage between 2004 and 2010 usage is largely explained by extended periods of drought affecting the majority of farming land across the nation throughout that period (Australian Bureau of Statistics 2011). Water supply to agricultural land in Australia is directly affected by climate. As such, extreme water-related climatic events such as drought and flood directly impact not only supply, but also availability of water, as runoff and seepage to surface and ground water resources (respectively) are both provided by rainfall. Therefore, agricultural water availability in Australia is immediately related to climate behaviour, which places direct stress on water sustainability with respect to usage and renewal in the agricultural sector. Increasing levels of population growth will demand more output from agriculture. Further, if population growth is intensive over a relatively short-time interval, as would be observed in the case of large-scale migration from Asian regions motivated by water scarcity, this will place additional stress on the capability of agriculture to provide demanded output whilst still consuming water at a rate that is sustainable in the long term.

In summary, consideration of the balance of the push and pull factors and conditions discussed above leads to a conclusion whereby Australia can be identified as a likely choice of target destination by migrants in the event of migration caused by water scarcity in the considered Asian regions. In this case, implications for Australia may become apparent in the form of short-term intensive population growth, urbanisation, and increased stress on the sustainability of water resources for the domestic, industrial and agricultural sectors.

9.5 Summary and Conclusion

The findings of this research show that water scarcity in Asia and associated long-term border security implications for Australia demands increased attention in terms of both qualitative and quantitative assessment. It is clear that if current levels of population growth, urbanisation, and industrial and agricultural development are maintained in certain regions of south and southeast Asia, these regions risk facing dire levels of water scarcity in the next 10–50 years. The majority of the regions selected for consideration in this study will face various levels of water scarcity within this time period if current water usage trends are maintained. Water scarcity will result in significant demographic, economic and environmental implications, many of which will immediately affect qualitative and quantitative aspects of the lifestyles of the population.

Forced and unforced degradations of lifestyle elements have been shown to be a motivating factor for the consideration of permanent migration at both individual and population levels. In the majority of nations considered in this research, it can be seen that demographic, economic and environmental implications directly related to water scarcity will likely result in some form of degradation of lifestyle for a considerably significant percentage of the population. Even when socio-economic conditions inherent to the individuals of a population are considered, the proportion of individuals that would be economically and/or financially capable of permanent migration remains considerably significant. The review of push and pull factors shows that in the event of degradation of lifestyle within a region, individuals generally seek a target habitat that displays improved qualities of lifestyle over a sustained period than that of their origin habitat.

In the event of large-scale migration over time from the Asian region due to prevalence of water scarcity, Australia's comparatively adequate status with respect to water usage and sustainability places it as a likely target habitat for a large percentage of the migrating individuals. Australia's developmental status, compared with those of the majority of nations from which the migration will occur, will also contribute to its attractiveness as a target habitat, as discussed in the review of push and pull factors. Projections of water usage in Australia up until the year 2026 (determined from domestic, industrial and agricultural usage indicators in recent years) show that levels of water usage will remain acceptable assuming minimal population increase. However, it has also been shown that current stress placed on water sustainability will worsen if a comparatively larger increase in population is observed over this time period. It would be reasonable to expect that large scale migration from the south and southeastern Asian regions due to water scarcity would contribute to a considerable increase in migrant intake in Australia, and thus, if the intake is significant, stress on water sustainability in Australia will correspondingly increase. Renewal of natural water resources in Australia has been shown to depend largely on climate behaviour, and as such, if usage increases as a direct result of population increase over relatively short-time period, implications for the domestic, industrial and agricultural sectors will arise. If the population increase is sustained over a longer time interval, this will exacerbate water sustainability issues observed in the short term, and associated stress on water usage and renewal may result in long-term implications for all sectors.

9.5.1 Implications and Limitations

Quantifying the magnitude of a large-scale migration due to natural resource depletion is critical in developing a rigorous understanding of the behavioural dynamics of such a migration, including the scale of population influx that would be observed in target habitats over certain time periods. Mathematical modelling of population dynamics is well studied with many properties translating directly to the historically recorded qualitative and empirical properties of human population dynamics. It is also noted that the conventional Fisher-Skellam population model may be applied to this specific context of dispersal but a number of parameters needed will be required for in-depth analysis of the various factors involved.

Whilst this report confirms that water scarcity in Asia will likely result in short- and long-term border security implications for Australia, further attention is needed to better understand not only the behavioural dynamics of dispersal motivated by such circumstances, but also the impacts of such a dispersal is likely to have on both target and origin habitats alike. Consideration of how the impact of the dispersal on target destinations will be affected due to socio-economic conditions amongst the migrating individuals is also needed. This will help develop a more rigorous understanding of the nature and magnitude of implications faced by the target habitats.

References

Adams J, Sluckin T, Steele J (1998) Modelling Paleoindian dispersals. World Archaeol 30(2):286–305

Ahmad M (2005) Living in the coast: urbanization', Government of the People's Republic of Bangladesh. Ministry of water resources, Water Resources Planning Organization (WARPO). http://www.warpo.gov.bd/rep/liv/living4.pdf. Accessed 20 Jan 2012

Alcamo J, Henrichs T, Rosch T (2000) World water in 2025—global modelling and scenario analysis for the world commission on water for the 21st Century. Report A0002, Center for Environmental Systems Research, University of Kassel, Kurt Wolters Strasse 3, 34109 Kassel, Germany

All J, Benjaminsen TA (2004) Fuelwood, timber, and deforestation in the Himalayas; the case of Basho Valley, Baltistan Region, Pakistan. Mt Res Dev 24(4):312–318

Australian Bureau of Statistics (2011) Water account, Australia 2009–2010, 2008–2009, 2004–2005, 2000–2001. http://www.abs.gov.au/. Accessed 5 Feb 2012

Australian Government Department of Agriculture, Fisheries and Forestry: Food Policy and Programs Section (2011). Australian food statistics 2009–2010, ISSN 1447-8358, ISBN 978-1-921575-11-2. http://www.daff.gov.au/__data/assets/pdf_file/0011/1910819/food-stats2009-10.pdf. Accessed 6 Feb 2012

Australian Government: Department of Immigration and Citizenship (2012) Statistical publications. http://www.immi.gov.au/media/publications/statistics/. Accessed 7 Feb 2012

Bauer A, Hasan R, Magsombol R, Wan G (2008) The World Bank's new poverty data: implications for the Asian Development Bank. In: Asian Development Bank—Sustainable Development Working Paper Series, No. 2, November 2008. http://www.adb.org/Documents/Presentations/New-Poverty-Estimates/Poverty-Data-Implications.pdf. Accessed 20 Jan 2012

Blood P (ed.) (1994) Pakistan: A country stud. http://countrystudies.us/pakistan/. Accessed 5 Jan 2012
Brooks N (2007) Imminent water crisis in India. http://www.arlingtoninstitute.org/wbp/global-water-crisis/606. Accessed 4 Jan 2012
Dzvimbo K (2003) The international migration of skilled Human capital from developing countries, World Bank—Human Development Network; Education (HDNED). http://siteresources.worldbank.org/INTAFRREGTOPTEIA/Resources/peter_dzvimbo.pdf. Accessed 25 Jan 2012
Fisher R (1937) The wave of advance of advantageous genes. Annals of Eugenics, vol 7. Cambridge University Press, Cambridge, pp 355–369
Gamble C, Sluckin T, Steele J (2000) Estimating the rate of PaleoIndian expansion into South America. In: Nicholson R, O'Connor T (eds.) People as an agent of environmental change, Oxbow Books, pp. 125–132
Gamble C, Sluckin T, Denholm D, Steele J (1996) Simulating the hunter-gatherer colonization of the Americas. Analecta Praehistorica Leidensia 28:223–227
Gleick P (1993) Water and conflict. Int Secur 18(1):79–112
Hays J (2008) Agriculture and organic farming in China. http://factsanddetails.com/china.php?itemid=348&catid=9&subcatid=63. Accessed 17 Jan 2012
Hays J (2008) Water pollution in China. http://factsanddetails.com/china.php?itemid=391&catid=10&subcatid=66. Accessed 18 Jan 2012
Hazelwood L, Steele J (2003) Spatial dynamics of human dispersals; constraints on modelling and archaeological validation. J Archaeol Sci. Elsevier Ltd, DOI 10.1016/j.jas.2003.11.009. Accessed 6 Jan 2012
Higgins J (2008) Push and pull factors. http://www.heritage.nf.ca/society/pfactors.html. Accessed 25 Jan 2012
Implications of population growth in Australia on urban water resources (2010) Water services association of Australia, occasional paper no. 25. https://www.wsaa.asn.au/Publications/Documents/201007%20WSAA%20Occasional%20Paper%2025.pdf. Accessed 7 Feb 2012
Kabir H (2011) Poverty rate now 31.5pc. Bangladesh Economic News. http://bangladesheconomy.wordpress.com/2011/05/04/poverty-rate-now-31-5pc/. Accessed 19 Jan 2012
Kainth SG (2009) Push and pull factors of migration: a case of brick kiln industry of Punjab state. Asia-Pacific J Soc Sci 1(1):82–116
Keeley B (2009) International migration: the human face of globalisation. OECD, Multilingual Summaries, ISBN 978-92-64-047280. http://www.oecd.org/dataoecd/19/61/43568302.pdf. Accessed 2 Feb 2012
Khan AE, Ireson A, Kovats S, Mojumder SK, Khusru A, Rahman A (2011) Drinking water salinity and maternal health in coastal Bangladesh: implications of climate change. Environ Health Perspect 119:1328–1332. http://dx.doi.org/10.1289/ehp.1002804. Accessed 19 Jan 2012
Kuzmin Y, Tankersley K (1996) The colonization of eastern Siberia: an evaluation of the paleolithic age radiocarbon dates. J Archaeoll Sci 23(4):577–585
Murray J (2002) Mathematical biology, 3rd Edn. Springer, New York
Napitupulu L, Gunawan D (2008) Economic impacts of sanitation in Indonesia: a five-country study conducted in Cambodia, Indonesia, Lao PDR, the Philippines, and Vietnam under the Economics of Sanitation Initiative (ESI). Water and sanitation program—East Asia and the Pacific (WSP-EAP), Research Report August 2008, World Bank, Jakarta, pp 1–94
Oglethorpe J, Ericson J, Bilsborrow R, Edmond J (2007) People on the move: reducing the impact of Human migration on biodiversity. World Wildlife Fund and Conservation International Foundation, Washington. http://www.worldwildlife.org/what/communityaction/people/phe/WWFBinaryitem9123.pdf. Accessed 29 Jan 2012
Olsson G (1965) Distance and Human interaction. A migration study. Geografiska Annaler. Series B, Hum Geogr 47(1):3–43
Parkins N (2010) Push and pull factors of migration. Am Rev Political Econ 8(2):6–24

Philibert J (2006) One and a half century of diffusion: Fick, Einstein, before and beyond. Mater Sci 4(6):1–19
Pradhan B, Gruendlinger R, Fuerhapper I, Pradhan P, Pradhanang S (2005) Rural communities' knowledge on water quality and water borne disease: the case of Bungamati locality in Kathmandu Valley, Nepal. Institute of Medicine, Utter Pradesh, pp 1–4
Raj A (2010) Water security in India: the coming challenge. Future Directions International—Independent Strategic Analysis of Australia's Global Interests, Pty. Ltd, Australia
Ravenstein E (1885) The laws of migration. J Stat Soc London 48(2):167–235
Reddy MS, Char NVV, Afzal N, Ayub QS, Basnyat DB, Karmacharya JL, Maniruzzaman Miah M, Mukherjee S, Nickum JE, Rahman K, Sajjadur Rasheed KB (2004) Water demand—Supply gaps in South Asia, and approaches to closing the gaps. WASSA Project Reports, 1, no. 1, South Asia Program/FPI—school of Advanced International Studies (SAIS), John Hopkins University, Washington
Schoorl J, Heering L, Esveldt I, Groenewold G, Erf R, Bosch A, Valk H, Bruijn B (2000) Push and pull factors of international migration: a comparitive report. Office for Official Publications of the European Communities, pp. 73–85. http://www.nidi.knaw.nl/Content/NIDI/output/2000/eurostat-2000-theme1-pushpull.pdf. Accessed 1 Feb 2012
Skellam J (1951) Random dispersal in theoretical populations. Biometrika 38(1):196–218
Steele J (2009) Human dispersals: mathematical models and the Archaeological record. Human Biol 81(2). http://digitalcommons.wayne.edu/humbiol/vol81/iss2/2. Accessed 7 Jan 2012
The World Bank (2012) http://www.worldbank.org/. Accessed 7 Jan 2012
Tularam GA, Ilahee M (2007) Environmental concerns of desalinating seawater using reverse osmosis. J Environ Monit 9(8):805–813 (Epub 2007 Jun 27)
Tularam GA (2010) Relationship between El Niño southern oscillation index and rainfall (Queensland, Australia). Int J Sustain Dev Plann 5(4):378–391
Tularam GA, Ilahee M (2010) Time series analysis of rainfall and temperature interactions in coastal catchments. J Math Statis 6(3):372–380
Tularam GA, Murali KK (2009) Water security issues of Asia and their implications for Australia. Environmental futures centre report, faculty of science, environment, engineering and technology [ENV], Griffith University, Brisbane, pp 1–22
Tularam GA (2012) Water Security Issues of Asia and their implications for Australia. In: Thangarajan M, Mayilswami C, Kulkarni PS, Singh VP (eds.) Proceedings of the IWGC-Fifth international groundwater conference, The assessment and management of groundwater resources in hard rock systems with special reference to Basaltic Terrain (pp. 288) held in Aurangabad, Maharastra, 18–21 Dec 2012
Tularam GA, Properjohn M (2011) An investigation into modern water distribution network security: Risk and implications Secur J 24:283–301
Turner J (2006) China's water problems. http://www.ocnus.net/cgi-bin/exec/view.cgi?archive=106&num=27251. Accessed 17 Jan 2012
Upali A, Giordano M, Liao Y, Shu Z (2005) Water supply, water demand and agricultural water scarcity in China: a basin approach. International Commission on Irrigation and drainage (ICID), New Delhi. http://www.iwmi.cgiar.org/assessment/files_new/research_projects/ICID_China_A_Basin%20Approach.pdf. Accessed 16 Jan 2012
Verhulst P (1838) Notice sur la loi que la population suit dans son accroissement. Correspondence Mathematique et Physique 10:113–121
Xia J, Liu M, Jia S (2005) Water security problem in North China: research and perspective. Web Sci Pedosphere 15(5):563–575, ISSN 1002-0160/CN 32-1315/P
Young D (2002) A new space-time computer simulation method for human migration. Am Anthropol 104(1):138–158
Young D, Bettinger R (1995) Simulating the global human expansion in the Late Pleistocene. J Archaeol Sci 22(1):89–92
Yusuf A, Francisco H (2009) Climate change vulnerability mapping for South East Asia. Economy and Environment Program for South East Asia—International Development Research Centre (IDRC); Swedish International Development Cooperation Agency (Sida);

Canadian International Development Agency (CIDA). http://web.idrc.ca/uploads/user-S/12324196651Mapping_Report.pdf. Accessed 29 Jan 2012
(2007) J Environ Monit 9(8):805–813 (Epub 2007 Jun 27)
Tularam GA, Ilahee M (2007) Environmental concerns of desalinating seawater using reverse osmosis. J Environ Monit 9(8):805–813 (Epub 2007 Jun 27)

Chapter 10
Water Diversion Projects in China

He Chen and Robert B. Wenger

Technical Terms and Definitions

B.C.E	Before the Common Era
EPBs	Environmental Protection Bureaus
ERP	Eastern Route Project
MEP	Ministry of Environmental Protection
MRP	Middle Route Project
MWR	Ministry of Water Resources
WRBs	Water Resource Bureaus
WRP	Western Route Project

10.1 Introduction

Water is a basic necessity for organisms living on the earth. Most organisms, including humans, consist mainly of water and live on a planet that is dominated by water. Yet, despite this global abundance of water and its renewable characteristics, as much as one-fifth of the world's population lives under conditions of water scarcity (Hering and Ingold 2012). This scarcity is due primarily to the heterogeneous distribution of freshwater in space and time. The availability of water is an important factor in determining the distribution of human populations, but historical, social, political, and institutional factors may contribute to situations where large groups of people suffer from shortages of fresh water.

H. Chen (✉)
School of the Environment, Beijing Normal University, Beijing, China
e-mail: chenhe@bnu.edu.cn

R. B. Wenger
Natural and Applied Sciences, University of Wisconsin-Green Bay, Green Bay, USA
e-mail: wengerr@uwgb.edu

R. M. Clark and S. Hakim (eds.), *Securing Water and Wastewater Systems*,
Protecting Critical Infrastructure 2, DOI: 10.1007/978-3-319-01092-2_10,

China is a country that faces many challenging water resource problems. According to the country's State Statistics Bureau, on average, the annual amount of available water is 2.8 trillion m^3 (98.9 trillion ft^3), a quantity that constitutes approximately 6 % of global water resources. China ranks sixth among all countries in the world in the amount of water received from rainfall. However, the average annual rainfall amount is 648 mm (25.51 in.), a figure that is 19 % less than the world average of 800 mm (31.5 in.). As the country with the largest population in the world, China has an average annual per capita water resource availability of 2200 m^3 (77.7×10^3 ft^3), an amount that is one-fourth the world average, placing it in the 109th position among the world's countries.

There is, however, a considerable variation in water availability across the country; in south China the annual per capita fresh water resource is 3000 m^3 (105.9×10^3 ft^3) or more, while in the north it is about 500 m^3(17.7×10^3 ft^3). The Haihe, Huaihe, and Yellow River basins in north China are particularly hard hit; in these areas the per capita volumes of water resource availabilities are only 290 m^3(10.24×10^3 ft^3), 478 m^3(16.88×10^3 ft^3), and 633 m^3(22.35×10^3 ft^3), respectively. The North China Plain, which encompasses much of these river basins, contains approximately 25 million hectares (61.78 million acres) of cultivated land—about a quarter of China's total—yet possesses only 10 % of the nation's water resources (Zhang 2009).

Because of the water shortage in northern China, groundwater in some locations has been pumped out at a very rapid rate, with the result that subsidence of ground surfaces has occurred. Over the past 40 years, the amount of water pumped from the land has exceeded the amount replaced from rainfall by 120 billion m^3 (4237 billion ft^3); as a result, the water table in the North China Plain has fallen steadily (Li 2010). Another perspective on the groundwater situation in northern China is provided by the results from a recent study conducted by Gleeson et al. (2012). Using a measure called groundwater footprint, which they define as the area required to sustain groundwater use and groundwater-dependent ecosystem services, the authors state that this measure for the North China Plain is approximately 7.9 times the aquifer area. This ratio, computed using data from China's Ministry of Environmental Protection, is considerably greater than 1, an indication of widespread stress on groundwater resources and groundwater dependent ecosystems in the North China Plain. Totaling results from large aquifers around the world, the authors go on to estimate that the global groundwater footprint is currently about 3.5 times the actual area of aquifers. Thus, other regions around globe have groundwater depletion problems similar to the one that exists in the North China Plain.

The national annual water shortage in China is estimated to be 40 billion m^3 (1, 412 billion ft^3), of this amount about 6 billion m^3 (211.9 billion ft^3) is ascribed to urban areas. Among a group of 668 large cities, more than 300 face water shortage problems, with 110 of these cities characterized as having severe water shortages.

China has a monsoon climate that varies considerably over its vast territory and complex domain. The rainfall associated with this climate is highly uneven over time and space; southern areas enjoy relatively abundant amounts while northern areas experience scarcity. River runoffs vary greatly across the country from year to year. The ratio of the intra-annual maximum runoff to the minimum runoff is less than 5 in the Yangtze River, but for the rivers in the northern part of the country this ratio is usually more than 10. The variation in runoff from year to year can cause wet years and dry years to appear in succession. Such changes clearly have negative impacts on people's lives and affect their productive capabilities. In addition to the irregular annual patterns, in most parts of the country the rainfall is also unevenly distributed over the months of a given year. In the Yangtze River area 60 % of the rainfall occurs between April and September. In the Yellow River area the rainfall is concentrated even more; 80 % occurs between June and September. Southwestern China receives 70 % of its rainfall from June to October. These heterogeneous spatial and temporal distributions are important causes of the frequent occurrences of drought and flood that plague China.

10.2 Water Diversion Projects in China

The rapid economic development that has occurred in China in recent decades, along with population growth, has led to an increase in water demand. Available water resources are well short of the amounts needed to meet this increased demand. As a result, water shortages in certain parts of the country have created obstacles to socio-economic sustainable development (Wei 2005). In an effort to address these water shortage problems, China has implemented a number of water diversion projects. These large infrastructure projects are designed to alleviate the conflict between supply and demand by diverting water from areas with plentiful water resources to areas with scarce amounts of fresh water. In the last 60 years, especially in the most recent three decades of this period, China has constructed, or is in the process of constructing, 20 large inter-basin water diversion projects. Others are in the planning stage. The most celebrated of these projects is the South-to-North Diversion Project (nanshui beidao gongcheng), but there have been a number of other lesser known diversion projects.

It is worth noting that China had undertaken water diversion projects well before this 60 year period. In fact, China has the honor of being the originator of water diversion projects with its construction of the famous Beijing-to-Hangzhou Grand Canal. The oldest parts of the Grand Canal date back to the fifth century B.C.E., although various sections were not finally combined until the Sui Dynasty period, 581–618 C.E. Later, in the fifteenth century, during the Ming Dynasty era, the Grand Canal was almost entirely renovated.

In addition to the South-to-North Diversion Project, other notable projects, all of which have been completed, are the Luanhe-to-Tianjin Water Diversion Project in Hebei Province and Tianjin Municipality, the Yellow River-to-Qingdao Water Diversion Project in Shandong Province, the Datong River-to-Qinwangchuan District Water Diversion Project in Gansu and Qinghai Provinces, and the Biliuhe River-to-Dalian Water Diversion Project in Liaoning Province. A design plan exists for another project, the North-to-South Water Diversion Project in northeast China, but the Chinese government has yet to decide when it should be constructed.

These projects are derived from important strategic policy measures that have been undertaken to address water shortage problems in north China. It is hoped that these infrastructures will assist in meeting water demand requirements that are necessary to support and promote the country's economic development goals and the needs of its citizens. However, many scientists have concluded that the South-to-North Water Diversion Project, along with the other diversion projects, will not, by themselves, be sufficient to solve North China's water shortage problems. The water diversion projects must be complemented with other initiatives, such as better management of groundwater resources. As a result, the Chinese government is calling for stepped-up monitoring and scientific advice on groundwater management. To save water, researchers have proposed improved water conservation, better water pricing policies, and more rational agricultural practices (Li 2010). Another issue is inefficiency of water use. Due to its water intensive industrial structure, outdated technologies, and low reuse rates, China's water consumption per unit of Gross Domestic Product is three times the world average (Liu and Yang 2012).

Even though water diversion projects are seen by many as an important piece of the puzzle in solving water shortage problems in North China and are needed to provide benefits to people's lives and to promote socio-economic development, negative impacts cannot be ignored. Water diversions change water flows, nutrient flows, and water chemistry, all of which contribute to alteration and fragmentation of riparian habitats in rivers and streams. Fragmented areas are often too small and disconnected to support mammals and birds that need large territories for food sources and reproduction requirements. Habitats for aquatic species are also fragmented, in this case by dams and other types of structures associated with water diversion projects. Other kinds of environmental impacts that may occur are the creation of anoxic conditions in hypolimnetic water, alteration of water temperatures, eutrophication, increased concentration of mercury in fish, intermixing of alien invasive aquatic species, increased erosion of downstream sediments and associated contaminants, and changes in metal concentrations in outflows (Das 2006).

Water diversion projects can also change downstream hydrological regimes. In a theoretical morphological analysis conducted by Wang et al. (2008), they show that large scale diversions like those along the lower Yellow River can, in the long-term, cause the development of a convex riverbed profile. Deposition will take place along the whole reach of the river, with an increasing deposition depth from downstream to upstream. This means that the slope of the river bed increases from upstream to downstream.

In addition to environmental and ecological impacts, the construction of water diversion projects often involves the relocation of large numbers of people. Thus, there are major social impacts as well. As a result of these types of impacts, and the very large construction costs, water diversion projects are not without controversy in China.

10.3 The South-to-North Water Diversion Project

The South-to-North Water diversion Project is the largest inter-basin diversion project ever undertaken in China or, for that matter, anywhere in the world. As early as 1952, Mao Zedong put forward the idea of diverting water from the southern part of China to the north when he said, "The south has lots of water, the north little. If possible, it is okay to lend a little water." In 1979 a project planning office was established in the Ministry of Water Resources and a number of environmental impact studies were carried out in the early 1990s. A number of locations were considered and studied over several decades and then, finally, in August 2002 the Chinese State Council approved a three part diversion project and created a limited liability company to oversee operations and management (Liu et al. 2011). The approved project was comprised of three components: the Eastern Route Project (ERP), the Middle Route Project (MRP), and the Western Route Project (WRP). By 2050, after the completion of all three components, the project is expected to divert 44.8 billion m^3 (1,581.9 billion ft^3) of water per year from the water rich Yangtze River and its tributaries to the North China Plain.

The ERP, constructed along the route of the Beijing-Hangzhou Grand Canal and its parallel rivers, is designed to divert water from the downstream portion of the Yangtze River. Beginning at Yangzhou, a city located on the north bank of the Yangtze River in Jiangsu Province, water is diverted to the north, connecting along the way with the Hongze, Luoma, Nansi, and Dongping Lakes, which serve as regulating reservoirs. Dongping Lake, located just to the south of the Yellow River is the highest point in the entire range. Its elevation is 40 meters (131.2 feet) higher than the Yangtze River at the point where the diversion begins. As a result, 13 stages of pumps are needed to lift the water to the height of Dongping Lake. After passing through a tunnel under the Yellow River, the water flows to the city of Tianjin by gravity. The length of the overall route is 1150 km (714.6 mi) (Wang et al. 2006).

The region receiving water from the ERP consists of three sectors: the area south of the Yellow River, the area north of the Yellow River, and the Shandong Peninsula. Portions of the Jiangsu, Anhui, Shandong, and Hebei Provinces are recipients of water from the ERP, as well as the city of Tianjin. From a river basin perspective, the ERP extends across the eastern Huai-Yellow-Hai Plain (Huaihe, Yellow, and Haihe Rivers) (Lu et al. 2006).

The first phase in the construction of the ERP was designed to provide water to municipalities in Shandong Province and to improve agricultural conditions in the northern part of Jiangsu Province and the southwestern part of Shandong Province. This construction was based primarily on the existing Northward Water Transfer Project in Jiangsu Province. The average annual discharge of water diverted from the Yangtze River at the completion of this phase was expected to be 8.9 billion m^3 (314.26 billion ft^3). The flow after crossing the Yellow River was expected to be in the range of 1.0–1.5 billion m^3 (35.31–52.96 billion ft^3) and a similar flow was to be supplied to the Shandong Peninsula. Construction of this phase of the ERP was started in 2002 and was completed in 2007.

The second phase of the ERP, which began at the completion of the first phase, focused on enlarging the project's scope. Parts of Hebei Province and the city of Tianjin were added as recipients of water from the eastern route and the overall diversion rate was increased to an annual average of approximately 10.6 billion m^3 (374.3 billion ft^3). The amount transferred across the Yellow River was in the range of 3.0–4.0 billion m^3 (105.9–141.2 billion ft^3), with Tianjin receiving about half of this amount. An amount in the range of 1.0–1.5 billion m^3 (35.31–52.96 billion ft^3) was transferred to the Shandong Peninsula. The construction under this phase was completed in 2010.

The goal of the third and final phase of the ERP is to achieve an additional increase in scale. The design plan for this phase calls for an annual diversion of 14.8 billion m^3 (522.6 billion ft^3). From this diversion it is expected that 6.0–8.0 billion m^3 (211.9–282.5 billion ft^3) will be channeled across the Yellow River and 1.5–2.0 m^3 (52.96–70.62 ft^3) will be transferred to the Shandong Peninsula. This part of the project is targeted for completion in 2030.

The MRP will divert water from the Danjiangkou Reservoir on the Han River—a tributary of the Yangtze River—and channel it to parts of Hubei, Henan, and Hebei Provinces in the western Huai-Yellow-Hai Plain and to a number of cities including Beijing and Tianjin. The route for the MRP traverses the Yangtze River watershed in the vicinity of Fangcheng, continues north to a point west of Zhengzhou where it crosses the Yellow River, and then is extended further following the Beijing-Guangzhou Railway on its way to the cities of Beijing and Tianjin (Wang and Ma 1999). The total length of the route from its intake point at the Danjiangkou Reservoir to Beijing is 1246 km (774.2 mi); 482 km (299.5 mi) of the route lies to the south of the Yellow River and 764 km (474.7 mi) to the north. The water, starting from a relatively high terrain, will flow by gravity and will be used to irrigate thirsty croplands, in addition to addressing the needs of urban areas. In addition to the construction of the canal, another part of the MRP is an expansion of the dam at the Danjiangkou Reservoir. The expansion will elevate the dam from a height of the original 162 m (531.5 ft) to 176.6 m (579.4 ft) and increase the dam's storage capacity from 11.6 billion m^3 (409.6 billion ft^3) to 29.05 billion m^3 (1025.8 billion ft^3) (Zhang 2009).

The construction of the MRP is divided into two phases. The first phase began in 2002 and was completed in 2010. The average annual output of water at the completion of this phase was projected to be 9.5 billion m^3 (335.4 billion ft^3); 4.0–5.0 billion m^3 (141.2–176.6 billion ft^3) from this diversion crosses the Yellow River, with 1.0 billion m^3 (35.31 billion ft^3) of this amount targeted for Beijing and Tianjin. The second phase was initiated at the completion of the first phase. When this phase is completed around 2030 the annual flow rate for the MRP is expected to be in the range of 13.0–14.0 billion m^3 (459.0–494.3 billion ft^3) (Wang and Ma 1999).

The WRP will be located in the Qinghai-Tibet Plateau in an area where the elevation is 3000 m (9843 ft), and in some places, considerably higher. The terrain is exceedingly mountainous and the proposed design involves construction of high dams, lengthy tunnels, and elevated canals that would be extremely costly and difficult to construct (Zhang 2009). Water is to be diverted from three rivers that are tributaries to the upper Yangtze River—the Dadu, Yalong, and Tongtian Rivers—to an upstream portion of the Yellow River. Preliminary plans call for diverting an annual total of 12.0–17.0 billion m^3 (423.7–600.3 billion ft^3) from these rivers with 3.0–5.0 billion m^3 (105.9–176.6 billion ft^3) to be taken from the Dadu River, 3.5–4.0 billion m^3 (123.6–141.2 billion ft^3) from the Yalong River, and 5.5–8.0 billion m^3 (194.2–282.5 billion ft^3) from the Tongtian River. Following the course of the Yellow River, the water would flow to several northwestern provinces and autonomous regions, including Qinghai, Shanxi, Shaanxi, Gansu Provinces and Ningxia and Inner Mongolia Autonomous Regions.

Due to the extremely difficult terrain, the formidable engineering challenges, and the severely cold winter climate, additional scientific research and studies are required before construction of the WRP can begin. Because time is needed to complete these studies, construction is not likely to begin anytime in the near future. Others have urged that the project never be undertaken because data on rainfall, geology, and receding glaciers, call into question estimates on water volumes in the upper Yangtze River that are contained in the government's blueprint for the WRP (Larsen 2009). If the project is constructed and water

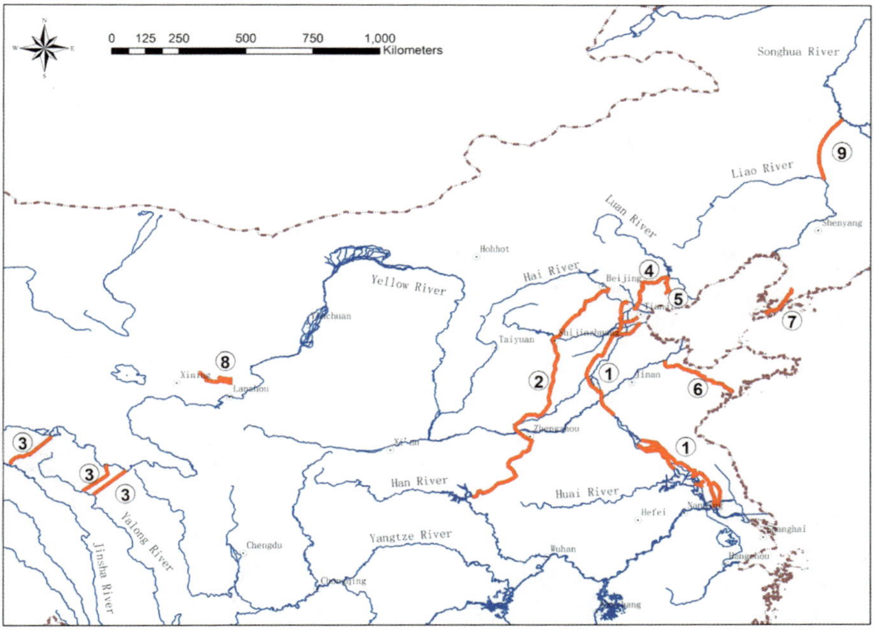

Fig. 10.1 Map of China's actual and proposed water diversion projects: *1* Eastern Route Project (ERP), *2* Middle Route Project (MRP), *3* Western Route Project (WRP), *4* Luanhe-to-Tianjin water diversion project, *5* Luanhe-to-Tangshan water diversion project, *6* Yellow river-to-Qingdao water diversion project, *7* Biliuhe river-to-Dalian water diversion project, *8* Datonghe river-to-Qinwangchuan district water diversion project, *9* North-to-South water diversion project

volumes fall short of predictions, the consequences could be disastrous for downstream communities that depend on Yangtze River water for agriculture, industry, and hydropower.

In the map above (Fig. 10.1), the location of the three routes in the South-to-North Water Diversion Project are shown. The other water diversion projects, which are described in the next section, are also shown on the map. A table with information about all of the diversion projects can be found at the end of Sect. 10.4.

10.4 Other Water Diversion Projects

The South-to-North Water Diversion Project overshadows all others in China in terms of size, cost, engineering demands, and length of construction time. But, as noted earlier, there have been a number of other diversion projects of note. In this section several such projects are described; the goal in choosing these specific projects for discussion is to portray a range of water resource needs that diversion projects are designed to address. The map shown below in Fig. 10.2 displays rivers that are involved in the diversion projects and identifies provinces in which they are located.

Fig. 10.2 Map of rivers and provinces in China

The need for water in the city of Tianjin has been long-standing. As a result, a water supply project was undertaken as early as the 1980s to address this need, at least in part. The Luanhe-to-Tianjin Water Diversion Project was designed to divert water from the Luanhe River Basin in Hebei Province to Tianjin. In addition to water diversion, other types of applications were part of the project, including water storage, water conveyance, and water purification. The water channel begins at the Panjiakou Reservoir on the middle reach of the Luanhe River in Hebei Province and ends at the Yuqiao Reservoir in Tianjin, covering a distance of 234 km (145.4 mi). On average, the amount of water diverted in a given year is 1.0 billion m^3 (35.31 billion ft^3). The construction of this project took place from May 1982 to September 1983.

The Luanhe River is the source of another diversion project that was constructed at an earlier time. In this project, one of the largest in northern China at the time, water was diverted from the Daheiting Reservoir on the Luanhe River to the Douhe Reservoir on the west branch of the Douhe River and then on to the city of Tangshan. Tangshan is located about 100 km (62.1 mi) northeast of Tianjin and is remembered as the city where a devastating earthquake occurred in 1976 that killed more than 200,000 people. The diversion project was started 2 years later in 1978 and was completed in 1984. The project is capable of supplying 0.6-0.8 billion m^3 (21.2–28.2 billion ft^3) per year to the Tangshan District and to the Huaixinghe and Douhe Rivers where it is available for agricultural use.

As described earlier, some of the water from the ERP in the South-to-North Water Diversion Project goes to Shandong Province. Another source of diverted water for the Shandong Province, specifically for the city of Qingdao located on the south side of the Shandong Peninsula and overlooking the Yellow Sea, comes from the Yellow River. The Yellow River-to-Qingdao Water Diversion Project carries water in a conveyance canal from the Dayuzhang intake sluice on the Yellow River to Qingdao, traversing a distance of 275 km (170.9 mi). Water is diverted into this channel from November 11 in a given year to March 20 of the following year; the average annual output is 0.55 billion m^3 (19.4 billion ft^3). The diversion takes place during this time period because the sediment load in the Yellow River water is lower at that time. This project was started in April 1986 and completed in November 1989.

Across the Bohai and Yellow Seas in a northerly direction from Shandong Province lies Liaoning Province. Dalian, the second largest city in Liaoning Province with a population of more than six million people, is another urban area that has had insufficient sources of water needed to sustain its economic development and the living requirements of its residents. As a result, the Biliuhe River-to-Dalian Water Diversion Project was constructed during the period from June 1995 to October 1997 with the design objective of diverting water to Dalian and smaller cities and towns along the way. The project consisted of two parts: the northern part, which carries water from the Biliuhe River to the Wazidian Reservoir, and the southern part, which continues the flow from the Wazidian Reservoir to the district of Dalian. The total distance of the channel is 68 km (42.3 mi) and its annual diversion capacity is approximately 0.33 billion m^3 (11.65 billion ft^3).

Moving to the western province of Gansu, one finds another diversion project, this one constructed in a discontinuous manner. Construction of the Datonghe River-to-Qinwangchuan Water Diversion Project began in 1976, was halted in 1980, and then was resumed in 1985 and finally completed in 1995. The purpose of the project was to improve the environmental and ecological conditions in the Qinwangchuan District by gradually enhancing the vegetation diversity there, and to form a green zone in the northern part of Lanzhou, the capital of Gansu Province. In this manner the goal was to bring significant economical, societal, and environmental benefits to the region. The project was designed to deliver 0.44 billion m^3 (15.53 billion ft^3) of water annually, a sufficient amount to irrigate a land area of about 58,000 km^2 (22,394 mi^2).

China's vast land mass and extensive topographic variation has allowed planners to consider a north to south diversion project, a potential counterpart to the well-known diversion in the opposite direction: the South-to-North Water Diversion Project. The north to south diversion, formally known as the Northeast China North-to-South Water Diversion Project, is a proposed plan that would divert water from the Songhua River and its tributaries in the far northeastern Heilongjiang Province to the Liao River. The purpose of this diversion would be to supplement the flows in the middle and lower branches of the Liao River and, along the way, to supply water to areas in the Inner Mongolia Autonomous Region. Part of this project would involve the construction of reservoirs to enhance water storage capabilities. One reservoir would be constructed on the Second Songhua River and another on the Nen River (the Nen and Second Songhua Rivers meet near the city of Songyuan to form the Songhua River). The design calls for a maximum runoff in the range of 400–500 m^3/s (14,124–17,655 ft^3/s) and, after completion of the project, a diversion of up to 7 billion m^3 (247.2 billion ft^3) each year. The water from this project would address water shortage problems in the industrial, agricultural, and domestic sectors in the Liaoning and Jinlin Provinces and the Inner Mongolia Autonomous Region. As noted above, no date has been set for the start of this project (Table 10.1).

Table 10.1 Information on Water Diversion Projects

Project	Construction period	Length of transfer (km)[b]	Annual diversion discharge(10^9 m^3)[c]
South-to-north (eastern route)	2002–2030[a]	1150	15
South-to-north (middle route)	2002–2030[a]	1246	13–14
South-to-north (western route)	Planned	450	12–17
Luanhe river to Tianjin city	1982–1983	234	1
Luanhe river to Tangshan city	1978–1984	51	0.6–0.8
Yellow river to Qingdao city	1986–1989	275	0.55
Biliuhe river to Dalian city	1995–1997	68	0.33
Datonghe river to Qinwangchuan district	1976–1995	87	0.44
Songhua river to Liao river	Planned	650	7

[a] 2030 is the completion date for the final phase; previous phases will be completed earlier
[b] To convert from km to mi multiply by 0.6214
[c] To covert m^3 to ft^3 multiply by 35.31

10.5 Environmental and Ecological Challenges

Water diversion projects bring with them many challenges. Some of these challenges occur during the construction phase; others arise after construction is complete and operation of the system has begun. The tremendous physical disruptions that water diversion projects impose on the earth's surface and on water bodies, as well as impacts upon the natural environment, inherently bring with them major environmental and ecological challenges. As noted in Sect. 10.2, water diversion projects influence and affect environmental flows; i.e., the provision of water within rivers, surface water bodies, and groundwater systems that is necessary for the maintenance of ecosystems and the benefits they provide. In China, diversion projects have contributed to a number of specific problems that arise from disturbance of environmental flows, including reduction in river flows, shrinkage of wetlands, deterioration of grasslands, and increased soil erosion. Ecological and environmental problems must be given full attention to ensure that the benefits from water diversion systems are not overridden by negative environmental impacts. In the remainder of this section we focus on environmental issues associated with the South-to-North Water Diversion Project.

A major issue that arises when water diversion projects have been completed and water is being diverted to needy cities, agricultural areas, and industrial facilities is the quality of the water itself. In fact, concerns about water quality are often present before diversion systems are constructed or have become fully operational. A directive approved by China's State Council, issued before construction of the ERP in the South-to-North Water Diversion Project had begun, required local authorities to ensure that the water quality is at least Grade 3, the minimum standard for water that is drinkable after treatment (China Daily 6 July 2010). This has been a difficult challenge. The construction of the ERP was halted for a time because authorities in cities destined to receive water carried through the system didn't want to pay for the privilege of drinking polluted water (Hao 2007). The eastern route traverses one of the most heavily developed regions of China. Inadequately treated domestic sewage and industrial wastewater was being released into the lakes that serve as storage facilities along the channel. A significant portion of the funds allocated for the first phase of the ERP were earmarked for the construction of wastewater treatment facilities. Nevertheless, in July 2010, the Director of the South-to-North Water Diversion Project under the State Council was quoted as saying that local authorities still had a long way to go before the eastern route would be transformed into a clean-water corridor (China Daily 6 July 2010).

In addition to pollutants arising from point sources, agricultural activities have been a source of nonpoint pollution along the eastern route (Zhang 2009). In a water quality modeling study focusing on pollution control in the eastern route, Wang et al. (2006) came to the following conclusions: urban point sources are the leading contributors to the pollutant load, nonpoint sources from agricultural areas take second place, and the internal contributions from watercraft are the least significant.

Recent reports indicate that the quality of the water in the eastern route is improving. Money invested in more than 400 pollution control projects over the past nine years has led to an 85 % decline in the average concentration of chemical oxygen demand and a 92 % decrease in ammonia nitrogen. Portions of the water body have reached the Grade 3 level, the required water quality standard for drinking water (People's Daily 2012).

Given the extant water quality challenges in the eastern route, people in places like Beijing and Tianjin and Henan and Hebei Provinces have eagerly anticipated receiving water from the Han River via the middle route. As noted earlier, the Danjiangkou Reservoir on the Han River is the major source of water for the MRP. This source was chosen in part because its water was deemed to be less polluted than that in most other reservoirs. But water quality issues are present here as well, not only with the water diverted into the northbound canal, but with the water that follows the course of the Han River downstream from the Danjiangkou Reservoir. During the early construction stages of the MRP Project, environmental authorities in Hubei Province raised concerns about potential difficulties in managing the lower reach of the Han River, which enters the Yangtze River at the city of Wuhan. The Han River supplies the 10 million residents of Wuhan, Hubei's capital, with its drinking water. With a decrease in flow caused by the diversion of water, the authorities feared a range of environmental and ecological problems would arise, including, among others, a decline in the river's ability to cleanse itself and a loss of breeding grounds for fish (Wang 2011). Hubei's concerns did lead to modifications in the project design resulting in a lowering of the planned quantities of water to be moved in the first stage of the middle route. But worries remain; a scheme to divert water from the Yangtze River to the Han River to compensate for reduced flows expected from the diversions have been discussed (Freeman 2011).

The Upper Han River Basin, the source of the water flowing into the Danjiangkou Reservoir, has an area of 95,000 km^2 (36,680 mi^2) and a population of 13 million people. In Shiyang, a city of 500,000 people, 90 % of the industrial wastes and 75 % of its sewage wastes are treated, but elsewhere in the Upper Han River Basin considerably smaller percentages of sewage wastes receive treatment. In addition, approximately 80 % of the population in the basin resides in areas devoted to agriculture. Nonpoint pollution arising from the agricultural sector has had a major impact on the water quality in the river network with the result that elevated phosphorus and nitrogen levels have led to algal blooms in the spring (Zhang 2009). Recent monitoring has shown that water quality standards are not fully compliant with the criteria specified by the Ministry of Environmental Protection in China. Specifically, the total nitrogen concentration of the water in the storage area of the Danjiangkou Dam failed to meet the standards required for a healthy water body (Hao et al. 2012). The origin of the problem, these researchers conclude, is runoff from agricultural areas in the Danjiangkou Reservoir basin. Thus, it is critical to reduce pollution from these nonpoint sources if the quality of the water diverted to the middle route of the South-to-North Water Diversion Project is to be improved. Given these problems, Zhang (2009) has urged the establishment of an integrated water resources management plan for the Upper Han River Basin that includes a centralized long-term monitoring network and a strategy for providing timely management information with the goal of protecting the ecological security of the MRP.

The proposed WRP brings with it very challenging ecological and environmental problems. The construction of this project would take place in a sparsely populated remote area, much of it at an elevation above 3000 m (9,843 ft). The mountainous landscape is fragile and some of the environment is in permafrost. In addition to impacts on riverine ecosystems in this region, the construction of dams and roads poses a danger of landslides and other types of geological problems. Water quality issues are not likely to be a pressing concern here because the water will be taken from tributaries in the upper Yangtze River where the population is sparse and few industries are in place. However, as noted in Sect. 10.3, available water quantities may prove to be shy of the amounts specified in the plans for the MRP. A shortfall of this nature would have major environmental impacts.

10.6 Institutional and Administrative Issues

In addition to engineering and environmental challenges, water diversion projects provide major institutional and administrative challenges. This is true of the water diversion projects in China, especially so in the case of the South-to-North Diversion Project. The routes for the ERP and MRP traverse several provinces and municipalities, thereby raising trans-jurisdictional water quantity and water quality issues. If the WRP is implemented, such trans-jurisdictional issues will be present there as well. County, township, and village governmental units are also impacted by the water diversion projects and represent constituents with water resource needs.

Administratively, the governmental units in China are organized in a hierarchical manner in the following descending order: Center, covering the entire country; Provinces; Cities and Counties; Townships; and Villages. Four cities—Beijing, Tianjin, Shanghai, and Chongqing—are an exception to this scheme. These cities are labeled as municipalities and, because of their large size, have ranks equivalent to that of a province. The State Council, the central administrative unit, directs the Ministry of Water Resources (MWR) and the Ministry of Environmental Protection (MEP). The MWR has primary responsibility for water quantity issues, while the MEP's main focus is on water quality problems. There are water resource bureaus (WRBs) and environmental protection bureaus (EPBs) at the provincial, municipality, and local levels that are counterparts to MWR and MEP, respectively. In addition to the MWR and the MEP, a number of other agencies are involved with water governance, including, among others, the Ministry of Agriculture, the State Forestry Administration, and the Ministry of Housing and Urban Development.

On the legal side, China's Standing Committee of the National People's Congress has established two major laws for water governance, the Water Law and the Water Pollution Prevention and Control Law. The Water Law was first enacted in 1988 and in revised in 2002. A number of the 82 articles in the Water Law contain provisions that apply directly to water diversion projects. For example, Article 22 states:

> For diversion of water across river basins, all-around planning and scientific demonstration shall be needed, overall consideration shall be given to the need of water by both the river basins where water is diverted from and the river basins where water is diverted to, and damages to ecological environment shall be prevented.

The Water Pollution Prevention and Control Law was passed in 1984 and revised in 1996. This law stipulates that the State Council shall establish national water quality standards. Provinces and municipalities may establish their own local supplementary standards for those items not specified in the national water quality standards, but, when doing so, are required to report the same to the MEP. The law specifies that EPBs of the people's government at all levels "shall be the organs exercising unified supervision and management of the prevention and control of water pollution". The law also states that individuals have the duty to protect the water environment and the right to provide information regarding any pollution or damage to the water environment. In addition, any individual or unit that has suffered direct damage from a water pollution hazard shall have the right to demand elimination of the hazard and receive compensation for the damage caused by the polluter. Article 2, one of 45 contained in the law, states:

> The law shall apply to the prevention and control of pollution of rivers, lakes, canals, irrigation channels, reservoirs, and other surface water bodies, and groundwater within the territory of the People's Republic of China.

Here, as was the case with the Water Law, the Water Pollution Prevention and Control Law contains provisions that apply to water diversion projects.

More recently, in January 2011, the Chinese government, in its annual "No. 1 Document," outlined a plan to expedite water conservancy development and reform and to achieve sustainable development and management of water resources within this decade (Liu and Yang 2012). This document, which reflects the government's top priorities, states that the government plans to quadruple its total investment in solving water problems in the next 10 years compared with the investment in the previous decade. The amount of this investment is four trillion Yuan (US $635 billion). In a supplement attached to their article, Li and Yang (2012) provide a translation of the document where they highlight the goal and the main objectives of the plan.

The overarching goal of the plan is to fundamentally reverse the backward situation of China's water conservancy development by 2020. There are four main objectives: (1) Establish a flood and drought control and relief system, (2) Establish a reasonable water resources allocation and a highly efficient water-use system, (3) Establish a system for protecting water resources and securing the health of aquatic ecosystems, (4) Establish a system and mechanism that facilitates the development of water conservancy. The plan focuses largely on engineering measures that address water quantity issues; water quality issues receive significantly less attention. A specific item under the second objective pertains to water diversion projects: enhance the capacity for water resources allocation and supply through water transfer projects and connecting different rivers, lakes, and reservoirs.

In addition to the Water Law, The Water Pollution Prevention and Control Law, and the Water Conservancy initiative, there are a number of other statutes that have been adopted by the Chinese government that apply to water diversion projects. These include the Law of Flood Control, the Law of Water and Soil Conservation, and the Law of Land Administration. The Law of Flood Control was enacted for the purpose of preventing and controlling floods, taking precautions against and alleviating calamities caused by floods and water logging, and maintaining the safety of people's lives and property. The Law of Water and Soil Conservation was formulated for the purpose of preventing and controlling soil erosion; protecting and promoting the rational utilization of soil and water resources; mitigating the disasters of flood, drought, and sandstorms; improving the ecological environment; and promoting production. The objective of the Law of Land Administration is to strengthen the administration of land use by ensuring a rational use of land and giving real protection to cultivated land to promote sustainable development of the socialist economy.

An additional administrative unit that has been created by the Chinese government is the Office of South-to-North Water Diversion Project under the State Council. The responsibility of this office is to provide governmental management oversight for the South-to-North Water Diversion Project and handle legal and contractual matters pertaining to the construction of the project (Feng et al. 2008). Among the responsibilities of the Office of the South-to-North Water diversion Project is to ensure that worker safety provisions as contained in the Work Safety Law of the People's Republic of China are followed. Within this law is a section, Provisions on the Administration of Work Safety in the Construction of Water Resource Projects, which spells out worker safety requirements in the construction of water resource projects. In the event of an accident, legal directives are given for investigating the accident, identifying liabilities, and implementing corrective measures.

It is clear that the leaders of China are aware of the challenging water quantity and quality problems that their country faces. As described above, they have enacted an impressive array of environmental laws, regulations, and initiatives that address water issues and problems. An extensive governmental administrative structure has been put in place, ranging from the Ministries of Water Resources and Environmental Protection under the State Council in Beijing down through Water Resource Bureaus and Environmental Protection Bureaus at the provincial, municipal, county, and township levels. In addition, China has become an active participant in the global environmental community. Yet, many environmental laws and policies, including those pertaining to water, are poorly or inadequately implemented.

There are multiple reasons for this conundrum, but at the heart of the matter is a failure by local authorities to enforce existing laws and to implement administrative decisions. The distribution of authority in the government system outlined above is such that officials in a local bureau report only to their corresponding local government, not to the bureaus or ministry above it (Liu et al. 2011). The result is a fragmented system whose core feature is the need for a tremendous amount of negotiation and consensus building for the implementation of environmental policy. Authority tends to be fragmented by function, by territory, and by rank (Lieberthal 1997).

In an earlier section we mentioned a halt that occurred in the construction of the ERP because of water quality concerns. This is an example of the difficulties that may arise because lines of authority are blurred. According to Hao (2007), nearly half of the $4 billion cost of the first phase of the construction was earmarked for improving the quality of the water. However, the central government provided only about 10 % of the needed funds; the rest was supposed to come from localities that were expected to benefit from the project. But localities bulked at building treatment facilities because nobody wants to clean up somebody else's dirty water.

10.7 Water Security in China

Water security, a concept that sometimes focuses primarily on the safeguarding of water and wastewater infrastructures, is now often viewed more broadly. An expanded perspective that has gained prominence in the last decade is the following: water security is defined as an acceptable level of water-related risks to humans and ecosystems, coupled with the availability of water of sufficient quantity and quality to support livelihoods, national security, human health, and ecosystem services (Bakker 2012). When considering human and biodiversity threats jointly, 80 % of the world's population is exposed to high levels of threat to water security (Vörösmarty et al. 2010). In the sense that water security is defined here, China is among the world's countries with high levels of water security threats. As we have stated earlier, this is particularly true in the northern part of the country.

The water diversion projects that have been constructed, or are being considered, are appropriately viewed as efforts by China to lessen incidents of water security threats. This can be seen in a number of ways: delivery of water to meet domestic needs, to promote industrial and economic development, to irrigate crops, and to restore and support ecosystems and the services they provide. The need to address impairments to ecosystems and biodiversity may not have received the attention that residential, agricultural, economic development issues have; nevertheless, research on the potential ecosystem benefits that can be derived from water diverted to areas that are deprived of water resources has been conducted. As an example, Lu et al. (2006) have provided an assessment of the potential ecological benefits to the water receiving areas of the eastern route of the South-to-North Water Diversion Project.

In their assessment, these researchers identified potential benefits to agricultural, forest, wetland, and city greenbelt ecosystems after each of the three stages of the ERP. In addition, they looked at benefits accruing from water logging control. They used market value and shadow engineering methods based on ecosystem service functions to assign monetary values to the eco-environmental benefits. Their calculations revealed projected benefits that would have a monetary value of approximately 50 billion Yuan (US $8 billion). The overall conclusion was that the ERP will result in enormous economic, social, and eco-environmental benefits within the water-accepting area and will help alleviate water scarcity in northern China. Although not explicitly stated, the researchers clearly assume that the delivered water will be of sufficient quality to provide the services that they assess.

Chen and Du (2008) point out that water transferred by the MRP is focused primarily on supplying water for industrial and domestic use. However, they see a potential for using this water for ecological benefits, particularly in the long term. In their view, this is a necessary objective because of the ecological degradation that has occurred in north China. After some of the more immediate needs are met and water shortages are no longer in a crisis mode, they advocate the construction of artificial niches along the channel that would provide an environment where aquatic plants could be grown, which, in turn, would create an environment for aquatic animals and amphibians. Further down the road, these two environmental scientists envision modifications to the entire channel that would transform it into a stream resembling a natural river. This could be accomplished by removing concrete layers on the channel bed. They believe the natural conditions would enhance biodiversity and improve the self-purifying capacity of the channel.

As noted earlier, the Chinese government in its January 2011 "No. 1 Document," outlined a plan to reverse the backward situation of the country's water conservancy development. This plan includes the following objective: establish a system for protecting water resources and securing the health of aquatic ecosystems. In spelling out details under this objective, attention is given to water quantity needs for healthy ecosystems, but the account doesn't stop there; water quality issues are addressed as well. Water quality aims include reductions in chemical oxygen demand, total nitrogen discharge, the emission of heavy metal pollutants. The advocacy for the inclusion of ecological benefits in the ERP and MRP, as delineated above, certainly appears to be consistent with the objective of securing healthy ecosystems as described in the "No. 1 Document."

It is noteworthy that "improvements in the efficiency of industrial water use" is one of the items listed under the healthy ecosystem objective. Specific targets of improving the water use per industrial added value by 8, 10, and 30 %, respectively, in comparison to 2010 are included in the document. This suggests the need for jointly designing policies for water use.

Clearly, if water diversion systems are to play a role in improving the water security of north China, initiatives like those suggested by Lu et al. (2006) and

Chen and Du (2008), and the objectives contained in the January 2011 "No. 1 Document", must be robustly pursued. Policies and management programs that focus on ecosystem principles and goals are required. In addition, integrated approaches and coordinated efforts are needed.

Another element of water security is the protection and safeguarding of water infrastructure systems. We have not seen any evidence that the Chinese government has formulated policies to address human or bioterrorist threats to water infrastructures, more specifically water diversion projects, but the need to repair deteriorating dams and reservoirs is a governmental concern. More than 46,000 of the 87,000 dams and reservoirs that have been built since the 1950s have surpassed their life spans, or will within 10 years. To avert potential disasters, the central government plans to repair and reinforce these structures by the end of 2015 (Liu and Yang 2012).

10.8 Concluding Remarks

Water diversion projects are designed to move water from one place to another, from a region with abundant water resources to one with a scarcity of water. But more than a singular or direct transfer of quantities of water is involved. As we have seen, water diversion projects in China inevitably intersect a number of other water related issues and systems, including water quality, water use efficiency, groundwater systems, and aquatic ecosystems. Therefore, if water diversion projects are to play an effective role in enhancing water security in China, their design, and policies related to their use, need to be embedded in integrated water resource management approaches. Such integrated and comprehensive approaches also need to take into account future trends. For example, the Intergovernmental Panel on Climate Change's projected climate scenario for China in 2050 indicates that north China will be wetter and south China will be dryer than at present. This predicted outcome should be considered when making a decision on whether to proceed with the western route of the South-to-North Water Diversion Project.

Given China's challenging water security problems and its fragmented environmental administrative structure, we concur with Liu and Yang (2012) in recommending that policies for water issues in China be jointly designed, implemented, monitored, and evaluated. Water diversion projects and water diversion systems placed in such a context would improve China's water security. Accomplishing this goal will require improved coordination among the various ministries that have responsibilities pertaining to water-related matters. In order to enhance policy implementation, better coordination is needed between the central and local governments. Integrated monitoring programs are needed to assess water quantity and water quality indicators so that proactive management measures can be implemented. Monitoring of social behaviors and people's attitudes about water usage is also required, so that effective programs can be designed to promote more efficient use of water.

Acknowledgments The authors gratefully acknowledge the assistance of Michael Stiefvater, GIS/Mapping Specialist at the University of Wisconsin-Green Bay, for contributing his skills in constructing Fig. 10.2. The map in Fig. 10.2 was created as part of the World Hydro Reference Overlay map service by HydroTeamRC, a unit of Environmental Systems Research, Inc. It was compiled from multiple public data sources and published by ESRI at http://184.72.212.114:6080/arcgis/rest/services/WorldHydroReferenceOverlay/MapServer.

References

Bakker K (2012) Water security research challenges and opportunities. Science 337(6097):914–915

Chen H-C, Du P (2008) Potential ecological benefits of the middle route for the south-north water diversion project. Tsinghua Sci Technol 13(5):715–719

China Daily (2010) Pollution still hinders water diversion in east China. http://news.xinhuanet.com/english2010/china/2010-07/06/c_13385886.htm

Das DK (2006) Environmental impact of inter-basin water transfer projects: some evidence from Canada. Econ Polit Weekly 41(17):1703–1707

Feng J, Zhou Y, Hong Y, Zhao S, Ren X (2008) Reliability model of organization management chain of south-to north water diversion project during construction period. Water Sci Eng 1(4):107–113

Freeman C (2011) Quenching the dragon's thirst: the south-north water transfer project—old plumbing for new China? Chin environment forum, Woodrow Wilson Center for Scholars

Gleeson T, Wada Y, Bierkens MFP, van Beek LPH (2012) Water balance of global aquifers revealed by groundwater footprint. Nature 488(9):197–200

Hao F, Zhang X, Wang X, Ouyang W (2012) Assessing the relationship between landscape patterns and nonpoint-source pollution in the Danjiangkou reservoir basin in China. J Am Water Resour Assoc 48(6):1162–1177

Hao X (2007) Pollution slows China's canal project. Science 317(5846):1846

Hering JG, Ingold KM (2012) Water resources management: what should be integrated. Science 336(6086):1234–1235

Larsen C (2009) On Chinese water project, a struggle over sound science. Yale Environment 360

Li J (2010) Water shortages loom as north China's aquifers are sucked dry. Science 328(5985):1462–1463

Lieberthal K (1997) China's governing system and its impact on environmental policy implementation. China environmental series. The Woodrow Wilson Center, Washington

Liu J, Yang W (2012) Water sustainability for China and beyond. Science 337(6095):649–650

Liu J, Dorjderem A, Fu J, Lei X, Liu H, Macer D, Qiao Q, Sun A, Tachiyama K, Yu L, Zheng Y (2011) Water ethics and water resource management. UNESCO Bangkok

Lu Z, Ma X, Sang W, Fan Y (2006) Ecological benefit assessment of the south-north water transfer project: an eastern route. Int J Sustain Dev World Ecol 13(3):221–227

People's Daily (2012) Water quality standard reached in 89% of water body because of pollution control over nine years

Vörösmarty CJ, McIntyre PB, Gessner MO, Dudgeon D, Prusevich A, Green P, Glidden S, Bunn SE, Sullivan CA, Liermann CR, Davies PM (2010) Global threats to human water security and river biodiversity. Nature 467:555–561

Wang C, Wang Y-Y, Wang P-F (2006) Water quality modeling and pollution control for the eastern route of south to north water transfer project in China. J Hydrodyn Ser. B 18(3):253–261

Wang K (2011) Drying up the Han. China Dialog 1 Feb

Wang L, Ma C (1999) A study on the environmental geology of the middle route project of the south-north water transfer. Eng Geol 51(3):153–165

Wang Z, Wang Z, de Vriend HJ (2008) Impact of water diversion on the morphological development of the lower Yellow River. Int J Sedim Res 23(1):13–27

Wei D (2005) Beijing water resources and the south to north water diversion project. Can J Civ Eng 32(1):159–163

Zhang Q (2009) The south-to-north water transfer project of China: environmental implications and monitoring strategy. J Am Water Resour Assoc 45(5):1238–1247

Chapter 11
Feasibility of Using Satellite Water Tanks for Protecting Drinking Water in Urban Communities in Developing Countries

Manish Shrestha and Steven G. Buchberger

Technical Terms and Definitions

BC	Benefit Cost
Cfu	Colony forming unit
CBS	Center of Bureau Statistics
CLOFs	Continuous rainfall and cloudburst
EPANET	A public sector hydraulic and water quality model developed by the USEPA
GDP	Gross domestic product
GLOFs	Glacial lake outburst floods
K_b	Bulk decay coefficient (day^{-1})
KUKL	Kathmandu Upatyaka Khanepani Limited
K_w	Wall decay coefficient (day^{-1})
LDOFs	Landslide dam outburst floods
Lpcd	Liters per capita per day
Lps	Liter per second
m	Meters
MIKE NET	A water distribution model developed by DHI
MIKE URBAN	A model integrating GIS and water modeling capability developed by DHI
MLD	Million liters per day
NWSC	Nepal Water Supply Corporation
RCC	Reinforced concrete cement
RWSSFDB	Rural Water Supply and Sanitation Fund Development Board

M. Shrestha (✉) · S. G. Buchberger
Environmental Engineering Program, University of Cincinnati, Cincinnati, OH 45221-0012, USA
e-mail: shrestmm@mail.uc.edu

S. G. Buchberger
e-mail: steven.buchberger@uc.edu

R. M. Clark and S. Hakim (eds.), *Securing Water and Wastewater Systems*, Protecting Critical Infrastructure 2, DOI: 10.1007/978-3-319-01092-2_11,

SWMM	Storm water management model
USEPA	United States Environmental Protection Agency
WHO	World Health Organization

11.1 Introduction

Nepal is one of the few countries in Southeast Asia that is uniquely rich in water resources according to a report prepared by Nepal's Water Energy Commission Secretariat (WRNCCS 2011). Nepal has approximately 6,000 rivers with a drainage area of 191,000 km^2 (73,745.1 mile^2), 74 % of which lie in Nepal alone. If properly developed these rivers could generate hydropower; provide water for irrigation, industrial uses; and supply water for domestic purposes. About 10 % of total precipitation in Nepal falls as snow, and 23 % of Nepal's total land area lie above the permanent snowline of 5,000 m (16,404 ft), with 3.6 % being covered by glaciers. Nepal's economy is largely based on agriculture which contributes about 40 % to the gross domestic product (GDP) and provides employment to two-thirds of the population. However, Nepalese agriculture is mainly rain fed so that agricultural areas are being badly affected by droughts, flooding, erratic rainfall, and other extreme weather events. Nepal was self-sufficient in food grain production until 1990 but due to drought conditions in 2005/06, production fell short by 21,553 metric tons (25,863.6 short tons) and by 179,910 metric tons (201,499.2 short tons) in 2006/07. Only about 72 % of the country's population has access to basic water supply and only 25 % of the population has sanitation facilities. In Nepal, devastating floods may be triggered by (WRNCCS 2011):

- Continuous rainfall and cloudbursts (CLOFs)
- Glacial lake outburst floods (GLOFs)
- Landslide dam outburst floods (LDOFs)
- Floods triggered by the failure of infrastructure
- Sheet flooding or inundation in lowland areas due to obstructions.

Availability of water resources is increasingly recognized as being sensitive to climate change impacts and this is especially true in Nepal. Water supply is challenging because water availability, quality, and stream flow are sensitive to changes in temperature and precipitation. Increased demand for water is caused by population growth, changes in the economy, development of new technologies, changes in watershed characteristics, and water management decisions are some of the other factors to be taken into consideration.

The present situation with regard to Nepal's water resources can be summarized as follows (Durbar 2002):

- 66 % of the population has access to safe water

- 41 % of irrigated land has year round irrigation
- Less than 400 MW of hydropower capacity is available
- Little consideration is being given to environmental requirements.

This chapter discusses the technical and economic feasibility of a technique that might materially increase the ability of Nepal's water managers to provide safe drinking water to consumers. It is based on the concept of using community based satellite tanks to provide reliable continuous supply of suitable quality water at a satisfactory pressure to a residential neighborhood where the existing water distribution system is intermittent. Under this concept, the supply to the network should be sufficient so that the tank does not run empty. The study is based on an existing intermittent water supply network in a neighborhood in Kathmandu, Nepal and explores the possibility of providing continuous and reliable drinking water using satellite tanks.

11.2 Water Supply Systems in Developing Countries

Water distribution systems in developed countries are typically designed to provide uninterrupted service to the end users. However, water supply systems in many developing nations are operated in an intermittent fashion due to the lack of proper hydraulic infrastructure and scarce water resources. Approximately, 350 million people in South Asia and about 50 million people in Latin America depend on intermittent water supplies which delivers water to the end users for limited hours and only on certain days (McIntosh 2003). However, the service provided is often unsatisfactory due to low network pressure. During non-supply periods, networks may experience negative hydraulic pressures (Marchis et al. 2010). This may result in the intrusion of contaminants including pathogens which may enter the distribution system through cracks and leaks in the pipeline (Lee and Schwab 2005).

In an effort to minimize vulnerability to the effects of intermittent supply, end users may rely on service pumps to collect water, private tanks to store water, and portable filters to treat water (Cobacho et al. 2008). Users nearest to the source can draw abundant water to fill tanks to be used during non-supply hours. However, users who are furthest from the source may have to wait for a dwindling supply which is often insufficient to fulfill their basic water demands (Batish 2003).

A potential solution that might reduce the adverse effects of an intermittent supply system is to install a community based satellite water tank as part of the distribution system. Satellite tanks are elevated structures, sized to balance the water supply and demand of the consumers and to provide sufficient hydraulic head to pressurize the downstream water distribution system (USAID 1982). Elevated satellite tanks in a community would help to maintain hydraulic pressure throughout a service zone and eliminate the need for individual household tanks and pumps. In addition, chlorine booster stations can be established alongside a satellite tank to maintain free chlorine residuals in the network (Tryby et al. 1999). Computerized hydraulic analysis of both the existing and the proposed water distribution system including a satellite tank was conducted. The economic benefits and costs of the satellite tank system were compared to the existing system.

11.3 Literature Review

Modeling an intermittent system is more challenging than modeling a conventional continuous system since an intermittent system is governed by "pressure-dependent demand". Under an intermittent scenario water supply is provided for a limited time and the amount of water that a user can collect depends on the available pressure at the service connection (Tzatchkov and Cabrera-Bejar 2009).

Vairavamoorthy et.al (2000), Battermann et.al (2001), and Petr Ingeduld et.al (2006) have developed models to simulate intermittent water supply. These investigators have applied their models to communities in developing countries.

Marchis et al. (2010) developed a FORTRAN-based numerical model for simulating the filling and emptying of private storage tanks in an intermittent water distribution system. The proposed model was successfully applied to a real monitored case study of Palermo, Italy.

Tzatchkov and Cabrera-Bejar (2009) proposed a model to simulate intermittent water supply using a combination of EPA SWMM and EPANET, which incorporated SWMM for modeling initial pipe network charging and EPANET for modeling the network intermittent operation. The proposed model was successfully applied to a real monitored case study of Guadalajara, Mexico.

Klingel et al. (2008) proposed a strategy to improve the inefficient intermittent water supply system which involved three steps: central data management of the network; deficiency analysis and restructuring of the network; and introduction of continuous supply. He demonstrated a pilot study of the water distribution system in Beni Abbes, Algeria.

11.4 Case Study Area

11.4.1 Study Area Description

The case study area is located in Kamaladi, a neighborhood in Kathmandu. The specific site was chosen because information about the water distribution system and the water demand of the local residents are known. The site has 25 bi-level houses, as shown in Fig. 11.1, along two branches of a water supply line. The population served is approximately 150 residents with a domestic water demand of 71 Lpcd (0.071 m^3 $capita^{-1}$ day^{-1} or 18.74 galcapita^{-1} day^{-1}). Water demand at the study site is expected to rise to 94 Lpcd (0.094 m^3 $capita^{-1}$ day^{-1} or

Fig. 11.1 A part of the study site (*Photo taken* December 2011)

24.82 galcapita^{-1} day^{-1}), an increase of 33 % above existing system demand when the water supply system changes to a continuous pressurized system (Joshi et al. 2003; CBS 2008).

Because the Kamaladi neighborhood is underserved by the current system (ADB 2010) almost all households are forced to supplement their supply with water from private tankers and bottled water sources. All of the houses at the study area have access to piped water supply but pipelines at the site are 50–60-year-old galvanized iron (Fig. 11.2). The Hazen–Williams roughness coefficient for all the pipes was assumed to be 150 and the total pipe length in the system main is approximately 750 m (2,460.63 ft) (personal communication, local resident).

The Kathmandu Upatyaka Khanepani Limited (KUKL) delivers water to the study area through a 150 mm (5.9 in) line at a rate of 2.6 Lps (0.67 gal/s) for 2.5 h for 3 days in a week (personal communication, KUKL personnel). Water costs were $0.75 (US) for the first 10,000 liters (2,640 gal) and $0.25 (US) per 1,000 liters (264 gal) for water supplied beyond that level (ADB 2010). There is no chlorine residual detected in the supplied water and the incidence of waterborne disease due to the untreated water supply is assumed to be 10 % of the population per year (Pradhan et al. 2005; CBS 2008). A summary of water quality parameters for the study area is shown in Table 11.1.

Services are either unmetered or do not generally function and local residents resort to direct pumping of water from the mains to augment their supplies and to

Fig. 11.2 Pipe connection at the study site (*Photo taken* December 2011)

Table 11.1 Water quality of private tap water at study site, 2005 (CBS 2008)

Parameters	Unit	Private tap water	WHO GL
pH		6.5–8.2	6.5–8.5
Temp	Celsius[a]	13–18	25
Iron	mg/l	ND–0.2	0.3–3.0
Chlorine	mg/l	ND	0.2
Chloride	mg/l	10–30	250
Ammonium	mg/l	ND–0.2	0.04–0.4
Phosphate	mg/l	0.1	0.4–5.5
Coliform bacteria	Source point	±	+
Coliform bacteria	Consumption point	+	−
E. coli	Cfu/100 ml	10–130	0

[a] To convert from degrees C to degrees F multiply by 9/5 + 32. *Note ND*, Not determined; *WHO GL*, World Health Organization guideline

Private pump Private water tank Water filter

Fig. 11.3 Current coping strategy of using private pumps, tanks, and water filter at the study site (*Photo taken* December 2011)

store in their private tanks (ADB 2010). Each household at the study site has a private rooftop storage tank with an average capacity of 3,400 l (897.6 gal) and a pump with an average pumping rate of 0.3 Lps (0.079 gal) (personal communication, local resident). Some other current coping strategies involving pumps, tanks and filters at individual houses are shown in Fig. 11.3.

These coping strategies could be eliminated if the piped water supplied were improved to provide sufficient, reliable and potable water (Whittington et al. 2005).

11.4.2 Current Water Supply Enhancement Programs

The Government of Nepal with the assistance of several international organizations initiated the "Melamchi Water Supply Project" in 2001 to divert water from the neighboring Melamchi, Yanri, and Larke rivers into the Kathmandu Valley. The project is designed to provide an additional 170 MLD (44.9 mil gal/day),

2.3 times more than currently available, into the capital by construction of a 26.5 km (16.47 mile) tunnel, a water treatment plant and other associated infrastructure. It is expected to be completed by the end of 2016 (ADB 2008).

11.5 Study Methodology

In order to understand the potential for the proposed satellite tank to provide continuous reliable water at the study site, the current and proposed systems were analyzed using hydraulic modeling. A computerized hydraulic analysis comparing the existing intermittent water supply system to a proposed continuous network including the satellite tank was conducted. All houses in the network were assumed to have the same elevation and same number of occupants.

11.5.1 Analysis of Existing Intermittent Supply System

11.5.1.1 Existing Intermittent Supply System

One of the distinct characteristics of an intermittent system is ‘network charging’ which occurs during the period when empty pipelines are slowly filling and eventually pressurized (Batish 2003). The water demand under an intermittent system has different characteristics as compared to a system with a continuous supply. The consumption pattern of users under an intermittent system is based on the quantity of water that can be collected during supply hours rather than the diurnal pattern of users (Vairavamoorthy et al. 2000; Marchis et al. 2010). Two distinct phases of water consumption occur under intermittent demand—first the private tanks are filled and then the water is consumed by users (Marchis et al. 2010). Filling of the private tanks depends on the pressure available in the network and the private pumping capacity. The distance between the inlet to the system and the users’ location in the system affects the first pattern. Similarly the consumption of water by users depends on the level of water in the private tank. Users tend to reduce their water consumption as the level of water in the private tanks begins to decrease.

11.5.1.2 Modeling the Existing Intermittent System

The existing intermittent water supply system was modeled using the USEPA Storm Water Management Model (SWMM 5.0) software. SWMM can simulate the transition of runoff from free surface flow to pressurized flow in network pipes and has been used to model initial distribution system charging and pressurized flow thereafter (Tzatchkov and Cabrera-Bejar 2009). An intermittent supply of

2.6 Lps (0.69 gal/s) for 2.5 h for 3 days per week is initially introduced at the inlet. This pattern was developed based on correspondence with KUKL personnel.

11.5.1.3 Flow of Water from the Network Pipes to the Private Tanks

Water is pumped into a private rooftop tank after the water reaches the user's connection node from the inlet. The pumping is based on the pump curve in which flow depends on the head difference between the inlet and outlet (Rossman 2008). The tank water level control is integrated with the pumping system to avoid overflow and the dimension of the private tanks are assumed as 1.5 m (4.92 ft) × 1.5 m (4.92 ft) × 1.5 m (4.92 ft).

11.5.1.4 Flow from Private Tank to the End Users

Consumption patterns in an intermittent supply system depend on the water level in the private tank rather than the normal diurnal water use pattern established by users having access to constant supply (Marchis et al. 2010). The flow from each private tank was set at 4.52×10^{-3} Lps (1.19×10^{-3} gal/s) (equaling the daily demand per house per day) when the depth of water in the tank is from 1.5 m (4.92 ft) to 0.5 m (1.64 ft). The 0.5 m (1.64 ft) level of water in the private tank was assumed to be the cut-off point below which the users start to reduce their water consumption gradually. Thus, the flow from the tank to the user is set to gradually decrease after the water level falls below 0.5 m (1.64 ft).

11.5.2 Analysis of Proposed Continuous Supply System

11.5.2.1 Proposed Continuous Supply System

The proposed system is assumed to be pressurized from the beginning and the satellite tank is assumed to be constructed of Reinforced Concrete Cement (RCC) with a minimum design period of 50 years (Washington State Department of Health 2009). The size of the satellite tank is determined based on USAID guidelines (USAID 1982). Future population is also based on USAID guidelines (USAID 1982). The elevation of the satellite tank above ground level is selected such that it meets the minimum residual pressure at which users receive water from the network. This standard is based on Indian government-approved standards since Indian standards are frequently used in Nepal. The minimum residual pressure according to the Manual of Water Supply for India is: 7 m (22.97 ft) for single-storey buildings and 12 m (39.37 ft) for two-storey buildings (Batish 2003). The water demand per person per day under a continuous system at the study area is estimated to rise to 94 Lpcd (24.82 galcapita^{-1} day^{-1}) (Joshi et al. 2003). The proposed water distribution system was modeled using EPANET.

11.5.2.2 Flow from Inlet into the Satellite Tank

A pump is incorporated into the system whose flow is defined by the head difference between the inlet and outlet pumps water from the inlet into the satellite tank. The tank water level control is integrated with the pumping system to avoid overflow. The dimension of the satellite tank is 4 m (13.12 ft) × 4 m (13.12 ft) × 4.5 m. The depth of tank is 4.5 m (14.76 ft).

11.5.2.3 Flow from the Satellite Tank to End Users

Flow from the satellite tank into the network pipes is regulated to meet the collective diurnal water demand pattern of the end users using a flow control valve. The diurnal pattern of water consumption of all houses is considered to be same. Since the actual diurnal pattern for the consumers is not known, the diurnal pattern for a house in an urban area is assumed.

11.5.2.4 Modeling of Leaks in the Existing and Proposed Water Distribution System

Round leaks caused by loss of a small portion of pipe wall as a result of corrosion have only been assumed in this analysis. These types of leak frequently occur in old iron pipes and the size of the opening remains fixed even when the pressure in the system changes on a daily basis (Lambert and McKenzie 2002; Shi 2006). The number of leaks for both-existing and proposed systems is derived from literature and ranges between 1.40 and 4.86 leaks/km (0.87–3.02 leaks/mile) in average (Al-Ghamdi 2011). The leaks are assumed to be randomly placed in the network for our analysis.

Orifices have been used to represent leaks in the existing system and emitters have been used to model the leaks in the proposed system. The rate of flow from an emitter is based on the Toricelli equation for flow through an orifice (Rossman 2002).

11.5.2.5 Analyzing Water Quality in the Proposed Water Distribution System

Chlorination disinfects water by preventing the growth of algae and other microorganisms and helps in maintaining the taste and odor control for water (Ministry of Urban Development-India 1999). The occurrence of waterborne disease—diarrhea, which accounts for most of the disease related to poor water quality, sanitation, and hygiene, is estimated to be reduced by at least 60 % if the minimum chlorine residual criterion of 0.2 mg/l set by World Health Organization for potable water is met (Hutton and Haller 2004; Semenza et al. 1998; WHO 2011). A chlorine booster station is assumed to be added at the satellite tank to augment the chlorine residual in the network and meet the minimum chlorine residual criterion. The chlorine residual content and water quality in the distribution system was analyzed using EPANET. The effect of bulk fluid and wall reactions on the chlorine residual was introduced by means of a first-order kinetics constant bulk decay coefficient (Kb) and wall decay coefficient (Kw), respectively (Rossman 2002). The Kb is taken to be 1 day^{-1} and Kw is varied between 1.8 and 3.0 day^{-1}. These values are valid for iron pipes where the flow rate in the pipes is between 0 and 0.18 Lps (0.05 gal/s) (Clark et al. 2010).

After determining the maximum dosage of chlorine from analysis in EPANET, the required feed rate of chlorine or chlorine compounds for chlorination is determined as (Arizona Department of Health Services 1978):

$$\text{Chlorine feedrate (mg/day)} = \frac{(\text{Max. chlorine dosage (mg/l)}) \times (\text{Max.flow (l/day)})}{\text{Percenatage of free chlorine available}} \quad (11.1)$$

11.6 Economic Analysis

A benefit cost ratio analysis was used to evaluate the economic feasibility of the proposed system. All costs and benefits were converted to present value.

11.6.1 Costs Under Proposed System

Capital and operating costs were calculated for the proposed system. The capital costs include:

- Land acquisition cost
- Construction cost of the satellite tank
- Cost of water meters
- Cost of the supply pump
- Cost of the chlorine booster pump.

Costs for the water meter and the land acquisition cost for the satellite tank at the study site were determined by direct communication with the local residents. The cost of constructing the satellite tank was derived by communicating with personnel from the Rural Water Supply and Sanitation Fund Board (RWSSFDB).

The RWSSFDB is active in Nepal in developing new sustainable and cost-effective water supply and sanitation systems in rural areas since 1996 (RWSSFDB 2006). The cost of the supply pump and the chlorine booster pump were derived from the company website of the pump manufacturer (Commercial Aquatic Supplies 2011; Tiwari 2008). Similarly the operating costs include:

- Operating and maintenance cost of the satellite tank and pump
- Chemicals needed for disinfection
- Leakage cost and the cost of increasing water supply.

The leakage cost is determined by multiplying the total leak loss with the water tariff set by the KUKL. Similarly, the cost of increasing water supply is determined by the product of increased amount of water supplied with the water tariff set by KUKL. The operation and maintenance of the satellite tank and pumps are assumed as 10 % of the capital cost of the satellite tank and 5 % of the cost of the chlorine booster pumps, respectively (Washington State Department of Health 2009). The design and contingency cost is taken to be 8 % of the total cost (ADB 1999).

11.6.2 Benefits Under Proposed System

The benefits from the proposed system include direct and indirect benefits. The direct benefits include:

- Gain in productive time
- Reduction in coping costs mainly in pumping and storage, filtration and use of alternative sources of water.

The details of the coping costs are discussed below.

11.6.2.1 Productive Time Cost

The time spent waiting to receive water and on constantly checking if the water has been pumped or not, during the supply hours, is converted into monetary terms by multiplying the total time spent with the average hourly wage of an individual in the household. The average time spent on checking the water being pumped is found to be 3 h per week (personal communication, local resident). The average hourly wage of a skilled worker in Kathmandu was found to be $0.50 (US) per hour in 2003 (ADB et al. 2009). The average hourly wage has been inflated to obtain the prevailing wage in 2011 (ADB et al. 2009).

11.6.2.2 Pumping Cost

Each household at the study site uses a pump to draw water during supply hours. The duration of pumping time is equal to the supply hours, i.e., 7.5 h per week. The average monthly cost for pumping per household at the study site is $1 (US) (personal communication, local resident).

11.6.2.3 Storage Cost

Users at the study site use private tanks and jars to store water collected during the supply hours. The average storage cost per household per month was found to be $1.33 (US) in 2003 (Whittington et al. 2005). The average storage cost per household has then been inflated to obtain the prevailing storage cost in 2011 with the inflation rates derived from the Central Bank of Nepal (Central Bank of Nepal 2007, 2011).

11.6.2.4 Filtration Cost

The treatment of water in houses at the study site involves boiling and filtering which was found to be $0.8 (US) in 2003 (Whittington et al. 2005). The average treatment cost per household has then been inflated to obtain the prevailing treatment cost in 2011.

11.6.2.5 Cost of Using Alternative Source of Water

Users at the study site often purchase water from private water vendors as the supplied water under the existing condition is unable to satisfy their water demand. The average purchase cost of an alternative source of water per household per month was found to be $20 (US) in 2003 (Whittington et al. 2005). The average purchase cost per household was modified using projected inflation rates.

The indirect benefit includes healthcare cost averted as a result of improved water quality under the proposed system. The healthcare cost has been limited to the cost related to waterborne disease—diarrhea as it accounts for most of the disease burden related with poor water, sanitation, and hygiene (Hutton and Haller 2004). The reduction in occurrence of diarrhea is estimated to be at least 60 % of the disease incidence after the minimum chlorine residual criterion of 0.2 mg/l as set by WHO is satisfied (Semenza et al. 1998). The healthcare cost is determined assuming 91.8 % of the people suffering from diarrhea are ambulatory and the rest 8.2 % are hospitalized (Hutton and Haller 2004). The detailed healthcare cost per hospitalized and ambulatory patient is shown in Table 11.2.

After determining the cost and benefits, the economic feasibility (i.e., benefit–cost ratio) of the proposed system and existing system with increased inflow was

Table 11.2 Annual healthcare cost in US$ per hospitalized and ambulatory patient (Hutton and Haller 2004)

Healthcare cost per hospitalized patient	64.55
Healthcare cost per ambulatory patient	21.31

analyzed by varying the number of leaks and percentage of disease incidence and reduction scenarios. The disease occurrence under the existing system is assumed to be a minimum of 10 % and a maximum of 90 % of the total population and the percentage of disease reduced with the proposed system is assumed to be a minimum of 60 % and a maximum of 90 % of the disease incidence for our analysis.

11.7 Results and Discussions

11.7.1 Disparity in Existing Water Distribution System

With no leakage and no water extracted from by water users in the network it was found from the analysis that the maximum recharge time for the network was 12 min. However, under the existing system, the recharge time increases to 53 min. The percentage of water demand satisfied gradually decreases as the distance between the user and inlet increases. For example, the closest user receives enough water (2,800 l–739.2 gals) to satisfy 100 % of their daily demand of 71 Lpcd (18.74 galcapita^{-1} day^{-1}) while the user furthest from the inlet received about 60 % (1,750 L or 462 gals) of their weekly demand. It was found that intermittent supply generates high inequality among end users with the amount of water received and waiting time largely depending on the distance between the inlet and end users.

11.7.2 Proposed System Versus Existing System

The existing water distribution system was converted to the proposed water distribution system using a satellite tank of suitable capacity and by increasing the intermittent inflow. In the proposed system it was assumed that all private tanks and residential pumps are removed. The 72,000 L (19,008 gals) satellite tank covers a land area of 32 m^2 (38.27 yard^2) and the elevation of the floor of the tank from ground level is 10 m (32.81 ft). Intermittent inflow of 4 Lps (1.06 gal/s), an increase of 50 % from the existing inflow has been used to meet the increasing water demand under a continuous system. The increase in supply is expected to be provided from the ongoing 'Melamchi Water Supply Project' that is expected to be completed by 2016 (ADB 2008). The water demand under a continuous system is estimated to be 94 Lpcd (24.4 galcapita^{-1} day^{-1}) (Joshi et al. 2003). Also to better understand the effect of satellite tank on the water distribution system at the study site, the existing water distribution system was also analyzed by increasing the water supply by 50 % without using the satellite tank.

Results show the water demand of 94 Lpcd for all users is completely (100 %) satisfied under the proposed system and the existing system with 50 % increased inflow while in the existing system none of the user's water demand is completely satisfied and the percentage rate of water demand satisfied gradually declines as the distance between the inlet and user increases. Results from the analysis also show that there is no weekly delay time to receive water for users under the proposed system as the system is always pressurized while under the existing system with 50 % increased inflow, the average weekly delay time to receive water is 33 min with a standard deviation of nearly 23 min. In summary, the proposed system completely eliminates the weekly delay period whereas the existing system with 50 % increased inflow only reduces the weekly delay time by 75 % in average per week.

11.7.3 Leak Loss Under Existing and Proposed System

The study shows that the amount of leak loss under the proposed system is five times the leak loss under the existing system due to the increased pressure. For example, consider the case with 5 leaks of 6 mm (0.24 in); the total leak loss is 9,100 L (2402.4 gals) per week under the existing system but the leak loss increases to 39,000 L (10,296 gals per week) under the proposed system. The fivefold increase in leak loss under the proposed system reflects the continuous leak loss under the pressurized system whereas under the existing case there is no leakage loss during the non-supply hours. The average leak loss per week under the existing system is 7,100 L (1,874.4 gals) and 34,500 L (9,108 gals) per week under the proposed system.

11.7.3.1 Water Quality Results

Under the proposed system the chlorine booster pump was set at 1.5 mg/l and chlorine was added at the satellite tank in order to meet the minimum World Health Organization (WHO) chlorine residual criterion of 0.2 mg/l at the extremities of the system. The dosage rate of 1.5 mg/l was determined from running water quality simulations in EPANET. While running the simulation, the initial chlorine residual content in water under the existing water supply system is taken to be zero and analyzed. Result shows that the concentration of chlorine residual for all users lies between 0.2 mg/l and 1 mg/l for the bulk decay coefficient (Kb) value of 1 day^{-1} and wall decay coefficient (Kw) values of 1.78 and 3.0 day^{-1}. Results indicate that the water quality under the proposed system satisfies the WHO water quality criterion.

11.7.4 Economic Feasibility Results

11.7.4.1 Costs and Benefits Under the Existing and Proposed System

Costs incurred include leakage cost, cost of increasing supply, productive time cost, pumping cost, storage cost, healthcare cost, and home treatment cost whereas the benefits include gain in productive time and reduction in use of alternative source of water supply. The present value multiplier to convert the annual cost to present cost is found to be 15.76 where the project life is taken to be 50 years and the interest rate as 6 %.

11.7.4.2 Costs and Benefits Under the Proposed System

The land acquisition cost at the study site is US$ 390.65 m^{-2} (326.63 $yard^{-2}$) and the cost of water meter is US$ 10.00 per unit (personal communication, local resident). The annual operation and maintenance cost of satellite tank and pumps are computed from their respective capital cost and are converted to the present value using the present value multiplier of 15.76. A summary of capital cost and operation and maintenance cost incurred under the proposed system is shown in Table 11.3.

The benefits and costs are listed in Table 11.4. The values listed in Table 11.4 are for a case with five 6 mm leaks and when the disease incidence is 10 % of the total population and percentage of disease reduced is taken to be 60 % of the incidence. The results in this case show that the BC ratio is 1.32 with the benefit exceeding the cost by nearly US$ 2,450 per household.

The BC ratio for the proposed system will rise as the disease reduction increases. The existing system with the increased inflow fails to meet the minimum chlorine residual criterion of 0.2 mg/l and is hence unable to reduce the disease incidence at all.

The result shows that the proposed system is economically feasible with a BC ratio above 1. Thus, the results demonstrate that the satellite tanks are highly

Table 11.3 Summary of capital cost and O&M cost of satellite tank and pumps per house in US$

Total Capital and O&M cost per 25 houses	1,47,765
Total Capital and O&M cost per house	5,910

Table 11.4 Summary of variables that effect benefits and costs under the proposed system

Benefits per household (US$)	10090.50
Costs per household (US$)	7652.99
Benefit cost (BC) ratio	1.32

feasible in areas where the incidence of waterborne disease is high as it has the potential to significantly reduce the occurrence of waterborne disease by chlorination.

11.8 Conclusion

An intermittent water supply system can lead to major problems in the quantity and the quality of water delivered to end users in the distribution network. This case study shows that a suitably sized satellite storage tank in conjunction with a moderate increase in supply can be a viable remedy for water quality and quantity problems that afflict intermittent water systems in developing urban areas. The computer code SWMM was used to model the existing intermittent water supply system, whereas the computer code EPANET was used to model the proposed continuous supply system with the satellite tank. The simulation results show that the intermittent water supply system is fraught with inequities among end users in the neighborhood. There is an uneven distribution of the already scarce water. There is an uneven distribution of the waiting time burden that users have to endure to receive water. There is an uneven distribution of the additional cost that end users must pay to secure adequate drinking water from other sources. Those living upstream near the entrance to the supply link generally are able to secure their full quota for water in a relatively short period. In sharp contrast, those living downstream near the terminal end of the supply link often are unable to secure an adequate water supply even after a relatively long waiting period. These users are forced to procure some of their water needs from other sources (e.g., bottles from a local water truck).

The picture changes dramatically with the proposed pressurized system. Here simulation results show that a pressurized system leads to equal distribution of supplied water and eliminates the waiting time burden. Further, the water quality under the proposed pressurized system is expected to be much better than the quality of the intermittent supply. The pressurized supply will meet the minimum chlorine residual criterion set by WHO. In return, the provision of high quality drinking water reduces the incidence of waterborne disease. One of the drawbacks of the proposed pressurized water distribution system is the increased leakage loss, projected to be five times greater than leakage from the existing intermittent system. The leakage problem can be minimized, however, through a neighborhood-wide water conservation program that uses meters (and other devices) to identify likely leakage locations under the streets and inside homes and, then, quickly repairs those pipes with confirmed leakage problems. In summary, this case study clearly demonstrates that the proposed pressurized distribution system with satellite storage tank and supplemental water supply can have a significant positive impact on the quality and quantity of water provided to residential consumers in Kathmandu.

References

Al-Ghamdi AS (2011) Leakage-pressure relationship and leakage detection in intermittent water distribution system. J Water Supply Res Technol AQUA 60(3):178–183

Arizona Department of Health Services (1978) Disinfection of water systems. Engineering Bulletin No. 8. http://www.azdeq.gov/function/compliance/download/eb8.pdf. Accessed on May 2012

Asian Development Bank (1999) Handbook for the economic analysis of water supply Projects. ISBN: 971-561-220-2

Asian Development Bank (2008) Nepal: Melamchi water supply project and Kathmandu valley water services sector development program. Project Nos. 31624 and 36609

Asian Development Bank, Department for International Development and International Labour Organization (2009) Nepal: critical development constraints. ISBN: 978-971-561-803-8. Publication Stock No. RPT090743

Asian Development Bank (2010) Nepal: Kathmandu valley water distribution, sewerage, and urban development project. Project feasibility study. Project No. PPTA 4893-NEP

Batish R (2003) A new approach to the design of intermittent water supply networks. In: Proceedings of world water and environmental resources congress 2003, Philadelphia, Pennsylvania, 23–26 June 2003

Batterman A, Mache S (2001) A strategy to reduce technical water losses for intermittent Water Supply Syst. http://sdteffen.de/diplom/thesis.pdf

Central Bank of Nepal—Research Department (2007) Inflation analysis and price situation, vol II, No. 1. http://www.nrb.org.np/red/publications/inflation/Inflation%20Analysis%20and%20Price%20Situation–2006-08__Mid-August_2006__Text.pdf. Accessed on March 2012

Central Bank of Nepal—Research Department (2011) Comparative analysis of inflation in Nepal and India. http://econpapers.repec.org/paper/esswpaper/id_3a4633.htm. Accessed on March 2012

Central Bureau of Statistics, Government of Nepal (2008) Environment Statistics of Nepal 2008. http://s3.amazonaws.com/zanran_storage/www.cbs.gov.np/ContentPages/24857902.pdf. Accessed on Jan 2012

Clark RM, Yang J, Impelliteri CA, Haught R, Schupp D, Panguluri S, Krishnan ER (2010) Chlorine fate and transport in distribution systems: experimental and modeling studies. J AWWA 102(5):144–155

Commercial Aquatic Supplies (2011) Chlorine booster pumps. http://www.commercialaquaticsupplies.com/chlorine%20booster%20pump.pdf. Accessed on May 2012

Cobacho R, Arregui F, Cabrera E, Cabrera, E Jr (2008). Private water storage tanks: evaluating their inefficiencies. Water Pract Tech 3 (1). doi:10.2166/WPT.2008025

Durbar S (2002) Executive summary. Water resources strategy, Nepal. His Majesty's Government of Nepal Water and Energy Commission Secretariat, Kathmandu, Nepal

Hutton G, Haller L (2004) Evaluation of the costs and benefits of water and sanitation improvements at the global level. World Health Organization. WHO/SDE/WSH/04.04

Ingeduld P, Pradhan A, Svitak Z, Terrai, A (2006) Modeling intermittent water supply systems with EPANET. 8th annual water distribution systems analysis symposium, Cincinnati, 27–30 Aug 2006

Joshi PS, Shrestha KB, Shrestha PL (2003) Household water use survey and research in Kathmandu valley. Final report. http://www.ngoforum.net/index.php?option=com_docman&task=doc_view&gid=7. Accessed on Jan 2012

Klingel P, Deuerlein J (2008) Rehabilitation strategy for water supply systems in developing countries case study IWRM Oasis Beni Abbes, Algeria. In: Van Zyl JE, Ilemobade AA, Jacobs HE (eds) Proceedings of the 10th annual water distribution systems analysis conference WDSA, Kruger National Park, South Africa, 17–20 Aug 2008

Lambert A, McKenzie R (2002) Econoleak—economic model for leakage management for water suppliers in South Africa. WRC Report TT 169/02

Lee JE, Schwab KJ (2005) Deficiency in drinking water distribution systems in developing countries. J Water Health 3(2):109–127

Marchis MD, Fontanazza CM, Freni G, Loggia GL, Napoli E, Notaro, V (2010) A model of the filling process of an intermittent distribution network. Urban Water J 7(6):321–333

McIntosh AC (2003) Asian water supplies: reaching the poor. ISBN 971-561-380-2. Publication Stock No. 070101

Ministry of Urban Development, India (1999) Manual on water supply and treatment, 3rd edn., pp 120–211. http://www.urbanindia.nic.in/publicinfo/o_m/Chapter%206-9.pdf. Accessed on Mar 2012

Pradhan B, Gruendlinger R, Fuerhapper I, Pradhan P (2005) Knowledge of water quality and water borne disease in rural Kathmandu valley, Nepal. Aquat Ecosyst Health Manage 8(3):277–284

Rossman L (2002) EPANET2-user's manual. Water Supply and Water Resources Division, US EPA. http://www.epa.gov/nrmrl/wswrd/dw/epanet.html. Accessed on Nov 2010

Rossman L (2008) Stormwater management model user's manual version 5.0. Water Supply and Water Resources Division, US EPA. http://eng.odu.edu/cee/resources/model/mbin/swmm/win/epaswmm5_manual.pdf. Accessed on Nov 2010

Rural Water Supply and Sanitation Fund Development Board (2006) Introduction of RWSSFDB. http://www.rwss.org/index.php?mainid=1. Accessed on July 2012

Semenza JC, Roberts L, Henderson A, Bogan J, Rubin H (1998) Water distribution system and diarrheal disease transmission: a case study in Uzbekistan. Am Soc Trop Med Hyg 59(6):941–946

Shi P(2006) Seismic response modeling of water supply systems. Ph.D. Dissertation, School of Civil and Environmental Engineering, Cornell University, Ithaca, NY

Tiwari BN (2008) Pricing of water and sanitation: the Nepalese experience central department of economics. Tribhuvan University, Nepal. ISBN 978-9937-524-00-1 pbk

Tryby ME, Boccelli D, Koechling MT, Uber JG (1999) Booster chlorination for managing disinfectant residuals. Am Water Works Assoc J 91(1):95–108

Tzatchkov VG, Cabrera-Bejar JA (2009) Intermittent drinking water supply: problems, solutions, modeling and experiences from Mexico. In: Proceedings of world environmental and water resources congress. Great Rivers, Kansas City, Missouri, 17–21 May 2009

USAID (1982) Water for the World: Designing an elevated storage tank. Technical Note No. RWS.5.D.3. http://www.watersanitationhygiene.org/References/EH_KEY_REFERENCES/WATER/Reservoirs/Designing%20an%20Elevated%20Storage%20Tank%20%28USAID%29.pdf. Accessed on Nov 2011

Vairavamoorthy K, Akinpelu E, Lin Z, Ali M (2000) Design of sustainable water distribution systems in developing countries. Technical report, South Bank University, London

Washington State Department of Health (2009) Water system design manual. DOH 331-123 (Revised 12/09)

Water Resources of Nepal in the Context of Climate Change (WRNCCS) (2011) Aryal RS, Rajkarnikar G (eds), Water Energy Commission Secretariat (WECS), P.O. Box 1340, Singha Durbar, Kathmandu, Nepal

Whittington D, Pattanayak SK, Yang JC, Balkumar KC (2005) Coping with unreliable public water supplies: averting expenditures by households in Kathmandu, Nepal. Water Resour Res 4(2):11

World Health Organization (2011) Guidelines for drinking-water quality, 4th edn. ISBN 9789241548151.43

Chapter 12
The Impacts of Global Climate Change on Water Treatment Design and Operations

Robert M. Clark

Technical Terms and Definitions

CDF	Cumulative probability distribution functions
DBP	Disinfection byproducts
D/DBP	Disinfectant/Disinfection byproduct
DOC	Dissolved organic carbon
EBCT	Empty bed contact time
GAC	Granular activated carbon
GCWW	Greater cincinnati water works
HAA5	Haloacetic acid five
IPCC	Intergovernmental panel on climate change
NASA	National Aeronautics and Space Administration
NOM	Natural organic matter
MC	Monte Carlo
RSSCT	Rapid small-scale column test
SUVA	Specific ultraviolet absorbance
TOC	Total organic carbon
TTHM	Total trihalomethanes
UK	United Kingdom
US EPA	United States Environmental Protection Agency
UVA	Ultra violet adsorption
WTP	Water treatment plant

R. M. Clark (✉)
Environmental Engineering and Public Heath Consultant, 9627 Lansford Drive, Cincinnati, OH 45242, USA
e-mail: rmclark@fuse.net

R. M. Clark and S. Hakim (eds.), *Securing Water and Wastewater Systems*, Protecting Critical Infrastructure 2, DOI: 10.1007/978-3-319-01092-2_12,

12.1 Introduction

There is a growing concern that climate change and related global warming will impact every aspect of society, ranging from ecosystems and infrastructure to public health and economic and national security. Agencies in the United States including the National Aeronautics and Space Administration (NASA), the National Climatic Data Center and the Hadley Centre in the United Kingdom (UK) have measured global surface temperature as shown in Fig. 12.1. According to the Intergovernmental Panel on Climate Change the Earth's surface has warmed by 0.74 °C in the period between 1906 and 2005 (IPCC 2007).

This heat, augmented by an enhanced greenhouse effect is stored in the world's oceans as shown in Fig. 12.2. Storage of this heat has resulted in increases in sea surface temperature, ocean heat content, and sea-level rise due to thermal expansion of water (Domingues et al. 2008). Figure 12.2 shows increases in ocean heat content for the upper 300 and 700 m (984–2300 ft) of the ocean between 1950 and 2003, along with increases in sea surface temperatures.

There are some major consequence associated with ocean warming including melting of ice sheets and thermal expansion of the oceans, which also lead to higher sea levels. It is estimated that the average sea-level has risen by 17 cm (6.7 in) during the 20th century, and that the rate of increase from 1961 to 2003 was 1.3–2.3 mm/year (0.05–0.09 in/year) (IPCC 2007). Satellite altimeters have shown a fairly steady increase in global mean sea level of around 3.2 ± 0.4 mm/year (0.13 ± 0.016 in/year) from 1993 to present (Cazenave and Llovel 2010; White and Church 2011).

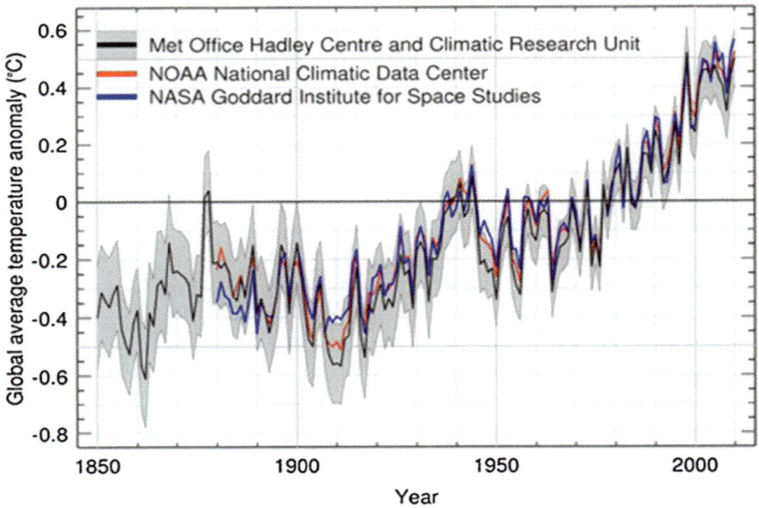

Fig. 12.1 Global-average temperature anomalies (°C), relative to the period 1906 to 2005 (IPCC 2007)

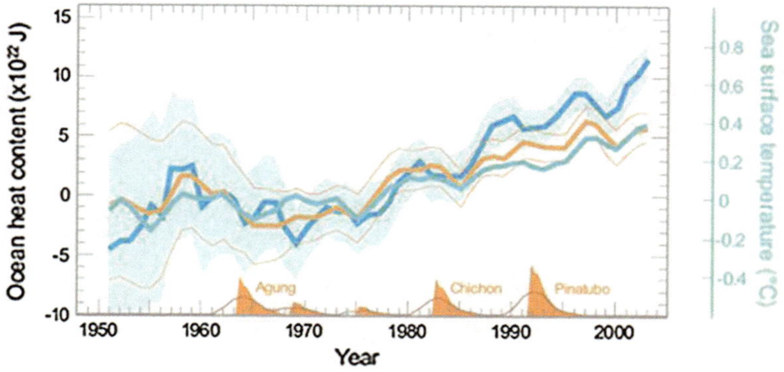

Fig. 12.2 Comparison of near global ocean heat content for the upper 700 m (*thick blue line*, *light blue* shading indicates an estimate of one standard deviation error) and upper 100 m (*thick orange line*; *thin orange lines* indicate estimates of one standard deviation error). The *brown curve* is a three-year running average of these values, included for comparison with the smoothed observations (Domingues et al. 2008)

Figure 12.3 shows increases in global annual average temperature (as measured over both land and oceans) and the carbon dioxide concentration from 1880 to present day. It demonstrates that this change in CO_2 levels is most likely due to increased greenhouse gas concentration (including carbon dioxide) in the atmosphere resulting in warmer average water and air temperatures. Red bars indicate temperatures above and blue bars indicate temperatures below the average temperature for the period 1901–2000. The year round average air temperature of the U.S. has already risen by more than 2 °F (1.1 °C) over the past 50 years with greater increases projected for the future. In the US, the intensity of severe

Fig. 12.3 Global annual average temperature and carbon dioxide concentration from 1880 to present day. To convert from Fahrenheit to Centigrade (Fx9/5-32). (NRC 2010)

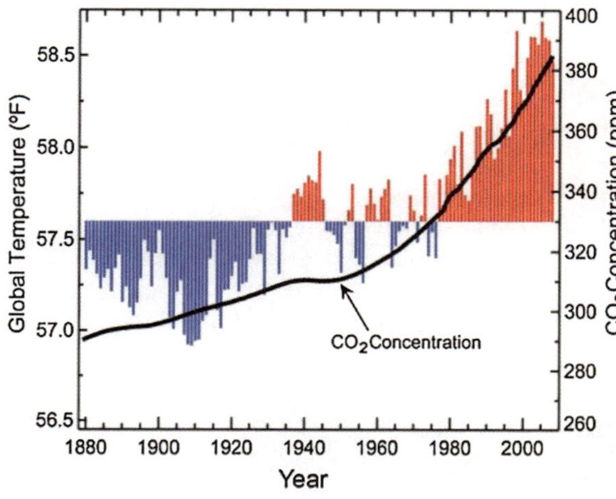

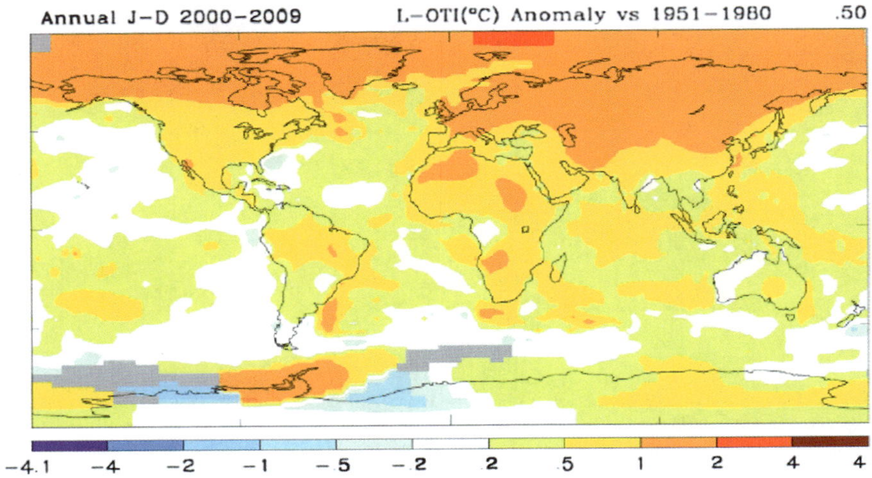

Fig. 12.4 Average surface temperature trends in Fahrenheit (degrees per decade) for the decade 2000–2009 relative to the 1950–1979 average. To convert from Fahrenheit to Centigrade (Fx9/5-32). (NRC 2010)

precipitation events has increased over the last 50 years, and continued increases in both frequency and intensity of precipitation are projected for the future.

It is unlikely that climate change impacts will be distributed evenly. For example, temperature increases in the past 10 years have generally been greatest in the northern latitudes as shown in Fig. 12.4. Warming has been more pronounced at high latitudes, especially in the Northern Hemisphere and over land.

12.2 Consequences of Climate Change

These climate change impacts will have consequences for the design and operation of water treatment plants and will influence decisions by water utilities as discussed in the following sections (Clark et al. 2011). Climate change is expected to adversely affect water quality in streams and other raw water sources and consequently impact the design and operation of existing and future drinking water treatment plants. Increases in precipitation and precipitation intensity will result in increased runoff and stream flows. These increases may result in problematic turbidity levels, increased levels of organic matter, potentially high levels of pathogens, and increased levels of pesticides in lakes, rivers, and streams. Climate change may affect both surface water and ground water quality. Although some areas may experience increases in runoff due to shifts in the spatial and temporal distribution of precipitation, some areas may experience droughts resulting in elevated levels of potentially toxic algae and high concentrations of organic matter, bacteria, etc. (Whitehead et al. 2009; Interlandi and Crockett 2003; Jacobs et al. 2001).

When assessing the impact of climate change it is important to understand that a weather event may actually consist of multiple events which occur simultaneously. For example, droughts coincide with high temperatures; storms, hurricanes, and tornadoes bring heavy rains and strong winds. Extended drought may be followed by heavy rains that negatively affect turbidity, contaminant concentrations, and organic matter in raw water supplies. Droughts affect water supply availability as well as water quality. As lake and reservoir levels decrease algal growth can occur in warmer stagnant waters, and water-use conservation enforced during drought events can increase water age in a distribution system, leading to increased disinfection byproducts (DBPs) in chlorinated systems or nitrification in chloraminated systems. Sea-level rise may increase saltwater intrusion in tidally influenced sections of rivers. Under low-flow conditions, higher sea levels tend to push saltwater farther upstream, temporarily affecting the quality of the supply (Wright et al. 2013).

Heavy rainfall can mobilize contaminants from a watershed (i.e., nonpoint source pollution) disturb sediments in streams and reservoirs, and result in sewer overflows. The most significant pollutant loads are usually associated with the first flush of rainfall after long, dry periods. Beyond heavy rainfall, flooding has its own negative consequences, including infrastructure damage and groundwater well contamination. In addition, alkalinity, which influences treatment plant processes, tends to decrease under heavy rainfall because the percentage of stream flow originating as groundwater decreases.

According to Wright et al. (2013) temperature extremes can result in significant challenges within a watershed and distribution system. Increased water temperatures can encourage algal growth, particularly toxic cyanobacteria, which prefer higher temperatures. High temperatures increase the potential for DBP formation in distribution systems and nitrification in chloraminated systems. As air temperatures rise, water temperatures increase and dissolved oxygen in water decreases.

Weather-related changes to source water quality can occur in a watershed or within reservoirs; changes to finished water quality can occur in the distribution system. Depending on upstream vegetation conditions and land-use types, nonpoint source pollution usually results in higher levels of nutrients, natural organic matter (NOM), pathogens, organic chemicals (e.g., pesticides and pharmaceuticals), and turbidity in watershed runoff.

Algae is a major culprit in taste-and-odor problems and is a source of NOM for DBPs. Although weather can influence algal growth, nitrogen and phosphorous are necessary components. The DBP and nitrification potential in chloraminated systems is closely correlated with increased water temperature and age.

An important aspect of water utility decision making that should not be overlooked is the ability to predict water demand. Accurate water demand projections help utility managers make short term operational decisions about water treatment plant operations, which include water production, balancing of supply and demand, and potential water use restriction during a drought. Demand projections assist managers make long-term decisions about acquisition of additional supply and treatment capacity, and investment in distribution system infrastructure. Haagenson et al. (2013) Illustrate these points by a case study focused on Aurora, Colo., a rapidly growing suburb of Denver with limited water supply in a semiarid region. Clearly climate change may have an adverse impact on the ability of drinking water utilities to meet US Safe Drinking Water Act regulations, therefore, requiring treatment systems to make major changes in their operations.

12.3 Impact of Climate Change-A Water Treatment Example

To illustrate some of the issues that need to be addressed regarding the impact of climate change on water supply Li et al. (in press) have developed a detailed example that attempts to assess the effects of climate of change on water treatment plant (WTP) design and performance. As discussed previously, WTP performance, design and operation, is affected by uncertainties in influent water quality and in water demand. The investigators studied the changes in finished water quality that will occur in response to changes in influent water quality (pH, TOC, ammonia, UVA, turbidity, bromide, etc.) under possible future climate change conditions. Their study should provide insight and guidance to water utility managers as to how to identify key source water quality parameters that can affect regulated water quality parameters such as disinfectant residual, pH, and total trihalomethanes (TTHMs). As part of the study the investigators examined the use of granular activated carbon as a possible treatment technology that is flexible enough to deal with extreme variations is source water quality. A tool developed by the US EPA called the US Environmental Protection Agency WTP Model (USEPA 2005) is used in the analysis.

The EPA model is based on empirical correlations to predict central tendencies of natural organic matter removal, disinfection, and DBP formation in a drinking water treatment plant. The model was designed to (1) identify and screen new treatment technologies that can assist in meeting both required levels of disinfection and the disinfection by products (DBP) levels regulated under USEPA's various rule making efforts, (2) evaluate the possible effects of source water or unit treatment process operations on DBP formation, and (3) assist regulatory agencies in assessing the potential impact of new regulations on treatment requirements over a large group of plants.

An important step in applying the WTP Model is validation. In order to perform this validation step the complete data set from USEPA's Information Collection Rule (ICR) in conjunction with ICR data obtained specifically for the Greater Cincinnati Water Works (GCWWs) Richard Miller Plant was utilized. The ICR was designed to obtain water quality, water treatment, and occurrence information needed for development of Safe Drinking Water Act regulations (Obolensky and Singer 2005; Obolensky et al. 2007). This data set was used to validate the WTP

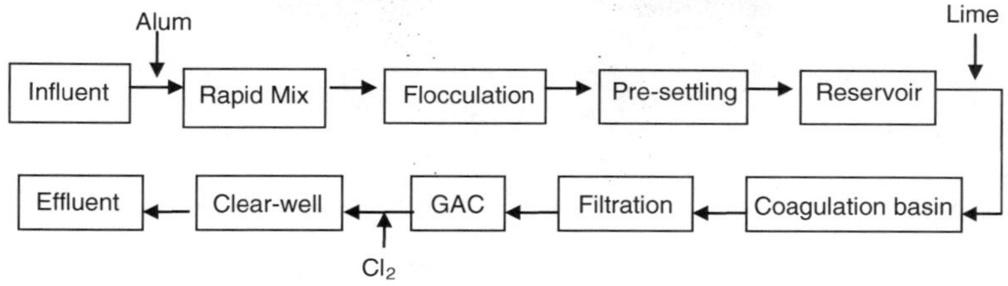

Fig. 12.5 Unit process treatment train for the miller plant

model and to characterize water quality and the changes in water quality in the Ohio River which is the source of supply for the Miller Plant.

The WTP Model was validated by simulating performance and operation of the Miller WTP. Individual influent water quality parameters were perturbed to identify the key influent water quality parameters that may significantly affect finished water quality. Source water quality in the Ohio River was simulated using ICR data based on a Monte Carlo simulation approach in conjunction with research conducted by Skjelkvale et al. (2005), Whitehead et al. (2006), and Cromwell et al. (2007). These steps are discussed below.

12.3.1 WTP Plant Validation

As discussed, validation was based on ICR data obtained specifically for the Greater Cincinnati Water Works (GCWW)'s Richard Miller Plant. Data sources include detailed information about treatment, plant design and operations from all large public water systems in the United States serving populations of 100,000 or more for 18 monthly monitoring periods from July 1997 to December 1998 (USEPA 2000).

Individual influent water quality parameters were perturbed to identify key influent water quality parameters that may significantly affect finished water quality. The Miller WTP unit process treatment train is shown in Fig. 12.5 and the design and operation parameters utilized in the simulation are listed in Table 12.1. In Table 12.1, the value for T_{10} refers to the time required for the effluent tracer concentration to reach 10 % of the influent tracer concentration in a step-dose tracer study. Inflow rate and chemical feed concentrations are given in Table 12.2 (Li et al. 2009).

Results from the WTP Model simulation and the validation data from the ICR database are compared in Table 12.3. They show close agreement for most water quality parameters including pH, alkalinity, total hardness, TOC, free chlorine residual, and TTHMs.

Table 12.1 Miller WTP unit process design parameters

Unit process	Volume (m^3)	T^*_{10} (min)
Rapid mixing	31.8	2
Flocculation basin	7342.9	14
Presettling	8440.6	14
Reservoir settling	1411805.0	1,728
Coagulation basin	98410.0	144
Filtration	9352.0	4
GAC contactors	9311.1	–
Clear-well	107115.5	77

Note: Data not available. (*Data source* USEPA ICR database)
* Time required for effluent concentration to reach 10% of influent concentration

Table 12.2 Inflow and chemical feed levels for the Miller WTP

Sampling period	10	13	16
Inflow rate, m3/s	4.41	5.27	5.76
Aluminum at RM, mg/L	0.87	1.82	0.87
OPC at RM, mg/L	1.50	1.50	1.30
Iron at COAG, mg/L	0.00	0.47	0.00
Lime at COAG, mg/L	6.73	7.92	4.62
Chlorine at CLR, mg/L	1.26	1.56	1.46
HFS at CLR, mg/L	0.63	0.63	0.55
NaOH at CLR, mg/L	3.00	5.00	5.50

Note: RM-rapid mixing; COAG-coagulation basin; CLR-clearwell
OPC-ordinary Portland cement
(*Data source* USEPA ICR database)

12.3.2 Water Quality Simulation

It is clear from the previous discussion that uncertainties in raw water quality may result from global climate change and result in a violation of drinking water standards. In order to characterize these changes, a Monte Carlo simulation model was developed using ICR source water quality data from six Ohio River WTPs during July 1997 to December 1998. Correlation analysis was used to identify whether relationships exist among the raw water quality parameters considered in the analysis.

12.3.2.1 Analysis of Raw Water Quality Data

Log-normal distributions are assumed to describe the probability distributions for all raw water quality parameters in the Ohio River. Figure 12.6 verifies this assumption by demonstrating the normal probability plot for logarithms of pH and TOC using ICR database for Ohio River. The log-normal distribution is used for raw water quality parameters in the Monte Carlo simulations.

Table 12.3 Comparison of sampled and modeled water quality results

Water quality parameter	Sampling period	Data type	Influent	Coag. basin	Filtration	GAC	Finished water	AVG1[a]	AVG3[b]
pH	10	Sampled	7.7	8.7	8.6	8.0	8.5	8.6	8.6
[–]		Modeled	7.7	9.4	9.4	9.4	9.1	9.1	9.1
	13	Sampled	7.6	7.9	7.8	7.8	8.2	8.4	8.5
		Modeled	7.6	9.2	9.2	9.2	8.9	9.0	9.0
	16	Sampled	7.7	8.3	8.1	8.0	8.4	8.3	8.6
		Modeled	7.7	8.8	8.8	8.8	8.2	8.2	8.2
Alkalinity	10	Sampled	56	59	58	58	58	60	60
[mg/L]		Modeled	56	64	64	64	62	62	62
	13	Sampled	63	56	59	58	57	64	68
		Modeled	63	72	72	72	69	69	70
	16	Sampled	75	77	80	77	81	81	82
		Modeled	75	81	81	81	78	78	78
Total	10	Sampled	113	128	120	119	120	121	115
Hardness		Modeled	113	122	122	122	122	122	122
[mg/L]	13	Sampled	98	106	107	108	106	115	120
		Modeled	98	109	109	109	109	109	109
	16	Sampled	164	162	166	164	169	164	165
		Modeled	164	170	170	170	170	170	170
TOC	10	Sampled	1.8	1.5	1.4	1.0	1.0	–	–
[mg/L]		Modeled	1.8	1.8	1.8	0.8	0.8	0.8	0.8
	13	Sampled	3.6	2.55	2.2	–	0.51	–	–
		Modeled	3.6	3.6	3.6	1.3	1.3	1.3	1.3
	16	Sampled	2.3	1.9	1.7	–	0.54	–	–
		Modeled	2.3	2.3	2.3	0.6	0.6	0.6	0.6
UVA_{254}	10	Sampled	0.069	0.028	0.024	0.012	0.010	–	–
[cm^{-1}]		Modeled	0.069	0.061	0.061	0.009	0.006	0.006	0.006

(continued)

Table 12.3 (continued)

Water quality parameter	Sampling period	Data type	Influent	Coag. basin	Filtration	GAC	Finished water	AVG1[a]	AVG3[b]
	13	Sampled	0.178	0.068	0.054	–	–	–	–
		Modeled	0.178	0.151	0.151	0.018	0.013	0.013	0.013
	16	Sampled	0.075	–	–	–	–	–	–
		Modeled	0.075	0.067	0.067	0.004	0.003	0.003	0.003
Free chlorine	10	Sampled	–	–	–	–	1.0	0.9	0.7
Residual		Modeled	0.0	0.0	0.0	0.0	1.0	0.7	0.5
[mg/L]	13	Sampled	–	–	–	–	1.2	0.8	0.7
		Modeled	0.0	0.0	0.0	0.0	1.2	0.7	0.5
	16	Sampled	–	–	–	–	1.3	1.0	0.7
		Modeled	0.0	0.0	0.0	0.0	1.2	1.0	0.9
TTHMs	10	Sampled	–	–	–	–	11.9	23.7	28.8
[μg/L]		Modeled	0.0	0.0	0.0	0.0	9	16	22
	13	Sampled	–	–	–	–	8.1	30.9	47.8
		Modeled	0.0	0.0	0.0	0.0	20	37	53
	16	Sampled	–	–	–	–	8.5	29.6	46.5
		Modeled	0.0	0.0	0.0	0.0	8	16	23

Note: [a] AVG1 refers to average retention time 1 day
[b] AVG3 refers to the maximum retention time, 3 days
–: Data not available

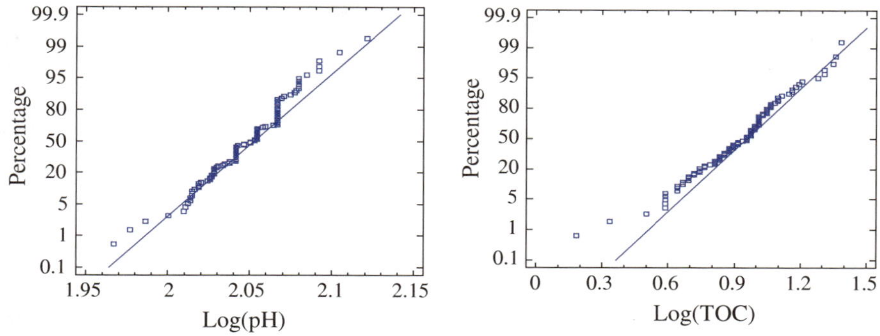

Fig. 12.6 Normal probability plots for logarithm of pH and TOC

A multivariate model by Salas et al. (1980) was adopted in this study as shown below:

$$c_i = \mu_i + z_i\sigma_i \quad i = 1, 2, \ldots, 9 \tag{12.1}$$

where μ_i, $\sigma_{i,}$ and z_i are the mean, standard deviation, and transformed correlation matrix of the other water quality variables and c_i is the mean value of the water quality variable under consideration. It is required that z_i is for normally distributed variables. Since the raw water quality parameters are log-normally distributed, the ICR data for the raw water quality are first transformed to a log normal form before computing the parameters used in Eq. (12.1).

12.3.2.2 Projection of Water Quality to 2050

Water quality was projected to 2050 in order to simulate the effects of climate change. These projections are based on Eq. (12.1) and research conducted by Skjelkvale et al. (2005), Whitehead et al. (2006) and Cromwell et al. (2007).

Skjelkvale et al. (2005) studied trends in surface water chemistry for the Appalachian Plateau located upstream of Ohio River were used to project values in alkalinity, total hardness, and dissolved organic carbon (DOC). The trends for alkalinity, total hardness, and DOC are equivalent to +0.036, –0.22, and +0.03 mg/L per year. Since DOC is usually the main component of TOC, the trend for TOC is assumed as the same as DOC.

Whitehead et al. (2006) investigated the impacts of climate change on ammonia using a dynamic modeling approach in the River Kennet of the United Kingdom (UK) for the period 1961–2100. An approximate 25 % increase of ammonia from 1998 to 2050 can be estimated from their study. It was also assumed that bromide levels pH, turbidity, calcium hardness, and UVA in 2050 are the same as baseline ICR data.

Cromwell et al. (2007) predicted that the temperature rise as consequence of green house gas emissions from 1990 to 2100, would increase from 1.1 to 6.6 °C. Therefore, it was estimated that the average temperature rise would be estimated as 2 °C over the baseline values for all seasons by 2050.

Using the ICR data discussed earlier, a baseline source water quality estimate was developed for the water quality as shown in Table 12.4. Note that means and standard deviations for the estimates are also included in Table 12.4.

Using the results from the development of Eq. (12.1) and the results from Skjelkvale et al. (2005), Whitehead et al. (2006) and Cromwell et al. (2007) the projected water quality for 2050 is shown in Table 12.5.

In order to eliminate population growth as a variable the flows to the Miller plant under the simulated conditions were assumed to be the same as that of the baseline. It was further assumed that the coefficients of variation for all quality parameters in 2050 are the same as those for the baseline data (Ratios of σ_0/μ_0in Table 12.4).

Monte Carlo simulations at the Miller Plant showed two important results: (1) TOC is a water quality parameter that dominates the levels of regulated DBPs such as TTHM and HAA5 in finished water and in distribution systems, i.e., TTHM requirements could be met if TOC compliance is met; (2) the TOC removal rate is always greater than 55 % owing due to the use of GAC. According to the U.S. EPA disinfectant/disinfection byproduct (D/DBP) rule, there are two important compliance criteria for TOC treatment when using surface water as source water. One criterion is treated water TOC level is less than 2.0 mg/L, calculated quarterly as a running annual average (based on Ohio Code, Section D 3745–81–77).

Figure 12.7 shows the average projected increase in mean TOC in the Ohio River between 2000 and 2050. Of note is the fact that the "one-standard deviation" line increases at a greater rate than the mean TOC line.

Table 12.4 Estimated baseline raw water inputs for the Miller WTP based on ICR data

		Spring		Summer		Autumn		Winter	
Parameter	Unit	μ_0	σ_0	μ_0	σ_0	μ_0	σ_0	μ_0	σ_0
pH	–	7.7	0.17	7.7	0.20	7.8	0.22	7.8	0.18
Alkalinity	mg/L	55.5	18.2	77.2	21.7	81.4	21.0	62.3	23.1
Turbidity	NTU	43.4	38.0	26.9	36.9	8.5	7.6	41.5	64.7
Ca hardness	mg/L	63.5	23.3	76.2	31.6	87.1	35.6	74.2	33.7
Total hardness	mg/L	110.4	18.6	140.3	26.1	161.3	31.1	133.	36.5
TOC	mg/L	2.3	0.6	2.9	0.6	2.6	0.3	2.5	0.6
UVA	cm^{-1}	0.12	0.06	0.11	0.06	0.08	0.02	0.09	0.05
Bromide	mg/L	0.03	0.01	0.05	0.02	0.10	0.04	0.07	0.04
NH_3_N	mg/L	0.29	0.41	0.20	0.11	0.18	0.10	0.18	0.10
Temperature	°C	12.4	–	25.7	–	20.8	–	9.8	–
Flow	m^3/s	4.75	–	5.01	–	5.75	–	5.30	–

Note: μ_0 is baseline value and σ_0 is standard deviation; –refers to not applicable

Table 12.5 Simulated raw water inputs for the Miller WTP in 2050

		Spring		Summer		Autumn		Winter	
Parameter	Unit	μ_1	σ_1	μ_1	σ_1	μ_1	σ_1	μ_1	σ_1
pH	–	7.7	0.17	7.7	0.20	7.8	0.22	7.8	0.18
Alkalinity	mg/L	57.3	18.9	79.1	22.1	83.3	21.7	64.1	23.7
Turbidity	NTU	43.4	38.0	26.9	36.9	8.5	7.6	41.5	64.7
Ca hardness	mg/L	63.5	23.3	76.2	31.6	87.1	35.6	74.2	33.7
Total hardness	mg/L	98.8	16.8	128.9	24.5	149.9	28.5	121.5	32.8
TOC	mg/L	3.8	1.0	4.4	0.9	4.1	0.5	4.1	0.9
UVA	cm^{-1}	0.12	0.06	0.11	0.06	0.08	0.02	0.09	0.05
Bromide	mg/L	0.03	0.01	0.05	0.02	0.10	0.04	0.07	0.04
NH_3_N	mg/L	0.36	0.50	0.25	0.13	0.23	0.13	0.23	0.13
Temperature	°C	14.4	–	27.7	–	22.8	–	11.8	–
Flow	m^3/s	4.75	–	5.01	–	5.75	–	5.30	–

Note: μ_1 is baseline value and σ_1 is standard deviation; –refers to not applicable

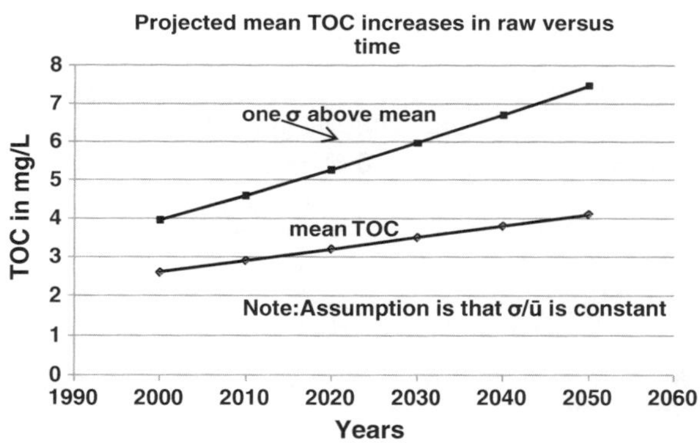

Fig. 12.7 Projected mean TOC increases in raw water versus time

12.4 Water Quality Projections Using the WTP Model

In order to explore the impact of climate change on water quality, the source code for the EPA WTP Model was modified to accommodate the algorithms introduced in the previous sections and to compute the quarterly running annual average for regulated TOC and TTHM in finished water and in the distribution system. In Monte Carlo (MC) simulations, the number of runs is an important criterion when attempting to obtain reliable results. Therefore, numerical tests were made to determine the number of runs that yielded stable mean, standard deviation, and skewness using running average of TOC in the finished water as the test case. The mean and standard deviation stabilize after 500 and 2,000 runs while the skewness does not stabilize until after 5,000 runs. Therefore, 5,000 runs were used in subsequent MC simulations.

12.4.1 Algorithms underlying the WTP Model

The following equations illustrate the types of algorithms used in the WTP model and which were utilized in this analysis. For iron and aluminum coagulated waters, the formation of TTHM was estimated with (USEPA 2005),

$$\mathrm{TTHM} = 23.9(\mathrm{TOC}*\mathrm{UVA})^{0.403}(\mathrm{Cl_2})^{0.225}(\mathrm{Br^-})^{0.141}(1.1560)^{(\mathrm{pH}-7.5)}(1.0236)^{(\mathrm{Temp}-20)}(\mathrm{time})^{0.264} \quad (12.2)$$

UVA in Eq. (12.2) is UV absorbance at 254 nm; Cl_2 is chlorine concentration in mg/L; Br^- is bromide concentration in mg/L; Temp is water temperature (°C); time is reaction time in hrs. It can be seen that formation of TTHM depends on and increases with increasing values of these parameters

According to the USEPA (2005), TOC can be removed by coagulation and GAC. In the Miller plant TOC is primarily removed by GAC. The expression for TOC removal by GAC in the EPA model is,

$$\mathrm{TOC_{eff}} = \mathrm{a} + \frac{\mathrm{a}}{\mathrm{d}\cdot\mathrm{D}}\ln\frac{1+\mathrm{b}\cdot\mathrm{e}^{-\mathrm{d}\cdot\mathrm{D}}}{1+\mathrm{b}} \quad (12.3)$$

Where, TOC_{eff} is effluent concentration from GAC processing; D is GAC reactivation period; $a = 0.682 \cdot TOC_{in}; b = 0.167pH^2 - 0.808pH + 19.086;$

$$\mathrm{d} = \mathrm{TOC_{in}}\{\mathrm{pH}[-0.0000058\mathrm{EBCT}^2 + 0.000111\mathrm{EBCT} + 0.00125] + 0.0001444\mathrm{EBCT}^2 + 0.005486\mathrm{EBCT} + 0.06005\}$$

In Eq. (12.3) TOC_{in} is the input TOC concentration to the GAC system; EBCT is empty bed contact time. Equation (12.3) shows that removal of TOC is primarily determined by influent TOC level and pH, GAC reactivation period and empty bed contact time.

12.4.2 Sensitivity Analysis of Raw Water Quality

Water utilities must be concerned about how their water treatment plants respond to changes in source water quality due to the uncertainties associated with global climate change. It is essential, therefore, for water utilities to be able to identify the water quality parameters that have significant effects on regulated contaminants such as TOC and TTHMs, and to seek effective treatment alternatives to control these parameters. Uncertainties in raw water quality may lead to increases in the risk of violating drinking water standards or compliance criteria for regulated contaminants.

The impact of climate change on performance of the Miller plant is evident by comparing the Cumulative Probability Distribution Functions (CDFs) for TOC, specific ultraviolet absorbance (SUVA), TTHM and haloacetic acids five (HAA5) in the finished water between the baseline (1998), and the climate change (future 2050) scenarios. These CDFs were calculated based on the 5,000 samples of quarterly running annual average data in the finished water. Based on the US EPA stage II D/DBP rule, the following alternative criteria for water treatment plant operations were used: finished water TOC $\leq$ 2.0 mg/L, SUVA $\leq$ 2.0 L/mg-m, TTHM $\leq$ 40 μg/L, and HAA5 $\leq$ 30 μg/L. Because GAC is in place under baseline conditions, the Miller plant always meets these performance criteria. Under the climate change scenario, SUVA and HAA5 levels increase but still meet the criteria. However, the Miller plant has 11.7 % of risk of exceeding the TOC performance criterion and has a 1 % of risk of exceeding the TTHM performance criterion. This is primarily due to projected increases of TOC in source water. TOC levels in the finished water at the Miller plant are usually in the range of 0.3 to 1.5 mg/L, based on field measurements from 2004 to 2010, in order to reduce DBP formation in peripheral regions with long residence time in the distribution network. If the TOC performance criterion was reduced to 1.5 mg/L, the Miller plant would have an 80 % of risk of exceeding this level (Li et al. in press).

12.4.3 Impact of Climate Change and Adaptation

The modified WTP model can also be used to assess potential engineering options. At the Richard Miller plant, there is a very low potential for violation of standards in the near future because the violation risk may be managed by operational adjustments because of the available GAC reactor capacity; for example, by reducing the GAC reactivation period. From USEPA (2005), the TOC removal rate by GAC is a function of the GAC reactivation period (t_p):

$$\frac{GAC\quad Effluent\quad TOC\quad Concentration}{GAC\quad Influent\quad TOC\quad Concentration} = a + \frac{a}{d \cdot t_p}\ln\frac{1 + b \cdot e^{-d \cdot t_p}}{1 + b} \tag{12.4}$$

where *a, b,* and *d* are empirical coefficients estimated using field measurements at the Miller plant: $a = 0.782$, $b = 8.191$, and $d = 0.029$ day^{-1} (Li et al. 2011).

Figure 12.8 demonstrates the effect of contactor reactivation period on finished water TOC concentration. The finished water TOC concentration decreases with decreasing contactor reactivation period. In adapting to climate change, TOC in the finished water and the resulting DBPs can be effectively controlled by modifying GAC operation at the plant.

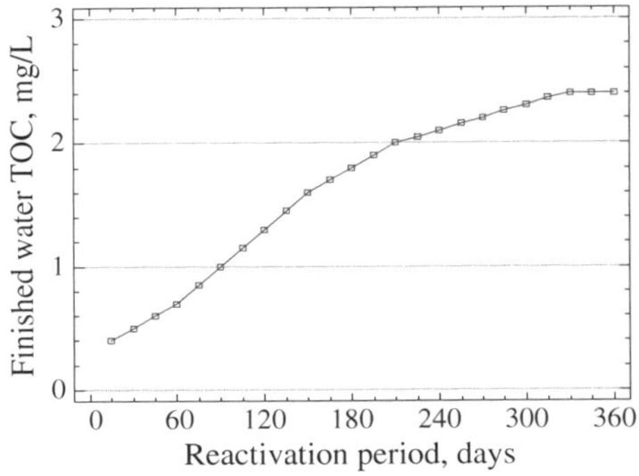

Fig. 12.8 Effect of contactor reactivation periods on finished water TOC

12.5 Cost Analysis for Future Operations

Costs associated with modifying the operation of the GAC unit process at the Miller plant were estimated using the general form of the cost models proposed by Adams and Clark (1988),

$$y = p_1 + p_2 x^c p_3^z \tag{12.5}$$

where y is the capital or operation and maintenance cost; x is the process design or operating variable, which is usually the total surface area of the GAC filter for contactors (total hearth area for GAC reactivation) or the total effective volume of the GAC unit for capital cost; p_1, p_2 p_3, and c are empirical parameters determined from nonlinear regression analysis, and z is either 0 or 1 for adjusting the cost functions for a range of x values, the values of these parameters are available from Adams and Clark (1988).

Costs were estimated by developing a curve that related the annual costs of GAC at the Miller treatment plant to the contactor reactivation period. The annualized costs include amortized initial GAC cost, cost to make up GAC loss due to reactivation, amortized contactor and furnace cost, and reactivation operating costs. The initial GAC cost is a one-time investment, based on total volume of contactors and GAC cost. The GAC make-up cost can be estimated by the GAC loss rate. The Miller plant has 12 down flow gravity contactors and two multi-hearth furnaces for onsite reactivation. Each of the Miller plant contactors has a volume of 595 m^3 (778.3 cu yds) and a surface area of 181 m^2 (216.5 sq yds). The overall GAC loss rate through the system is 7–8 %. The carbon loading rate is 482 kg/day (1063.3 1b/day) of GAC per square meter of hearth area in GAC reactivation. To estimate the annual cost, the capital recovery analysis assumed a return period of 20 years with an interest rate of 5 %. Because parameters in Eq. (12.5) are based on cost level in 1983 all costs were converted to 2011 using the Producers Price Index (US BLS 2008). Figure 12.9 shows the estimated annualized costs for GAC at the Miller treatment versus reactivation period. The annual cost rises when the reactivation period is reduced.

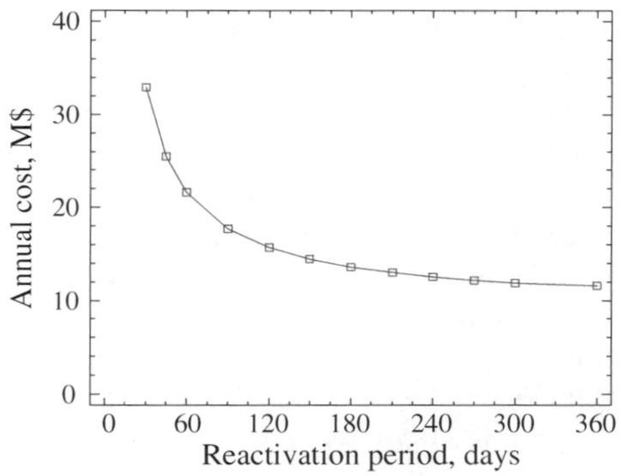

Fig. 12.9 Cost curve of GAC unit for GCWW's miller plant

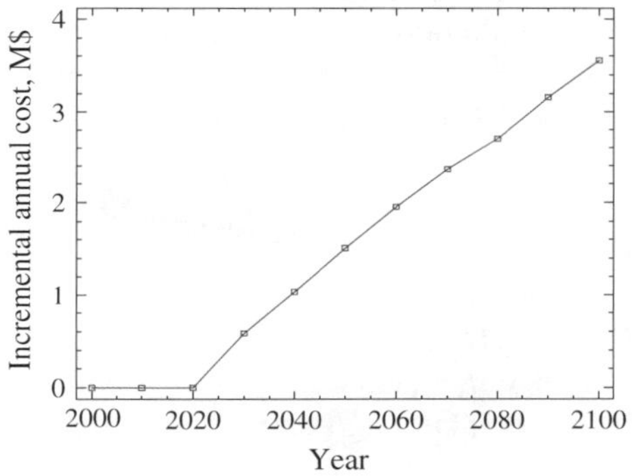

Fig. 12.10 Projected incremental annual cost from 2000 to 2100

To illustrate the cost impact associated with operational adaptation, a Monte Carlo simulation generated 5,000 samples for the baseline and future scenarios, respectively. GAC was assumed to be reactivated when the TOC concentration in the finished water exceeded 2.0 mg/L. As indicated the costs are minimal for the near future but as shown in Fig. 12.10 the resulting incremental annual cost (IAC), where IAC is defined as the difference between the baseline cost and the increased cost associated with climate change beyond 2050. After 2020, the IAC increases almost linearly with time. Therefore, planning decisions can be drawn from Fig. 12.10 in practice to control TOC concentration in finished water in order to minimize the adverse impact of climate change on drinking water treatment.

12.6 Extensions to the Water Quality Modeling Effort

It is clear that GAC treatment provides a very effective and flexible approach to deal with declining water quality and even more important the associated uncertainties with quality. GAC has been proposed to reduce organic contamination in water supplies since the early 1970s (Roberts and Summers 1982).

Logistic models for simulating TOC breakthrough behavior in a GAC column have been demonstrated (Clark et al. 1986; Clark 1987a, b). The logistic model has been well studied and verified but has never been applied to simulate an operational full-scale GAC treatment system. Full-scale systems can be subject to nonideal conditions including seasonal changes in influent concentration and inflows. Li et al. (2012) explored the logistic model to (1) establish a robust procedure for estimating the parameters of the logistic model describing TOC breakthrough in a field-scale GAC contactor, and to (2) apply the calibrated logistic model for predicting blended effluent TOC concentrations from multiple field-scale GAC contactors operated in various parallel configurations, and to (3) discuss the importance of a properly calibrated logistic model for GAC unit process operation.

A logistic function describes the characteristic 'S' shape breakthrough curves seen in a plot of TOC concentration versus GAC runtime. To model experimental breakthrough behavior of a *single* GAC column, a modified dimensionless version of the logistic function (Chowdhury et al. 1996) was applied in this study (Eq.12.6),

$$f(t) = \frac{Effluent \quad TOC \quad Concentration}{Influent \quad TOC \quad Concentration} = \frac{a}{1 + be^{-d \cdot t}} \tag{12.6}$$

where, $f(t)$ is defined as TOC fraction remaining; t is total continuous GAC runtime; a, b, and d are model parameters determined experimentally by a best fit to the breakthrough data. The parameter a represents the asymptotic steady-state value of $f(t)$. Parameters a and b govern the intercept of the logistic curve while the parameter d affects the "slope" of the logistic curve.

In a full-scale system, multiple GAC contactors can operate in parallel, with staggered reactivation cycles, and the effluent is blended prior to disinfection. Assuming identical flow rates through GAC contactors of equal size, the blended TOC fraction remaining can be the arithmetic average of the TOC breakthrough curves for each individual contactor operated in parallel (Roberts and Summers 1982):

$$\overline{f(t)} = \frac{Blended \quad Average \quad of \quad Effluent \quad TOC \quad Concentrations}{Influent \quad TOC \quad Concentration} = \frac{1}{m}\sum_{i=1}^{m} f_i(t) \tag{12.7}$$

In Eq. (12.7), m is number of contactors operating in a parallel system; $\overline{f(t)}$ is the average TOC fraction remaining in the blended effluent from m contactors; $f_i(t)$ is TOC fraction remaining in the effluent from active contactor i at time t.

A modified Gauss–Newton method was used to estimate the three parameters a, b, and d of the logistic function given in Eq. (12.7) (Hartley 1961). This method is an iterative nonlinear regression algorithm based on least square analysis between modeled TOC fraction remaining and corresponding field observations.

In this study, TOC breakthrough datasets were obtained from the Greater Cincinnati Water Work's (GCWW) Richard Miller Treatment Plant. There are 12 GAC contactors in the GAC system. As inflows vary throughout a year with water demand, contactors are brought in or taken out of service in order to maintain a relatively constant empty bed contact time (EBCT). Plant inflow and influent TOC concentrations follow an annual cycle with seasonal changes. For the 76-month period from January 2004 to April 2010, a total of 87 individual TOC breakthrough datasets were identified. Parameters for Eq. (12.7) were estimated using the nonlinear regression algorithm for each of the 87 datasets, and then were used in the revised GAC multiple contactor model. Because not all sets of model parameters can replicate the TOC effluent values observed in GCWW's historical operations for the 87 datasets, each of the datasets was evaluated against the operations guideline for GAC pilot studies (USEPA 1996). The datasets met the EPA criteria were defined as "developed". The rest of the datasets were defined as "under-developed". In all, 46 datasets were identified as developed.

An experimental technique, rapid small-scale column test (RSSCT) is commonly used to quantify breakthrough behavior. The RSSCT method uses mass transfer models to downsize the full-scale contactor to a small-scale column in order to reduce the GAC column runtime. Two sets of RSSCT experimental data were obtained for the GCWW Miller plant. The RSSCT datasets produced parameter estimates which for *a* and *d* are closer to the under-developed datasets than to the developed datasets. The RSSCT data yielded the lowest estimates for parameter *a* among all three datasets.

12.6.1 Results and Discussion

To assure the calibrated logistic model produces accurate and reliable simulations, internal validation was used to test the effectiveness of the calibrated model. To further inspect the influence of model parameter determination in calibration, the simulations results for the blended average TOC fraction remaining are compared in Fig. 12.11 among the logistic models based on the developed and RSSCT

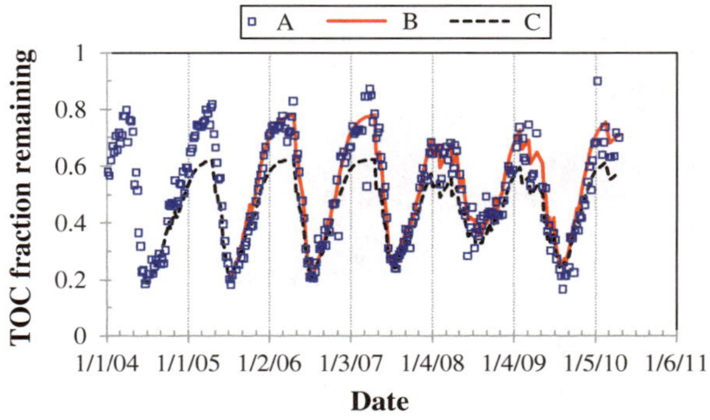

Fig. 12.11 Comparison of performance of logistic models with full-scale field measurements (A-field measurements; B-model based on developed datasets; C-model based on RSSCT datasets)

datasets against the full-scale measurements. Model B is based on developed datasets. Model C is based on the RSSCT datasets. As number of contactors in operation changed seasonally with water demands, was used to compute the blended average TOC fraction remaining for the two models. In both cases, the exact historical sequencing and corresponding runtimes of active GAC contactors at the Miller plant were followed.

Figure 12.11 shows that temporal concentration variation for models B closely replicate the performance history of the full-scale system, which validate the use of developed datasets and the modified GAC algorithm as the basis for the GAC simulation. Model residual sum of squares (RSS) of differences between the simulated fraction remaining and the corresponding field measurements is 0.60. In contrast, model C fails to replicate the performance of the multiGAC contactor bank when the TOC fraction remaining is high. The RSS value is high at 2.24. The poor model performance directly results from underestimation of parameter *a*, leading to a systematic model bias.

The accuracy of the logistic model parameter values depends upon the runtime length of the TOC breakthrough curve. To assess the minimum runtime length requirement in GAC model calibration, this study examined the effect of GAC runtime on the estimation of logistic model parameters based on segmenting each of the 46 developed datasets into runtime strings. Additionally, a numerical verification was conducted to study the effects from fluctuating influent TOC concentration and the random errors possibly introduced during the sampling or monitoring in practical operation using the calibrated GAC model. The results illustrates that adequate GAC contactor runtime is essential to achieve proper calibration of the logistic breakthrough model, and corroborates that the USEPA guideline for GAC pilot studies provides a reference to determine the contactor runtime. This insight has important implications for utilizing RSSCT experimental data.

12.7 Summary and Conclusion

Uncertainty analysis indicates that raw water quality parameters may be reasonably well described by log-normal distributions. Multivariate analysis based on Monte Carlo simulations demonstrates the correlation relations among the raw water quality parameters. The EPA WTP Model was utilized to evaluate the effect of deteriorating water quality due to climate change on water treatment performance.

If some raw water quality parameters deteriorate, as reported in the literature, due to climate change, the results show that the existing treatment at the Miller WTP is subject to high risk of violation for TOC compliance criteria. A validation study was conducted to show that the WTP Model provides a reasonable replication of the Miller plant. A sensitivity analysis using the WTP Model indicates that finished water TOC is sensitive to raw water TOC and pH and can be tracked from raw water through to finished water. Raw water UVA, alkalinity, and total hardness also show they have new significant effects on regulated TOC through correlation with either raw water pH or TOC or both.

The risk level of TOC violation can be controlled by reducing the GAC reactivation period. The results showed that TOC compliance can be met by modifying existing WTP operations by reducing the GAC reactivation period. However, the net annual cost also increases with reduced GAC reactivation period.

It has been shown that the logistic model for TOC breakthrough can be calibrated with data from full-scale GAC contactors using nonlinear parameter estimation algorithm. It can be validated internally through the accurate replication of historical data on TOC removal. The runtime length for the breakthrough dataset is critical for proper estimation of the three logistic model parameters. The guidelines for GAC pilot studies by USEPA (1996) provide good estimates for the runtime needed for screening developed field breakthrough datasets.

References

Adams JQ, Clark RM (1988) Development of cost equations for GAC treatment systems. J Environ Eng 114(3):672–688

Cazenave A, Llovel W (2010) Contemporary sea level rise. Ann Rev Mar Sci 2:145–173

Chowdhury ZK, Solarik G, Owen DM, Hooper SM, Summers RS (1996). NOM removal by GAC adsorption: implications of blending. In: Proceedings of the AWWA annual conference, Toronto

Clark RM, Symons JM, Ireland JC (1986) Evaluating field scale GAC systems for drinking water. J Environ Eng 112(4):744–756

Clark RM, Li Z, Buchberger SG (2011) Adapting water treatment design and operations to the impacts of global climate change. Front Earth Sci 5(4):363–370. doi: 10.1007/s11707-011-0197-7

Clark RM (1987a) Modeling TOC removal by GAC: the general logistic function. J. AWWA 79(1):33–37

Clark RM (1987b) Evaluating the cost and performance of field-scale granular activated carbon systems. Environ Sci Technol 21(6):573–580

Cromwell JE, Smith JB, Raucher RS (2007) Implications of climate change for urban water utilities. The Association of Metropolitan Water Agencies report. http://www.amwa.net/galleries/climate-change/AMWA_Climate_Change_Paper_12.13.07.pdf

Domingues CM, Church JA, White NA, Gleckler PJ, Wijffels SE, Barker PM, Dunn JR (2008) Improved estimates of upper-ocean warming and multi-decadal sea-level rise. Nature 453:1090–1093

Haagenson E, Rajagopalan B, Summers RS (2013) Projecting demand extremes under climate change using extreme value analysis. J Am Water Works Assoc http://dx.dol.org/10.5942/jawwa.2013.105.0013

Hartley HO (1961) The modified gauss newton method for the fitting of nonlinear regression functions by least squares. Technometrics 3(2):269–280

Interlandi SJ, Crockett CS (2003) Recent water quality trends in the Schuylkill river, Pennsylvania. USA: a preliminary assessment of the relative influences of climate, river discharge and suburban development. Water Resour 37:1737–1748

IPCC, 2007: Summary for policymakers. In: Solomon S, Qin D, Manning M, Chen Z, Marquis M, Averyt KB, Tignor M, Miller HL (eds) Climate change 2007: The physical science basis.

Contribution of working group I to the fourth assessment report of the intergovernmental panel on climate change Cambridge University Press, Cambridge

Jacobs K, Adams DB, Gleick P (2001) Chapter 14, Potential consequences of climate variability and change for the water resources of the United States. In: Jerry M, Anthony J Thomas K (eds) Climate change impacts on the united states–the potential consequences of climate variability and change, Cambridge University Press, New York

Li Z, Buchberger SG, Clark RM, Yang YJ, Swertfeger J (2012) Evaluation of logistic model for GAC performance in water treatment. J Am Water Works Assoc 104(9):E489–E500 http://dx.doi.org/10.5942/jawwa.2012.104.0120

Li Z, Clark RM, Buchberger SG, Yang YJ (2009) Assessing the impact of climate change on drinking water treatment plant design and operation. In: Proceedings of the ASCE/EWRI 2009 annual conference, Kansas, pp 52–60

Li Z, Clark RM, Buchberger SG, Yang YJ (in press). Assessing the impact of climate change on drinking water treatment plant design and operation. Accepted Publ J Water Res Plann Manag ASCE

National Research Council (NRC) (2010) Adapting to the impacts of climate change, America's climate. Progress report of the interagency climate change adaptation task force–2010

Obolensky A, Singer PC (2005) Halogen substitiution patterns among disinfection byproducts in the information collection rule database. Environ Sci Technol 39:2719–2730

Obolensky A, Singer PS, Shukairy H (2007) Information collection rule data evaluation and analysis to support impacts on disinfection by-product formation. J Environ Eng (ASCE) 133(1):53–63

Roberts PV, Summers RS (1982) Granular activated carbon performance for organic carbon removal. J. AWWA 74(2):113–118

Salas JD, Delleur JW, Yevjevich V, Lane WL (1980) Applied modeling of hydrologic time series. Water Resources Publications, Littleton, p 484

Skjelkvale BL, Stoddard JL, Jeffries DS, Torseth K, Hogasen T, Bowman J, Mannio J, Monteith DT (2005) Regional scale evidence for improvements in surface water chemistry 1990–2001. Environ Pollut 137:165–176

US BLS (Bureau of Labor Statistics) (2008) "BLS handbook of methods, chapter 14 producer prices-background". http://www.bls.gov/opub/hom/homch14_a.htm

US Environmental Protection Agency (1996) ICR Manual for Bench and Pilot-Scale Treatment studies. EPA 814-B-96–003, Office of Water, Washington, D.C.

US Environmental Protection Agency (2000) ICR Treatment Study Database, Version 1.0. EPA 815-C-00–003, Office of Water, Washington, D.C.

US Environmental Protection Agency (2005) Water Treatment Plant Model Version 2.2 User's Manual. Office of Ground Water and Drinking Water, Cincinnati, Ohio

White NJ, Church JA (2011) Sea-level rise from the late 19th to the early 21st century, Surv Geophys, centre for Australian weather and climate research CSIRO marine and atmospheric research, GPO Box 1538, Hobart, TAS 7001, Australia

Whitehead PG, Wilby RL, Battarbee RW, Kernan M, Wade AJ (2009) A review of the potential impacts of climate change on surface water quality. Hydrol Sci 54(1):101–123

Whitehead PG, Wilby RL, Butterfield D, Wade AJ (2006) Impacts of climate change on in-stream nitrogen in a lowland chalk stream: An appraisal of adaptation strategies. Sci Total Environ 365:260–273

Wright B, Stanford B, Weiss J, Debroux J, Routt J, Khan S (2013) Climate change: How does weather affect surface water quality? Opflow. Am Water Works Assoc, 10–15

Chapter 13
Integrated Control and Detection of Accidental Occurrences in Water Distribution Networks

Slobodan Macan and Edo Macan

Technical Terms and Definitions

3G	Third generation of mobile telecommunications
AMI	Advanced Metering Infrastructure
AMR	Automatic Meter Reading
CRO	Control-Regulation Objects
DMA	District Metering Areas
EDGE	Enhanced Data GSM Environment
GPRS	General Packet Radio Service
GSM	Global System for Mobile Communications
HSDPA	High-Speed Downlink Packet Access
ISDN	Integrated Services Digital Network
M-Bus	Meter-Bus Protocol (European Standard for remote reading)
PSTN	Public switched telephone network
SCADA	Supervisory Control and Data Acquisition
SMS	Short Message Service
TCP/IP	Transmission Control Protocol/Internet Protocol
UMTS	Universal Mobile Telecommunications System
VHF	Very high frequency

S. Macan (✉)
Coprogram d.o.o, Zagreb, Croatia
e-mail: s.macan@coprogram.hr

E. Macan
Coprogram Inc, Oakland, CA, USA
e-mail: e.macan@coprogram.com

R. M. Clark and S. Hakim (eds.), *Securing Water and Wastewater Systems*,
Protecting Critical Infrastructure 2, DOI: 10.1007/978-3-319-01092-2_13,

13.1 Introduction

This chapter focuses on the protection of the distribution network itself as contrasted to the supporting peripheral facilities such as pumping stations, storage reservoirs, etc. Since these facilities are at concentrated points of the system they are easier to defend in the case of deliberate threats and easier to monitor in the case of accidents. Most distribution systems are protected by supervisory control and data acquisition (SCADA) systems, video surveillance and automatic alarms, and frequently by physical protection. It is considerably harder to protect the network itself, which is often spread over very large areas and contains many highly vulnerable points. Most distribution networks, even in large cities, do not have continuous monitoring which may make them the most vulnerable part of a water supply system.

The possibility of an accident (due to a number of potential causes), in water distribution networks due to various vulnerabilities cannot be excluded in advance. Network protection should therefore be based on developing effective systems for the rapid detection of such accidents and the interpretation of specific control-regulation objects (CRO) that will allow system managers to quickly adapt to all new conditions and situations. A system is discussed in this chapter that can detect potential accidents (defects, or damage such as pipe breaks) in real time. A key factor in recognizing such situations is the detection of changes in hydraulic operational parameters (flow, pressure, water leaks, and energy losses). This recognition can be enhanced by the use of integrated systems of telemetry that include remote Automatic Meter Reading/Advanced Metering Infrastructure (AMR/AMI) and water leak detection systems.

The expansion of mobile telephones, with global system for mobile communications (GSM) and third generation mobile telecommunications (3G), (general packet radio service (GPRS), enhanced data GSM environment (EDGE), universal mobile telecommunications system (UMTS), high-speed downlink packet access (HSDPA)) has opened up new possibilities for development of a low-cost, simple, and effective system of real-time telemetry and monitoring of distribution networks. All three components of those systems (telemetry, AMR/AMI, and leak detection) integrated into a complete system, are complementary and provide full control of all hydraulic operational parameters, at all times and in all points in the distribution network. Since an approach is new, but these types of systems are being intensively developed and perfected in several countries (for example in the Republic of Croatia). They are important for successfully controlling normal water supply system operations and are also a prerequisite for the successful control and functional safety of such systems in emergency and accidental situations.

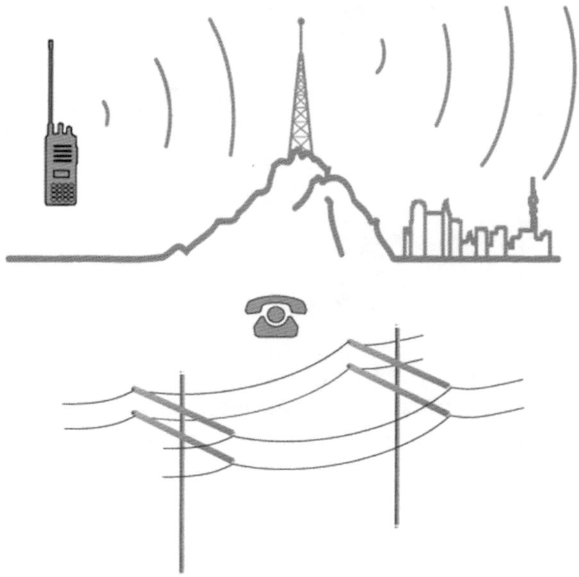

Fig. 13.1 VHF, PSTN, and ISDN

13.2 Communication Systems

13.2.1 VHF, PSTN, ISDN

Until recently, transmission of data from remote sites to a common telemetry, monitoring and operations center (and vice versa) has been done wirelessly with radio very high frequency (VHF), or using analog phone public switched telephone network (PSTN) or integrated service digital network (ISDN) communications (Fig. 13.1).

Each of these communication systems has very significant drawbacks. VHF radio systems require very large investments for construction, and require permission from the relevant institutions, and their effectiveness also depend on topographic and meteorological conditions. In addition, data transmission is slow and insecure. PSTN and ISDN communications are also slow, and for large telemetry networks additional cable networks need to be upgraded, which is a problem, especially where such networks do not already exist. In brief, all VHF, PSTN, and ISDN systems require large investments, licenses, and only authorized specialized institutions have been allowed to construct them. However, their efficiency is modest, with regard to speed, quality, and safety of data transmission. In addition to their high cost and accompanying problems in their development and usage, these communication techniques are not fast enough and safe enough to use as the basis for real-time monitoring and/or control system.

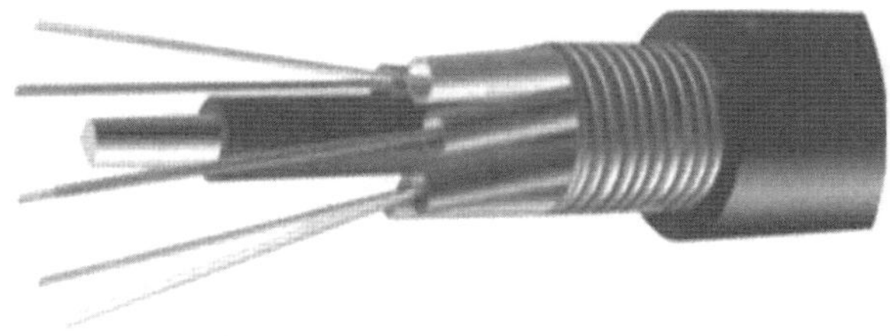

Fig. 13.2 Fiber Optic Cable

13.2.2 Fiber Optic Cable

A telemetry system may also be based on fiber optic cables (Fig. 13.2).

In this system, communication channels are available at all times (full time transmission channels), and the system is completely independent from other communications and from all external influences, and it enables fast and reliable transmission of a large amount of data. It is a communication technology, which could provide the bases for a real-time monitoring and/or control system. However, for obvious reasons, it can be used for only a small number of key facilities, not for a network of numerous dispersed measuring sites in a system.

13.2.3 GSM

The GSM system of mobile communications is suitable for use in dispersed networks. These systems do not require large investments (the existing infrastructure of the GSM provider is used), nor do they require special equipment or a special license from the relevant institutions. More than a sufficient transmission speed of short message services (SMS) messages/data is available (Fig. 13.3).

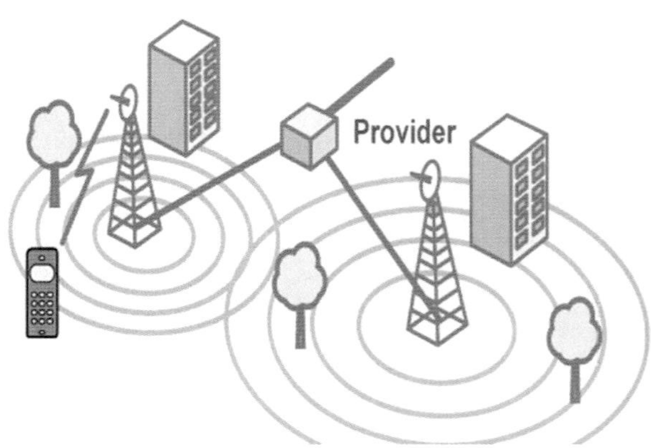

Fig. 13.3 GSM

Nevertheless, GSM also has significant deficiencies. Since it is a public (as opposed to an isolated private) network, it is vulnerable to external influences and therefore less reliable than a dedicated communications network. Communication is unidirectional, not bidirectional, and may be periodically interrupted. This communication protocol is not intended for transmission of large amounts of data. The potential for efficient real-time monitoring and/or control systems to be based on this type of communication protocol is limited.

13.2.4 GPRS : EDGE, UMTS, HSDPA

The general packet radio service (GPRS) protocol retains all the advantages of GSM technology, but has none of its disadvantages (Fig. 13.4).

Use of the GPRS protocol opens the possibilities for developing completely closed and reliable private transmission control protocol (TCP/IP) networks. This technology provides a constantly available communication channel, and the possibility of charging for the amount of data transferred rather than for the duration of the connection. This technology also opens-up the possibility of fast and reliable two-way transmission of large amounts of data at long distances and over large areas, especially in countries where GPRS coverage is already large. It is a communication technique on which efficient and reliable near-real-time monitoring and/or control systems could be based. Systems are being converted to GPRS communications (possibly in combination with fiber optic cable) technology, and provides a step toward the utilizing 3G capability (EDGE, UMTS, and HSDPA) which is already highly developed in many countries. Transmission of data using GSM protocol (SMS messages) is therefore becoming obsolete. In the future, GSM technology will most likely only be applied to some of the simplest functions or systems in which there is no possibility for the development of GPRS-EDGE, UMTS, HSDPA real-time monitoring, and control systems.

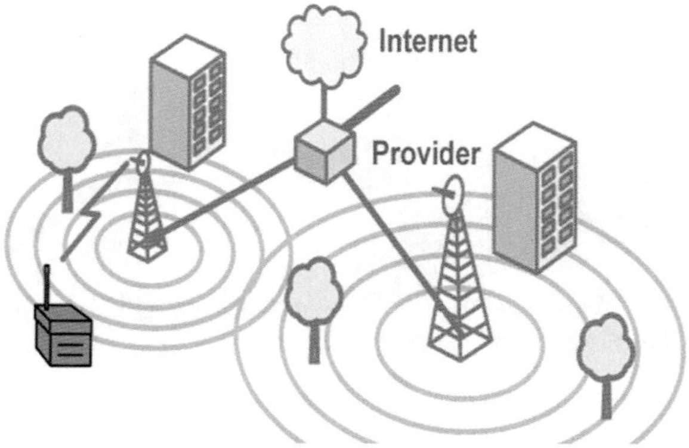

Fig. 13.4 GPRS-EDGE, UMTS, HSDPA

Another special advantage of GPRS/EDGE/UMTS/HSDPA systems is the capability of unidirectional connections with data bases and web servers. This technology enables the development of data web portals offering online access, at any time and at any place, for any authorized user. Real-time accessibility to data has uses beyond system operation. Although system operation is important, system operation does not necessarily contribute significantly to maintaining the long-term efficiency or maintenance of the system. An operator is generally a technician, with basic knowledge and skills in water supply system control and operators help maintain service. But these data must also be accessible to engineers and system managers who can optimize systems operations.

13.3 Telemetry Systems for Monitoring Water Distribution Networks

GPRS technology has the potential for wide system coverage, high-speed data transmission and reliability of bidirectional data transmission capability, as well as low prices for the necessary communications equipment and data transmission. These are reasons for selecting GPRS (EDGE, UMTS, HSDPA) communications networks as suitable for developing telemetry networks (Stanley, 2009). The availability of such communications networks and their possibilities is changing the water industries approach to the development of telemetry, monitoring, and control systems in water supply (Lin et al., 2008) (Fig. 13.5).

Traditional approaches to monitoring which focus on the measurement and monitoring of pump stations and reservoirs only are inadequate for water supply control. Using these traditional approaches, water supply networks have been operated without a real understanding of actual operational effects (flow, pressure, water consumption) in the network itself.

Lord Kelvin said that unless something is measured, it is not seen, and if it is unseen, it cannot be controlled, nor maintained, nor repaired. Translated into the language of water supply professionals, this would mean that one cannot see what

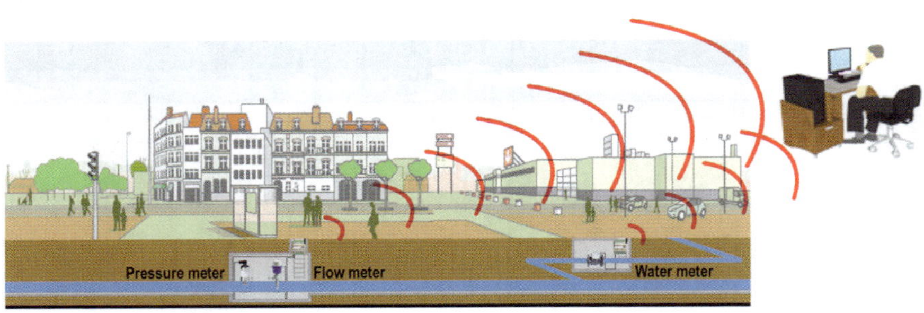

Fig. 13.5 Telemetry network-Telemetry of pressure, flow, greater water consumptions and greater accidental water and energy losses from water distribution networks

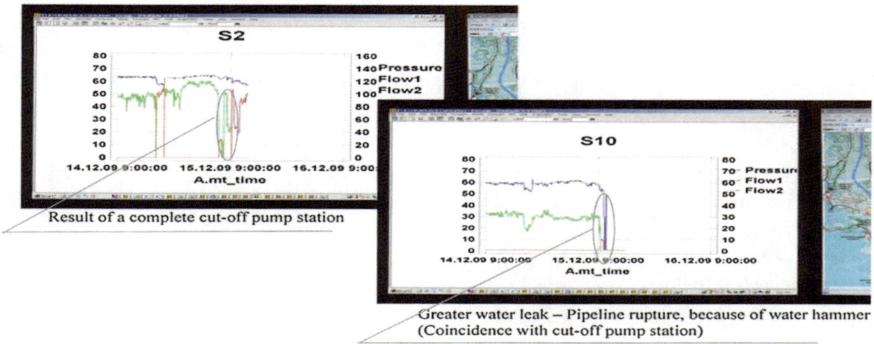

Fig. 13.6 Changes in pressure and flow at the measuring site: Site 2 and Site 10

is really happening in the water distribution network without direct measurements in the network itself. If these measurements cannot be made, then the distribution system cannot be successfully controlled, maintained, nor protected from accidental occurrences.

This is why development of real-time telemetry and monitoring of water systems in real time is needed. Telemetry of hydraulic parameters (e.g., flow, pressure, water consumption), water quality parameters (e.g., pH), and acoustic effects (e.g., noise of water leaks) in pump stations, reservoirs, at monitoring points in the network, as well as representative industrial and residential water use is a prerequisite for efficient monitoring and efficient management of networks. These data are important for efficient system operation, reducing water leaks, and maintaining and protecting the water supply system. Monitoring points in the system and network can be best established using a mathematical model of the water supply system. Data from these points can also be used to calibrate the model and for understanding the behavior of the water supply network/system (Machell et al., 2010) (Fig. 13.6).

Large losses of water and pressure/energy (as a result of malfunctioning in the network, natural disasters, or terrorist attacks) can be detected and identified using hydraulic measurements (flow, pressure, water consumption). Small water leaks can be identified by the detection of water leak noise, in the water distribution network, before they become major leaks.

13.4 AMR/AMI Systems and Water Leak Detection

Understanding the hydraulic performance, readiness and operational conditions in a distribution network is incomplete if there is no or little information on water consumption or on water and energy losses in the network.

Information on water consumption is an aspect of water supply control that can be supplied by a telemetry network. Automatic meter reading/advanced metering infrastructure (AMR/AMI) networks can provide information for billing of water consumption and can provide the basis for the development of water consumption zones (or district metering areas (DMA)). Telemetry and AMR/AMI networks are mutually complementary and therefore they need to be planned, designed, and developed integrally, as part of a complete network and system for monitoring and control of water supply.

An AMR/AMI system is not limited to monthly readings for billing of water consumption. It can be a system for consumption forecasting, as well as providing a system for detection of water leaks on the consumer side of water meters. This is why data from shorter, hourly intervals in real time might be required. This is also consistent with the requirements of mathematical models and integrated systems for the control of hydraulic performance and protection against accidental occurrences in water distribution networks.

It is recommended that an AMR/AMI network be a fixed private TCP/IP network, based on GPRS (EDGE, UMTS, HSDPA) communications, because it is a safer, faster, and cheaper way of communication than GSM/SMS communication. An AMR/AMI network should also be connected with a data base and web server. AMR/AMI system data would then be accessible via Internet, using a data web portal, at any time and at any site, to any authorized customer of such systems.

According to the laws governing municipal services in many countries, operators of water distribution systems (as well as gas, electric energy, and heat energy networks) have an obligation to their end-user customers to offer devices for measuring consumption in any part of a building which represents an independent unit. This would allow for access to information about current consumption, actual unit prices, total cost of consumption, and their water leaks through the Internet. These devices could be incorporated into newly constructed or existing buildings (Kenna, 2008) (Figs. 13.7, 13.8).

In many cities and water supply systems, the process of developing AMR systems has already begun. Unfortunately, they are based mostly on local M-Bus only, or possibly on PSTN/ISDN or GSM communications. A small number of these systems are based on up-to-date 3G communications (GPRS, EDGE, UMTS, HSDPA) systems and were incorporated into existing telemetry systems. These 3G communication systems, as well as telemetry systems based on these communications, have only been implemented recently. However, it is likely that will eventually be integrated in such communications networks. Fortunately, the adaptation of M-Bus, PSTN/ISDN and GSM/SMS AMR systems, and their transition into GPRS (EDGE, UMTS, HSDPA) networks and their integration with telemetry systems into these communications networks should not present any larger problem.

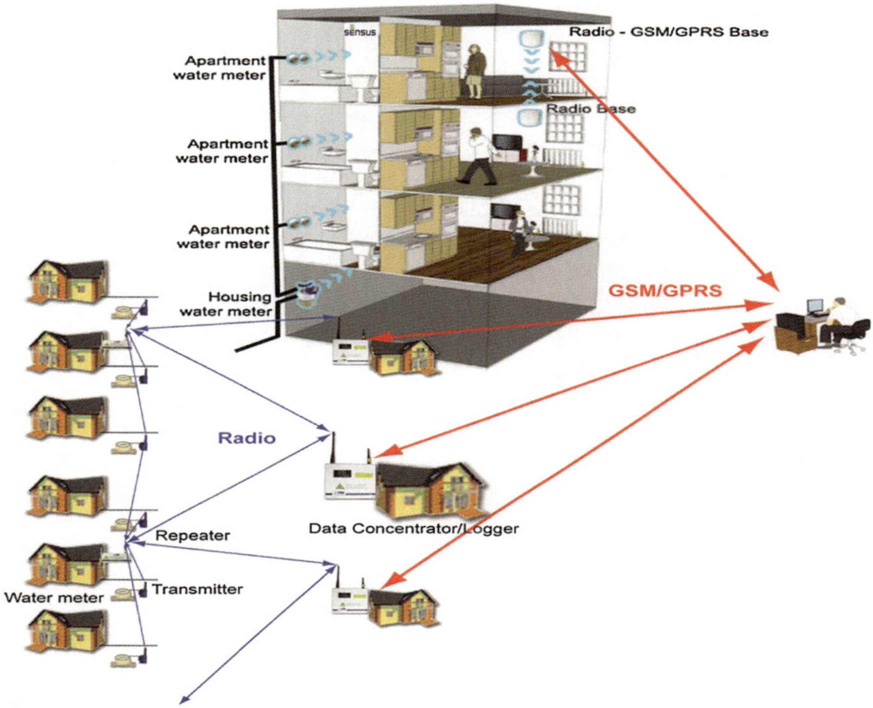

Fig. 13.7 Representative blocks and characteristics zones of water supply in telemetry and AMR/AMI system of water supply

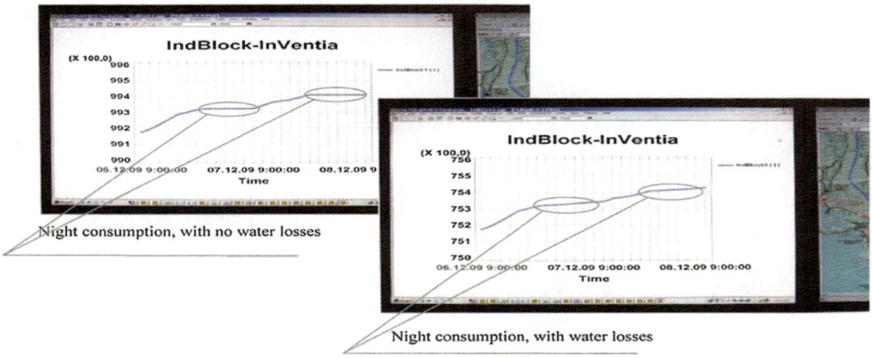

Fig. 13.8 Water consumption in Industrial blocks: IndBlock 1 and IndBlock 3

13.5 Detection and Localization of Water Leaks

There are three known techniques for investigating leaks in water distribution networks: (i) detection and localization of water leaks (using a fixed network of water leak noise sensors/loggers), (ii) detailed location (pinpoint) of water leaks (correlation of water leaks noise), and (iii) confirmation of the location of water leaks (by listening for water leak noises at the defined locations). Those are, not however alternatives. For a successful reduction of water leaks it is necessary to include all of these methods. However, in order to detect possible threats to the network water loss detection is of great significance, because it creates a high density of sensors in the telemetry system.

Due to the underdevelopment and inadequacy of available communication systems, fixed networks of water leak noise sensors/loggers are rarely employed. Therefore, remote systems for the detection of water leaks are also rarely developed. Instead teams with noise detectors are sent out immediately, relying on experience, intuition or any empirical standards, to the site of the supposed water leaks. Time is wasted on looking in the areas where there are in fact no water leaks, consequently, only after moving to another zone/area, leaks in the previous zone can be activated. Thus, the search is often reduced to an almost spontaneous never-ending walk around the network, without any significant success in detecting and reducing water leaks. Therefore, if an accident occurs in the distribution network, action is taken only if/when someone happens to reports it.

Research into the reduction of water leaks must be based on a permanent and continuous process of detecting and localizing them, as opposed to a one-time or even multiple campaigns after the event occurs. That is only possible with a fixed network of water leak noise sensors/loggers, combined with transmission of data from the network to the operations center. The development of GSM/GPRS (EDGE, UMTS, HSDPA) communication technology has simplified and reduced expenses for the development of such detection networks, so water leaks and any accidents and intrusions into the distribution network can now be very reliably detected and localized in time (Hunaidi and Wang, 2000).

Using these detection devices, teams with equipment for pinpointing and listening to leak noises can be on-time sent to the identified general locations. This approach will be significantly more successful in determining the exact location of the leak and make it possible to verify and reduce or eliminate water-energy losses or the consequences of any accidental occurrences in the water distribution network. (Stojanov et al., 2007) (Figs. 13.9, 13.10).

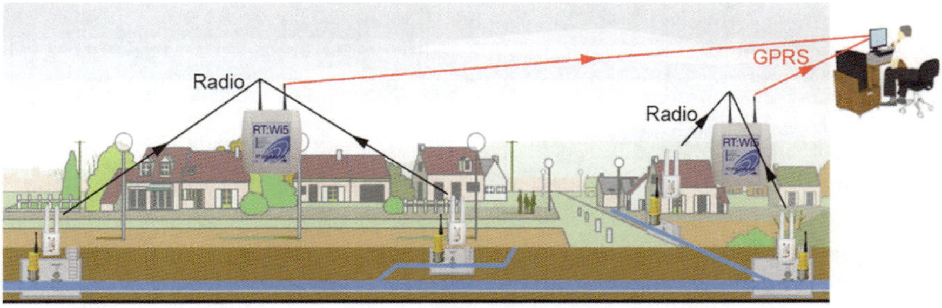

Fig. 13.9 Fixed network of water leak noise sensors/loggers

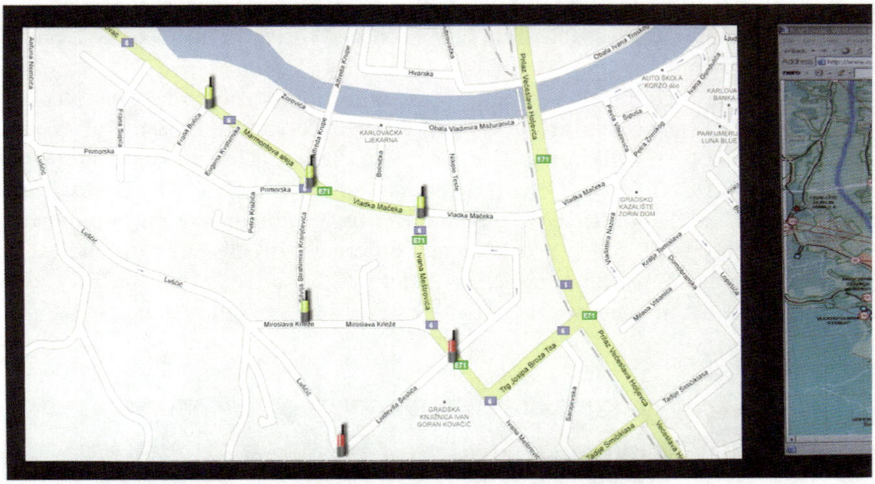

Fig. 13.10 Monitoring, detection, and localization of water leaks and accidental occurrences in water distribution networks water distribution networks

13.6 Integrated Telemetry, AMR/AMI, and Water Leak Detection Systems

From previous descriptions, it is evident that telemetry, AMR/AMI, and systems for detection of water leaks are interactive complementary systems, which can be integrated into a comprehensive system (Coprogram Ltd., 2010). The development of GPRS (EDGE, UMTS, HSDPA) communication technologies has simplified and has reduced the cost of developing such integrated monitoring systems, thus even smaller waterworks can now implement them. Differences to-date in the development of these systems does not represent an obstacle for such continued development. All this is more or less compatible, and can often be quite easily integrated, transformed, and upgraded into an integrated, holistic system.

Providing data just to the data center itself, is not an adequate protection against problems in the network. Frequently, the staff at the data center is composed of computer specialists, with insufficient knowledge of water systems. Limiting the availability of the data to staff in the data center would therefore be a relatively small contribution to the efficiency of the control and management of water systems. Data should also be available to water supply experts, because they will use these data in their analysis and to make decisions about the operation, maintenance, and development of the water supply system. Some of these data should also be made available to the waterworks consumers/users. Water users need to see data concerning their current consumption, and the costs and the possibility of water leaks on their side of water meters. The most suitable mechanism for transmitting this information is providing access to the Internet and the best way to do that is through a data web portal, available to everyone in accordance with their authorization, at any time and from any location where the Internet is available (www.coprogram.com) (Figs. 13.11, 13.12).

Since this is a comparatively new communication concept and suggests new solutions in telemetry, in AMR/AMI and water leaks detection, integral systems such as the one described in this chapter are in the initial stages of development. That is why there are only a few projects that might illustrate the importance of such systems and demonstrate the efficiency of their implementation. A number of such data web portals developed for many different systems can be found on the Internet. Interested parties are hereby referred to one that has initially been developed not for the needs of a particular water supply system alone, but for the

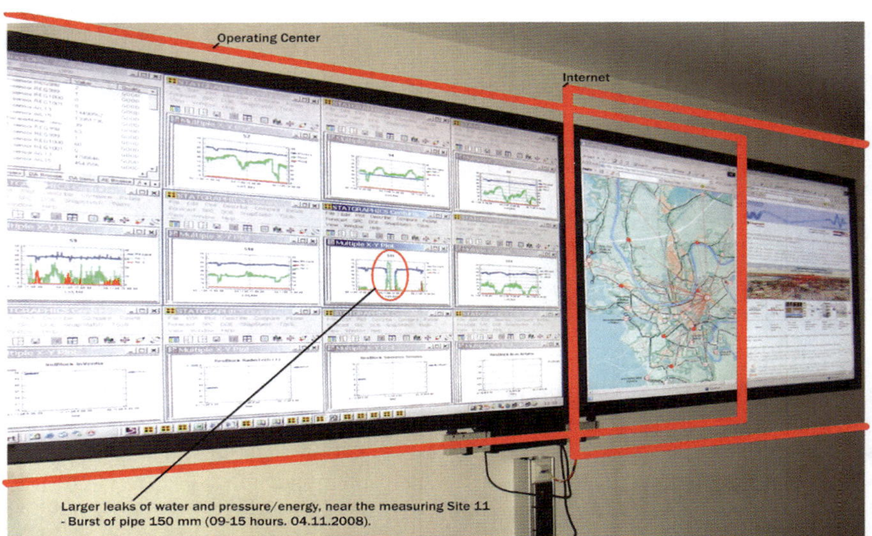

Fig. 13.11 Operating center of integrated telemetry, AMR/AMI, and system for detection of water leaks and accidental occurrences in water distribution networks

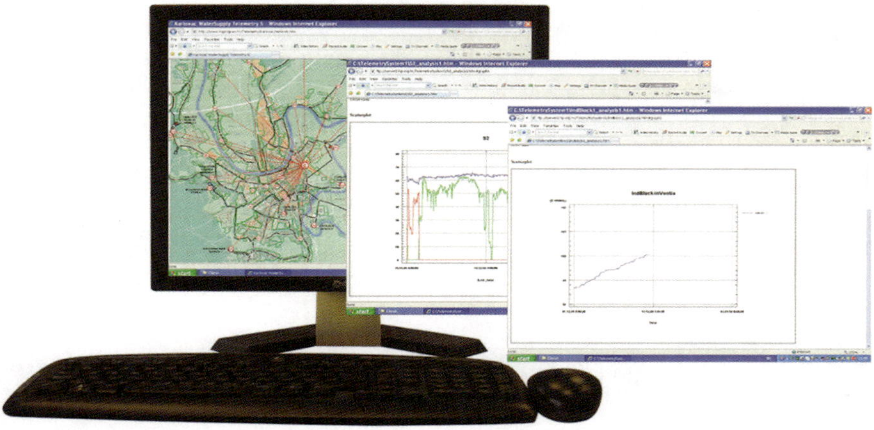

Fig. 13.12 Data web portal of integrated telemetry, AMR/AMI, and system for detection of water leaks and accidental occurrences in water distribution networks

needs of testing new components of such integrated system.[1,2] This system is composed of telemetry, AMR/AMI, and systems for the detection of water-energy losses and accidental occurrences, in water distribution networks.

References

Coprogram Ltd., (2010). "Why integrated Telemetry, AMR/AMI and Leaks Detection system". www.coprogram.com/Documentations.html

Hunaidi O, Wang A (2000) PC Multimedia Based Leak Detection System for Water Distribution Networks (LeakFinder – New Pipeline Leak Detection System). National Research Council, Ottawa, Canada

Kenna B (2008) Water Metering and Revenue Protection. University of Southern Queensland, AU

Lin M, Wu Y, Wassell I (2008) Wireless Sensor Network: Water Distribution Monitoring System. Computer Laboratory, Cambridge University, UK

Machell J , Mounce SR , Boxall JB (2010) "Online modeling of distribution systems: a UK case study". Department of Civil and Structural Engineering, University of Sheffield, UK

Morgan W (2006) "Managing water loss". Jour. AWWA, Vol 98, JAW_0062621

Stanley L (2009) "Overcoming IP address Issues with GPRS Remote Monitoring and Alarm System". Moxa Technical Writing Center (TWC), Moxa Inc., USA

[1] Experts of the Development Division of the manufactures of implemented equipment cooperated directly in the development of Coprogram GPRS Telemetry, AMR/AMI, and Water Leak Detection system.

[2] Companies Itron. Inc. and Sensus, GmbH. are given a permissions to use their graphics illustrations, in the background of Figs. 13.5, 13.7, and 13.9.

Stojanov I, Nachman L, Madden S (2007) PIPENET: A Wireless Sensor Network for Pipeline Monitoring. Imperial College, London, UK, Intel Research; MIT CSAIL, Cambridge, USA

Sensus Inc., USA; Itron-Actaris Inc., USA; Inventia Sp.zo.o., Poland; RadioTech Ltd., UK; FCS Inc., USA; Siemens GmbH, Germany; and Relay GmbH., Germany, (2010). Technical documentation

Chapter 14
Plan, Prepare and Safeguard: Water Critical Infrastructure Protection in Australia

Dave Birkett and Helena Mala-Jetmarova

Technical Terms and Definitions

ASIO	Australian Security Intelligence Organisation
BCP	Business Continuity Plans
CBRN	Chemical, Biological, Radiological, Nuclear
CFA	Country Fire Authority
CHOGRM	Commonwealth Heads of Government Regional Meeting
CI	Critical Infrastructure
CIAC	Critical Infrastructure Advisory Council
CIP	Critical Infrastructure Protection
CIPMA	Critical Infrastructure Protection Modelling and Analysis
CMT	Crisis Management Team
COAG	Council of Australian Governments
CSE	Crisis Simulation Exercise
DHS	Department of Human Services
DSE	Department of Sustainability and the Environment
EMA	Emergency Management Australia
IAAG	Infrastructure Assurance Advisory Group
IMT	Incident Management Team
NCTC	National Counter-Terrorism Committee
SCADA	Supervisory Control and Data Acquisition System
SES	State Emergency Services

D. Birkett (✉)
Global Terrorism Research Centre, School of Political and Social Inquiry, Monash University, Melbourne, 3800 Victoria, Australia
e-mail: dbirkett@crisisleaders.com

H. Mala-Jetmarova
School of Science, Information Technology and Engineering, University of Ballarat, Mt Helen Campus, University Drive, Ballarat, 3353 Victoria, Australia
e-mail: helenajetmarova@students.ballarat.edu.au

R. M. Clark and S. Hakim (eds.), *Securing Water and Wastewater Systems*, Protecting Critical Infrastructure 2, DOI: 10.1007/978-3-319-01092-2_14,

TISN	Trusted Information Sharing Network
TRG	Tactical Response Group
VicPol	Victoria Police
WSIAAG	Water Services Infrastructure Assurance Advisory Group

14.1 Introduction

Within the historical progression and the very nature of human beings, there exists an inherent and almost instinctive need to work and live in close proximity to large population clusters to grow, produce, manufacture goods and provide social support mechanisms. These ancillary support mechanisms are often viewed as part of the government structural and social contract of service provision within the complex political framework, whether in the private or public domains. As a consequence of the global increase of populations, the observed international phenomenon of the growth of mega-cities has emerged (Van der Ploeg and Poelhekke 2008). Although Australia has not yet attained the identified ranking of other recognised global mega-cities such as Tokyo, Guangzhou and Seoul as indicated in Table 14.1, there is an obvious tendency for the concentration of large population clusters close to the coast such as Sydney, Melbourne, Brisbane, Perth, Adelaide, Hobart and Darwin. In consideration of a national population of just over 22 million, most Australians prefer to live close to the coastal fringe, possibly due to the moderating effect of climate, the availability of infrastructure, services and above-mentioned social support mechanisms.

Table 14.1 Global mega-cities compared with Australian cities. *Source* Brinkhoff (2012)

City	Country	Population size	Mega-city (population >10 million)
International			
Tokyo	Japan	34,300,000	Yes
Guangzhou	China	25,200,000	Yes
Seoul	South Korea	25,100,000	Yes
Shanghai	China	24,800,000	Yes
Delhi	India	23,300,000	Yes
Mumbai	India	23,000,000	Yes
Mexico City	Mexico	22,900,000	Yes
New York City	U.S.A.	22,000,000	Yes
Australian			
Sydney	Australia	3,641,422	No
Melbourne	Australia	3,371,888	No
Brisbane	Australia	1,676,389	No
Perth	Australia	1,256,035	No
Adelaide	Australia	1,040,719	No
Hobart	Australia	128,577	No
Darwin	Australia	68,694	No

The Australian population clusters create a significant reliance on the extensive infrastructure catering for increasing demands. It is possible that there is an expectation by the Australian urban inhabitants that infrastructure services are to be provided on an uninterrupted basis. Moreover, there appears to be a perception that takes these supplies, which are usually invisible from view within underground services and often obvious to the ordinary person only at the tap or toilet, for granted.

Adequately treated and distributed water and collected wastewater are services essential for human health. As population centres have expanded, the criticality of these services and related infrastructure has increased. Subsequently, this contributes to a corresponding elevation of the consequences related to infrastructure or service failure threatening secure continuity of supply or safe water quality by various means.

Documented historical evidence (Gleick 2004) demonstrates that water and wastewater infrastructure have been impacted by multiple and variable external incidents over the past 2,500 years (Gleick 2006). From the broad perspective, these events may be defined as two designated groups, dependent on their source or origin. The first group represents incidents originating from 'all hazard', which are results of sudden and adverse natural or weather occurrences, infrastructure failure in its life cycle or accidental systemic errors. The second group includes incidents in the form of external human interventions such as organised transnational and national terrorism. This group is considered more dangerous as those events are *"extremely difficult to predict or provide early warning in most cases"* (Birkett and Mala-Jetmarova 2011) rendering them very challenging to mitigate and plan for.

14.2 Threats to Australia

In Australia, there is a documented history of both all hazard incidents and external human interventions. All hazard incidents such as bushfires, drought and floods create part of the national history over the past 200 years (Writer 2011), whereas terrorist incidents have been reported spasmodically in Australia since the 1970s (Australian Government 2010). Considering the fact that Australia has been subjected to a broad history of such incidents, the protection of infrastructure becomes a significant national objective.

14.2.1 All Hazard Events

All hazard events or incidents are often referred to as disasters. The term 'disaster,' according to Writer (2011), derives from the Greek term 'dus' and 'aster' meaning the term 'bad star', translated as an event which causes significant destruction affecting people, property and the environment. Australia has a documented history of disasters on the continent caused by bushfires, drought, floods, earthquakes, cyclones and epidemics. Flynn and Naseby (2011) observe that the Australians display an endemic national strength of character in the face of hardship to rebuild and reform in the face of adversity, which has been identified and documented over two centuries.

The consequences of all hazard events in Australia vary and are dependent upon impacted geographical locations. In sharp contrast to the large urban population clusters at the coasts with a high level of consequence, there are low densely populated rural areas spread over large geographical distances, where a lower level of consequence may be assumed.

14.2.2 External Human Interventions

External human interventions address the areas of human error, sabotage, vandalism, criminal activity and terrorism, with the focus of this section on acts of terrorism. Within the National Guidelines (Australian Government 2011), a 'terrorist act' is defined as: *"an act or threat intended to advance a political, ideological or religious cause by coercing or intimidating an Australian or foreign government or the public, by causing serious harm to people or property, endangering life, creating a serious risk to the health and safety of the public, or seriously disrupting trade, critical infrastructure or electronic systems"*.

In terms of a counter-terrorism perspective, Australia is in a unique position in the global environment with respect to its continental isolation, which provides some control over people entering and leaving the country. Nonetheless, terrorist incidents have occurred in Australia, and the terrorism threat and its high impact is still evident with terrorism signalling a new era of conflict for the foreseeable future (Australian Government 2004c).

The first modern observed and documented terrorist incident within Australia was the bombing of the Yugoslav General Trade Agency in Sydney in 1972, followed by the Sydney Hilton Hotel bombing in February 1978 at the venue for the meeting of the Commonwealth Heads of Government Regional Meeting (CHOGRM). Crown (1986) indicates a state of political confusion and overreaction related to the CHOGRM incident. Furthermore, there was the assassination of the Turkish Consul General also in Sydney in December 1980 (Crown 1986) and the bombing of the Israeli Consulate and Hakoah Club in Sydney in 1982 (Australian Government 2010). These incidents heralded a new emerging political environment and provided an alert message to Australia, which previously appeared immune from the global effects of terrorism occurring in other countries.

More recently, statements attributed to Al Qaeda with reference to the Bali bombings in October 2002 and Australia's support in Afghanistan, have specifically identified Australia as being a potential terrorist target (Australian Government 2004c). Indeed, Australia is defined as a terrorist target on a dual basis, as a Western

Nation and as a country which, as an international community member, is advancing an active foreign policy reflecting energetic and supportive global human rights. This active and energetic activity supports religious freedoms around the globe in the cause of democracy and freedoms of expression and action (Australian Government 2004c).

The authority assessing the terrorist threats to Australia is the Australian Security Intelligence Organisation (ASIO), the national security intelligence body. As one of its primary functions, the ASIO is designated with the responsibility for national threat assessments, which indicate levels of threat against, and probable nature of terrorism, espionage and politically motivated violence (Irvine 2012). These threat assessments are produced for specific events, facilities, people or sectors, and are separate to the national terrorism public alert levels. The current national terrorism public alert level related to a potential terrorist attack in Australia is advised by the Commonwealth Government as 'Medium' or that a terrorist attack is 'likely' to occur (Australian Government 2011).

14.3 Critical Infrastructure

The Australian Commonwealth Government classifies the 11 sectors identified under the category of critical infrastructure (CI) being (i) food supply, (ii) health, (iii) energy, (iv) utilities, (v) transport, (vi) essential manufacturing, (vii) communications, (viii) finance, (ix) emergency services, (x) government services, (xi) icons and public gatherings (O'Sullivan 2008), with an accepted definition for CI as (TISN 2004): *"physical facilities, supply chains, information technologies and communications networks which, if destroyed, degraded or rendered unavailable for an extended period, would significantly impact on the social or economic well-being of the nation, or affect Australia's ability to conduct national defence and ensure national security".*

Those 11 sectors of CI, in consideration of the increasing size of cities and complexity of urban development, have become more interconnected and interdependent on each other in relation to the adequacy of individual service provision. Nowadays, due to the complexity and inter-relativity of the various sectors of CI, an inter-utility reliance perspective is more closely associated with urban densities and population growth combined with the increasingly technical nature of utility business.

14.3.1 Dependency and Interdependency

Rinaldi et al. (2001) clarify the definitions of 'Dependency' and 'Interdependency' as applied to CI as follows:

- Dependency represents *"a linkage or connection between two infrastructures, through which the state of one infrastructure influences or is correlated to the state of the other"*.
- Interdependency represents *"a bidirectional relationship between two infrastructures, through which the state of each infrastructure influences or is correlated to the state of the other. More generally, two infrastructures are interdependent when each is dependent on the other"*.

According to Gillette et al. (2002), as city infrastructure networks grow, increase in complexity and develop an extended dependency on other forms of CI, the consequences and impact of infrastructure failure to society as a whole increases. This creates an elevated potential that a rather minor and routine disturbance can cascade into a significant outage involving multiple CI (Rinaldi et al. 2001). Indeed, Gillette et al. (2002) advise that the failure to comprehend how individual disruptions to one area of infrastructure may cascade to others can exacerbate response and recovery efforts. They suggest that this may extend to common cause failures leaving planners, operators, and emergency response personnel unprepared to deal effectively with the consequences of such disruptions. Subsequently, new assurance challenges are emerging which can only be met by a partnership among utility owners, operators and government at all levels.

Gillette et al. (2002) advise that mitigations relative to the issue of interdependencies require new modelling, simulation approaches and tools. These suggested approaches and tools would provide an assessment and analytical base to concisely review the technical, economic and national security implications of technology and policy decisions. These may also potentially impact and provide additional insight into interdependency relationships between areas of CI.

In Canberra, Australia, the aspect of interdependencies is currently being examined as part of the Critical Infrastructure Protection Modelling and Analysis (CIPMA) programme, under the leadership of Geoscience Australia as the specialised technical team (TISN 2007). The primary focus of the CIPMA project is to model and simulate the behaviour and dependency relationships across all sectors of CI; in order to analyse the effects of any disruption within and across CI sectors, and provide some assurance of reliability and security related to interdependent CI situations (TISN 2007). More specifically, the objectives are to determine what are the potential economic and social effects, how and which populations are to be affected and by what level of impact and consequence. The simulation additionally incorporates predictions related to the estimated time period of the disruption, the areas affected and how the various affected infrastructure behaves and recovers from a variety of disruption sources (TISN 2007).

14.3.1.1 Water Critical Infrastructure

Water and wastewater CI are considerably interdependent on other CI in numerous ways. Historically, according to Gillette et al. (2002), interdependencies of this CI have been considered to be either physical or geographic. Currently, however, Gillette et al. (2002) identify these four general categories of water and wastewater CI interdependencies as being (i) physical, (ii) cyber, (iii) geographic and (iv) logical. They also discuss the dimensions of infrastructure interdependency creating spatial, temporal and system representation complexities, which make analysing the water and wastewater CI particularly challenging.

The interdependency risk gap for water and wastewater is mutually reflective with other CI (Gillette et al. 2002). A typical example of a mutually reflective physical interdependence is that the water supply infrastructure depends on electric power to operate its distribution pumps while, at the same time, the electric power infrastructure requires water to produce steam for electricity generation and to cool the generating equipment. Conversely, water geographic interdependencies arise when infrastructure components, for example, water pipelines, transmission lines, gas pipelines and telecommunications cables share common service corridors; thus, increasing their individual and combined vulnerabilities.

In the Australian context due to the extensive distances between capital cities and regional locations across the continent, dependency of water and wastewater CI on other CI is compromised in consideration of the 'tyranny of distance'. As such, water organisations rely significantly on the transport sector and extended supply chains for essential chemicals for water treatment and disinfection. Another example of water CI dependency is the impact of electricity failure on water treatment, disinfection and distribution. Although this aspect may be mitigated by the use of dedicated emergency electricity generating units, the application of this mitigation would tend to be more prevalent in large urbanised centres such as Sydney, Melbourne and Perth, where the consequences are more significant.

Nevertheless, electricity failure is more likely to have an onerous impact on the transport and treatment of wastewater in geographically flat areas, where there are pumped wastewater collection systems, in lieu of a gravity system. These multiple and geospatially diverse pumping units are usually linked into conventional residential electricity distribution systems with a mixed pattern of electrical distribution feeders. As a consequence, they are more prone to spasmodic interruptions, which would display increased difficulty in supplementing portable electricity generating systems due to the number and locations of these pumping units.

To conclude, the aspect of dependency and interdependency of water CI is a complex issue, which requires an improved understanding in order to assist in determining the system's vulnerability.

14.3.2 Vulnerability of Water Critical Infrastructure

Various forms of threat exposures are apparent across the broader areas of water CI on a more universal basis. Perceptions of these threats are mainly linked to the concept that water and wastewater systems were originally designed for ease of maintenance access, with limited consideration for external adverse intervention.

Moreover, facilities such as water storages, water distribution and wastewater collection systems, and treatment plants cover a wide functional and geographic area to protect, and are potentially at risk from all hazard incidents as well as external human interventions.

Currently and internationally, most water security measures are primarily concerned with detecting and responding to water contamination (Bahadur et al. 2003; Ostfeld et al. 2008) from both all hazard incidents such as naturally occurring water pathogens and parasites, and deliberate interventions. Bahadur et al. (2003) identify the need to develop software for responding to malevolent and deliberate contamination of water distribution systems. Nevertheless, to adequately understand the risk exposures within the spectrum of water and wastewater CI, these systems require to be analysed from the broader perspective to ensure adequate risk mitigation.

The combined all hazard threats and external interventions or impacts in relation to water and wastewater CI may be interpreted in a typical example of a perception of a terrorist attack representation. To be more concise, individual water and wastewater CI system functions require to be classified in consideration of their system significance and potential consequence of their failure. As illustrated in Fig. 14.1 (Birkett and Mala-Jetmarova 2012), the graded threat sectors may be considered and divided into three groups, primary, secondary and tertiary. The primary group incorporates water systems, either drinking or non-drinking for human consumption and use. The secondary group represents the spectrum of wastewater systems. The tertiary group consists of other ancillary support system functions such as skilled human resources, buildings and facilities, vehicles and equipment, information technology and financial systems. As indicated, the physical and Chemical, Biological, Radiological, Nuclear (CBRN) attacks are the most probable, with cyber attack having an elevated risk in relation to financial and control systems.

Within Fig. 14.1, the illustrated radially designated significance of individual system elements of potential threats to water and wastewater CI may be classified as primary, secondary and tertiary targets to an all hazard event or deliberate human intervention. This reflects on an interpretation of the perception of consequences in terms of potential damage to the service to the consumer. For example, water supply, distribution, treatment and control systems are considered as possible primary targets, upon which an attack or external all hazard incident could potentially maximise damage to the final product of the service to the consumer. Conversely, consequences or damage to vehicles or equipment, finances, buildings and facilities or the specialised human resources, although potentially affecting services, are not considered to impact at the same significant level in relation to the final product.

It is also important to note that the global implementation of computers into industry and society has introduced innovative opportunities for advancement, business innovation and a reduction in labour. On the other hand, the introduction of cyber connectivity and control systems into CI has raised vulnerabilities to deliberate external unplanned interventions by saboteurs, vandalism, criminal activity and terrorism (US Army TRADOC 2006). This is particularly evident in

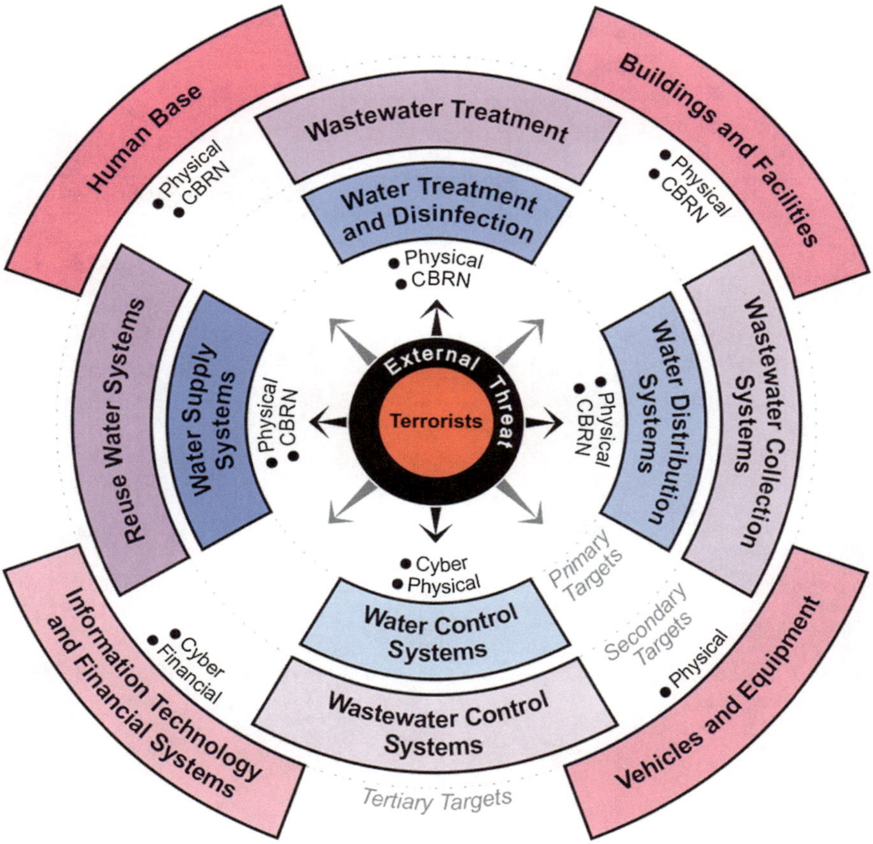

Fig. 14.1 Radially designated significance water and wastewater CI related to terrorism risk exposure. *Source* Birkett and Mala-Jetmarova (2012)

the water and wastewater sector where Supervisory Control and Data Acquisition (SCADA) systems control and monitor water distribution and treatment including system operation and water quality.

14.4 Planning for Water Critical Infrastructure Protection

14.4.1 Historical Background

The current overarching strategy of CI protection (CIP) in Australia has its origins in an announcement by Prime Minister John Howard in November 2001 to form a 'Business—Government Task Force' on CI, which first met in March 2002

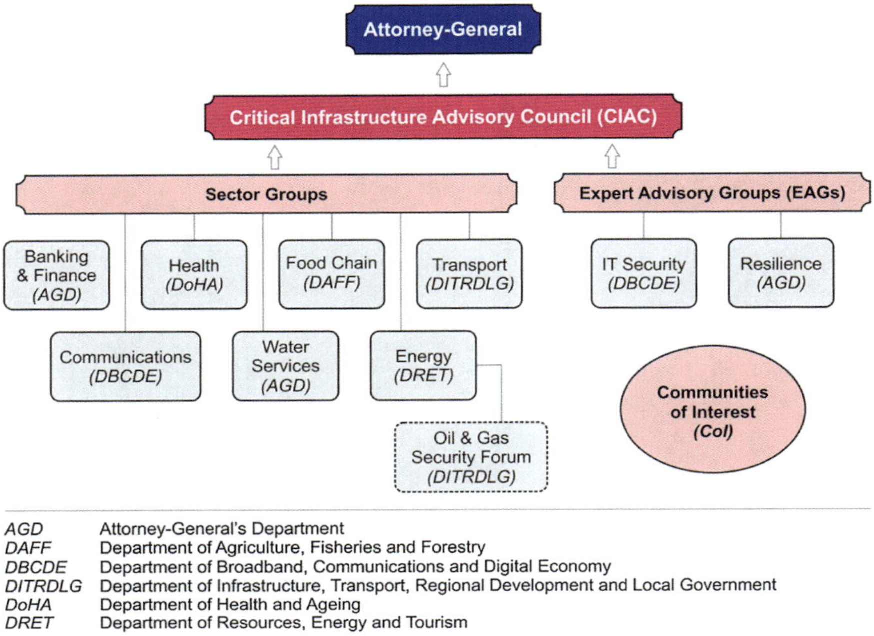

Fig. 14.2 Australian government structure of the trusted information sharing network (TISN). *Source* TISN (2012)

(TISN 2004). This action was subsequent to the Prime Minister's visit to the United States of America (U.S.), where he was present at the time of the terrorist attack on the Twin Towers and the Pentagon on 9th September 2011 (9/11). Though not specifically documented as such, but intimated by Prime Minister John Howard (Australian Government 2004b), it is feasible that this terrorist act 9/11 may well have initiated the framework leading to the protection of CI in Australia.

As a consequence of the Business—government Task Force, as the head agency at a Commonwealth level, the concept of a 'Trusted Information Sharing Network' (TISN) emerged in 2002 (Australian Government 2004b). This concept, which is shown in Fig. 14.2, was launched at the National Summit on CIP on 2nd April 2003. Prior to the TISN launch, guidelines were also endorsed by the Council of Australian Governments (COAG) based on developments by the Australian National Counter-Terrorism Committee (NCTC) in December 2002 (TISN 2004). These guidelines included the establishment of criteria to analyse and define CI sectors with appropriate security measures. Due to the elevated presence of the private sector in infrastructure ownership and operation, the COAG further identified and promoted the need to establish a more cooperative approach between Government and industry.

Although the primary trigger for the concept of CIP may well have had its birth from concerns of the potential effects of a terrorist action on CI, it has extended to an all hazard approach. This is due to a perception that CI may also be damaged, destroyed or disrupted by a wide range of natural external impacts, in addition to deliberate adverse human intervention of a terrorist act (TISN 2004).

14.4.2 Planning: Victorian and Queensland Perspective

In the area of CI protection, planning is a critical aspect which aims to ensure that an organisation is prepared to 'manage a crisis' in a form of an unpredictable interruption or adverse incident. As a result of such planning and preparedness, there is a level of assurance that the consequences of the potential incident are mitigated, which plays a particularly important role in the larger population areas across Australia.

The Victorian Government has led the way by incorporating the water sector in 2006 as an essential service into an Act of Parliament identified as the Terrorism (Community Protection) Act 2003 (Victorian Government 2003). Under this Act, it is now a mandatory requirement for water organisations to (Victorian Government 2003):

- Have a risk management plan which includes some provisions for a terrorist act.
- Audit this plan on an annual basis.
- Test this plan with an annual exercise, referred in this paper to as a Crisis Simulation Exercise, which is assessed for compliance by the Victoria Police and the Department of Sustainability and Environment.

From 2006, all 20 water organisations in Victoria are required to comply with the above mandatory requirements. In relation to their preparedness, they need to also demonstrate a continual improvement framework to further gradually strengthen the protection of water and wastewater CI and services provided. The Victorian Community Protection Act hardens the industry not only against acts of terrorism, but also illustrates an all hazards approach thereby increasing the overall resilience of the water industry across Victoria. This progressive action by the Victorian Government has ensured that water and wastewater organisations display an active stance to assess and adequately manage the risk related to the CI under their control.

Although in Australia, there is a central Commonwealth Government providing an umbrella of training, support and information, the operational aspects of CIP in Victoria are enhanced with the cooperation among State and Local Governments, the private sector, voluntary organisations and individual households. Other Australian States are observing the Victorian approach with interest and may well adopt this progressive step in the future.

In Queensland, within the range of facilities, networks and assets comprising Queensland's complex water CI base, it is considered that the potential interruption or cessation of water services would significantly impair the ability to maintain Queensland's way of life and cause considerable potential economic loss. Therefore, the Queensland Government is committed to the protection of all CI sectors and has examined major water infrastructure organisations with identification of individual CI assets. Subsequently, the Queensland Government has implemented a plan and programme to ensure protection of CI assets against acts of terrorism.

14.4.3 Risk Management

The Australian Commonwealth and State Governments encourage all risk management processes related to water and wastewater CI assessments to be conducted in accordance with the international risk management standard of AS/NZS ISO 31000:2009 (Standards Australia 2004). This standard illustrates a comprehensive framework which identifies not only the categories of prevention and preparedness, response and recovery (Australian Government 2009), but also represents a standard guiding tool to provide a consistent framework for water and wastewater risk assessments. Under this standard, risk is defined in terms of the effect of uncertainty on objectives with more explicit principles to achieve effective risk management in contrast to the former Australian standard AS 4360. Additionally within this standard, there is much greater emphasis and guidance on how risk management should be implemented and integrated into organisations through the creation and continuous improvement of a framework.

In relation to risk assessments of water and wastewater CIP, the Australian Government conducts risk assessments at a Commonwealth and State level to achieve a national industry criticality and threat analysis. These assessments are classified to provide governmental knowledge and awareness of levels of vulnerability for the specific sector, with individual infrastructure assessments conducted at the local industry and water utility level to evaluate local risk and mitigations required (Standards Australia 2004).

14.5 Preparation for Water Critical Infrastructure Protection

The Australian all hazard model of CIP, including protection against external human interventions, relies upon the strong cooperative, coordinated and consultative relationships from the Australian Commonwealth Government via TISN to the State and Local Governments. This is inclusive of the various Government Departments and agencies represented in the public, private sector and voluntary

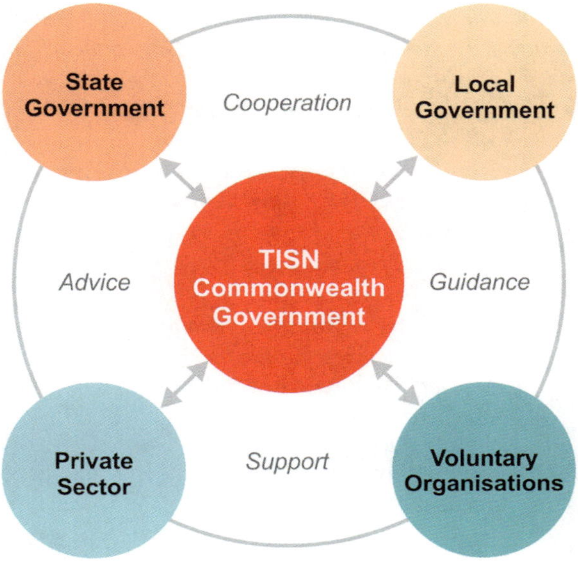

Fig. 14.3 Commonwealth coordination for water critical infrastructure protection via trusted information sharing network (TISN)

organisations (Fig. 14.3) such as State Emergency Services and Red Cross. Such a relationship across government, the private sector and voluntary organisations forms an effective business government partnership to assist with communication and support.

14.5.1 Role of Australian Government

The role of the Australian Government is allocated with a reflection of Government structure; more specifically, Commonwealth Government, State and Territory Governments and Local Government. Each tier of Government has an important and specific function in disaster and major incident management.

14.5.1.1 Commonwealth Government

The Commonwealth Government displays an umbrella role and stance in relation to the broad arena of emergency and crisis of CI in Australia. It adopts a lead national role in coordinating national research in disaster and emergency management, information, mitigation policies and practice (Australian Government 2009). This provision of national leadership ensures a risk and financial reduction of disasters to the nation, mobilisation of resources, with national support for disaster relief and community recovery (Australian Government 2004a). The Commonwealth Government has important and specific roles in (Australian Government 2004a):

- *"coordinating national strategic emergency management policy, in collaboration with the State and Territory Governments and Local Government,*
- *undertaking natural disaster research of national significance,*
- *identifying national priorities for natural disaster mitigation, in collaboration with other levels of government,*
- *providing support for disaster risk assessment and mitigation measures, in conjunction with the States, Territories and Local Government,*
- *providing operational support for disaster response to the States and Territories where their individual resources are insufficient,*
- *providing a national disaster relief and recovery framework and resources on a cost-sharing basis with the other levels of government, and*
- *providing vital information services such as meteorological, hydrological, geophysical and other geo-data services that support warnings and disaster management."*

The Commonwealth Government also has a continuing role in (Australian Government 2004a):

- *"providing national leadership on mitigation strategies and assessment,*
- *providing financial assistance to States, Territories and Local Government for cost-effective, priority disaster risk management,*
- *providing financial assistance to States, Territories and Local Government to assist them in meeting their disaster mitigation responsibilities, leading to an overall reduction in damage and costs, thereby benefiting all Australians and all levels of government."*

Furthermore, the Commonwealth Government under the Attorney-General's Department provides extensive training via the Mount Macedon, Emergency Management Australia (EMA), located in regional Victoria. The Commonwealth Government also produces a large range of resources, advice, guidance and educational publications through the publicly accessible TISN network on line (www.tisn.gov.au) to obtain advice and a structural framework for the protection of CI.

14.5.1.2 State and Territory Governments

In emergency management, State and Territory Governments assume a primary responsibility with a central objective for safety and well being of local communities, and have a focus within their own jurisdictions for disaster and major incident management. They also provide a forum and various support for the government business units and the private sector to ensure that the CI is adequately protected. The State and Territory Governments have important and specific roles in (Australian Government 2004a):

- *"developing, implementing and ensuring compliance with comprehensive disaster mitigation policies and strategies in all relevant areas of government activity, including land use planning, infrastructure provision, and building standards compliance,*
- *strengthening partnerships with and encouraging and supporting Local Governments, and remote and Indigenous communities, to undertake disaster risk assessments and mitigation measures,*
- *ensuring provision of appropriate disaster awareness and education programmes and warning systems,*
- *ensuring that the community and emergency management agencies are prepared for and able to respond to natural disasters and other emergencies,*
- *maintaining adequate levels of well equipped and trained career and volunteer disaster response personnel,*
- *ensuring appropriate disaster relief and recovery measures are available, and*
- *ensuring that post-disaster assessment and analysis is undertaken."*

In some States and Territories, legislative compliance is also structured to ensure that essential water and wastewater services are adequately protected and that effective mitigations are in place for the full range of all future emergencies which are likely to occur.

14.5.1.3 Local Government

In coordination with States and Territories, Local Government supports the safety and well being of local communities for disaster and major incident management. In most circumstances, the Local Government has important and specific roles in (Australian Government 2004a):

- *"ensuring all requisite local disaster planning and preparedness measures are undertaken,*
- *ensuring an adequate local disaster response capability is in place, including local volunteer resources,*
- *undertaking cost-effective measures to mitigate the effects of natural disasters on local communities, including routinely conducting disaster risk assessments,*
- *systematically taking proper account of risk assessments in land use planning to reduce hazard risk,*
- *undertaking public education and awareness, and ensuring appropriate local disaster warnings are provided,*
- *ensuring appropriate local resources and arrangements are in place to provide disaster relief and recovery services to communities,*
- *representing community interests in disaster management to other levels of government and contributing to decision-making processes, and*
- *participating in post-disaster assessment and analysis."*

14.5.2 Ownership and Responsibility

Although the responsibility for the protection of CI is a shared and cooperative relationship among Commonwealth, State and Local Government agencies, there is an issue with the ownership. On a broad industry basis, CI is owned by both Government and a range of independent, mostly commercial, organisations. While each of these individual organisations assumes security measures to protect their individual interests and those of their shareholders, it would be unrealistic to expect that they would invest in measures to secure those aspects that are of more relevance to the nation state.

Unlike other sectors of CI, water and wastewater CI are mostly owned and operated by Government in all States. Hence, a more standard regulatory approach is achievable with the appropriate supporting government legislation. As a result, the water sector in Australia is heavily regulated on a national and State basis. This regulatory impost is also partially a legacy of the emergence from a recent 12 year drought in Australia from 1997 to 2009, with associated regulatory controls necessary to conserve water.

In general, CI owners and operators in Australia are provided with the Commonwealth Government umbrella of support, training and response in adverse circumstances subsequent to a disaster or major incident. An example of a Commonwealth support initiative, introduced by the Commonwealth Attorney-General's Department, is the formation of Infrastructure Assurance Advisory Groups (IAAGs). These groups provide a partnership forum between the multiple government business units and private sector organisations which own and/or operate CI.

Moreover, water CI owners and operators in Australia are supported by 'Mutual Aid' agreements and Guidelines between States to provide various forms of physical assistance horizontally between States in the event of a major disaster. These Mutual Aid agreements were initiated by the Water Services Infrastructure Assurance Advisory Group (WSIAAG).

14.5.3 Coordination

Water CIP is coordinated under the guidance of the TISN under the Commonwealth Government Attorney-General's Department. The TISN membership includes owners and operators of CI, and representatives from Australian, State and Territory Government agencies and peak national bodies. As illustrated in Fig. 14.2, the TISN consists of (i) Sector Groups such as banking and finance, health, food chain, transport, communications, water services and energy including oil and gas security forum, (ii) Expert Advisory Groups and (iii) Communities of Interest.

The TISN groups aim to enable CI owners and operators to share information on threats and vulnerabilities, and collaborate on appropriate measures to mitigate risk and boost resilience. These groups are overseen by the Critical Infrastructure Advisory Council (CIAC) and supported by a designated Australian Government agency, usually the agency which has portfolio responsibility for that particular sector.

The TISN also formed the WSIAAG as a consequence of a Council of Australian Government (COAG) action in 2003 (TISN 2004). The WSIAAG consists of representatives from the owners and operators of water CI and meets on a 4-monthly basis in various capital city locations around Australia. The main focus of the WSIAAG is to share knowledge and initiatives within the context of risk reduction on an all hazard and human intervention basis, which has further extended to the promotion of a strong resilience model for water CIP.

14.5.4 Emergency Services: Victorian Perspective

In Victoria, the State Government structure of the emergency services is represented across a diverse combination of State Government agencies and Departments. As displayed in Fig. 14.4, these include the Department of Premier and Cabinet (DPC), Local Government (LG), the Department of Human Services (DHS), the Victoria Police (VicPol), the State Emergency Services (SES), the Country Fire Authority (CFA) and the Department of Sustainability and the Environment (DSE). These agencies maintain processes, which contribute to an effective reduction of risk related to any potential future emergency affecting any community service.

The DPC assumes a high level strategic function, marshalling resources from a national level and potentially interstate if required, with government to government discussions and liaison. In consideration of the close links and knowledge of community life, the Councils, Shires and organs of LG provide an essential linkage between other emergency services and the affected areas, with significant identification of potential social and material requirements by communities. The DHS coordinates the support and recovery functions, inclusive of health, medical facilities and counselling. It also acts as a conduit to feed relevant funding into areas of the State to restore the damaged structure of the community back to normalcy and functionality.

The VicPol liaises directly with the Commonwealth Government as the lead agency in relation to most emergencies. The VicPol includes the Tactical Response Group (TRG) and the significant VicPol CIP Branch. The TRG acts as the trained rapid response group to react and deal with physical external threats to CI. The CIP Branch works closely with the DSE in relation to the compliance, assessment and overview of all CI sectors across Victoria in accordance with the terrorism provisions of Part 6 of the Terrorism (Community Protection) Act 2003 (Victorian Government 2003). Specifically, the DSE Water Division is responsible for the coordination of water and wastewater CIP for the total 20 urban and rural water organisations in Victoria.

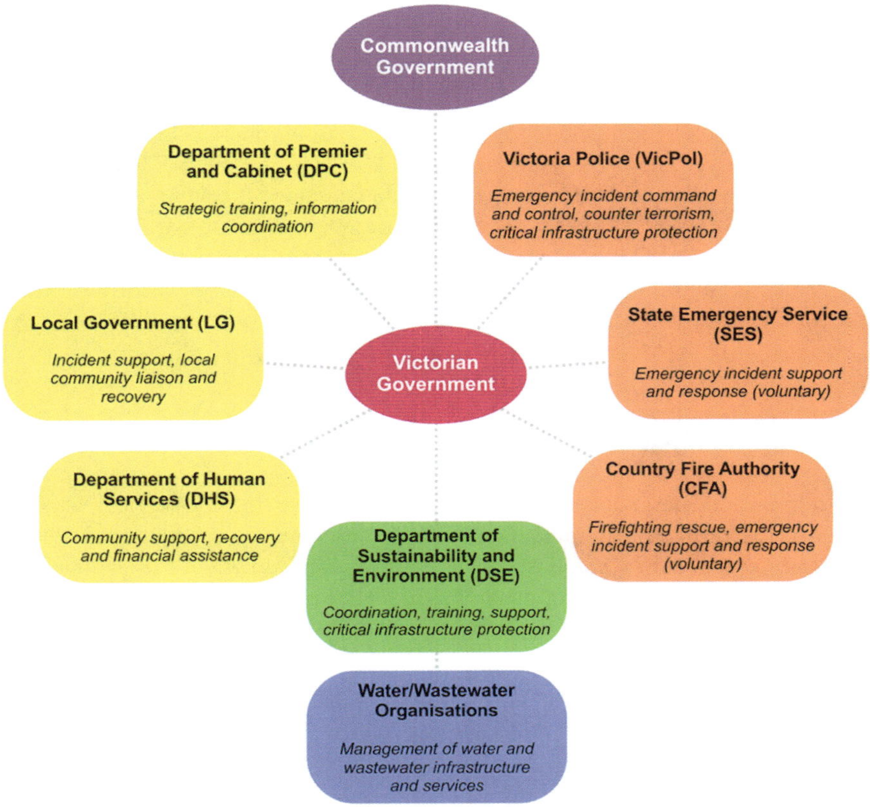

Fig. 14.4 Victorian state government structure of emergency services

The SES and CFA are community volunteer organisations, which are highly trained to react to and provide immediate emergency support during any incident affecting the Australian community such as flood, fire, earthquake and other forms of disaster. The CFA has a primary role of fighting and controlling bushfires across Victoria and other States under a variety of identities. During an emergency situation, both the SES and CFA provide emergency response to resolve the immediate perceived threat and alleviate any potential suffering, which may take the form of a rescue, extracting injured people, repairing facilities or restoring basic community services.

14.6 Protection of Water Critical Infrastructure

Australia's CIP arrangements are cooperative and coordinated across National and State lines, with protection translating into education and training. The important role in this arrangement is undertaken by WSIAAG, which ensures and maximises protection and risk reduction for the water sector. These arrangements progress and translate into a broader process of protection as illustrated in Fig. 14.5.

When an incident or emergency occurs there is little capacity of time to plan effectively, marshal resources, and establish relationships with key stakeholders and emergency services (Evans et al. 2003). These essential steps are an integral ingredient of the Australian planning cycle of identification and protection of CI, prior to an event occurring. In other words, the emphasis is on preparation, assessments, planning and effective mitigation of potential sources of emergency or disaster prior to the event. As indicated in Fig. 14.5, the cycle of identification and protection related to CI is a continual process, undertaken by water organisations mainly on an annual basis, and is dependent on the identified risk issues and size of population served by the organisation.

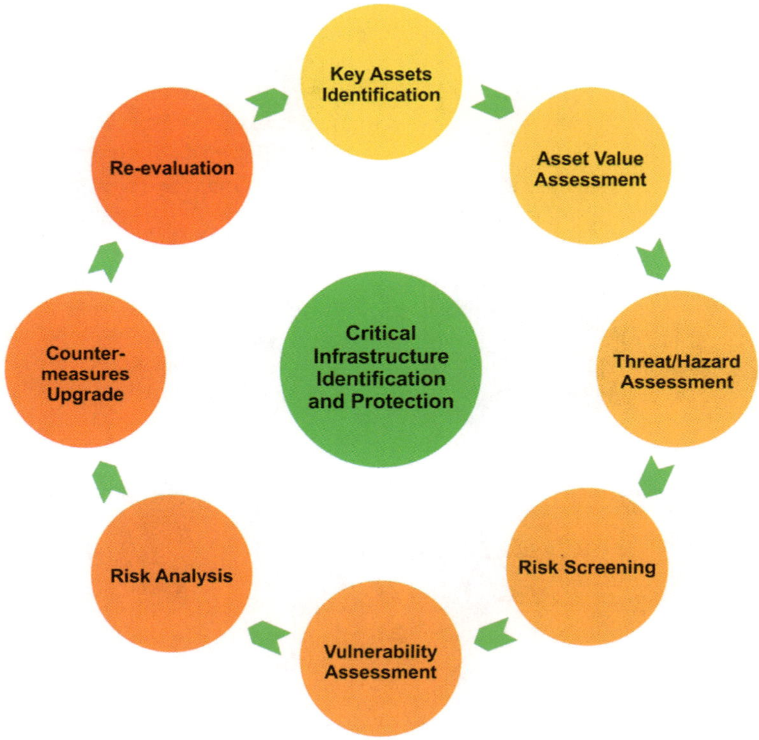

Fig. 14.5 Cycle of identification and protection of CI. *Source* Bennett (2007)

Because a major emergency or incident related to water CI can have a catastrophic effect on large populations and the economy by disruptions to business, the potential incident requires to be evaluated from both a 'tactical' and 'strategic' perspectives. The tactical component considers the impact on life, property and the environment, whereas the strategic component cascades to impact reputation, litigation and the corporate business continuity (Birkett et al. 2011). The evaluation of tactical and strategic perspectives dictates the approach towards any potential future incident. The outcome of such an approach may subsequently lead to a more accurate judgement or opinion on the planning decisions, demonstrating a more realistic outcome of public praise or condemnation affecting the business reputation, potential litigation or the longer term continuity of the business.

14.6.1 Threat Assessment

All hazard events such as adverse weather occurrences and infrastructure failure in its life cycle are theoretically predictable, with tangible indicators to guide the proactive planner. When conducting an infrastructure threat assessment for those events, some known and identified aspects can be included. Regarding external human interventions such as human error, sabotage, vandalism and acts of terrorism, however, threat assessments are not so clearly definable and are subject to external and unknown influences. Bennett (2007) suggests that the threat can be expressed in a mathematical representation as displayed in Fig. 14.6.

Methods of threat analysis include the 'CARVER' analysis tool (Birkett et al. 2011) and red cell analysis, which are known to be adapted and applied by various terrorist groups to accurately identify and compile a financial budget for the proposed attack. In the future, terrorists may engage a process of target selection related to attacks on water and wastewater CI. Indeed, Hillary Clinton, the US Secretary of State, recently warned that future wars will be over water as a depleting global resource and terrorists will attack water systems as a future attack strategy (AFP 2012).

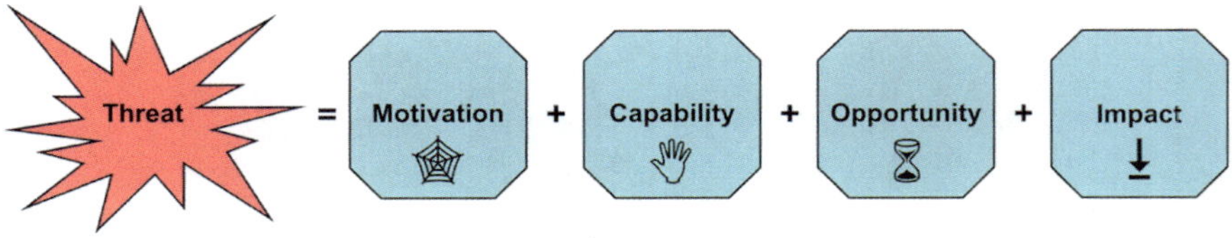

Fig. 14.6 Threat formula. *Source* Bennett (2007)

14.6.2 Risk Assessment

In a broad Australian water industry view, a general risk analysis indicates the existence of 'soft' and 'hard' targets, or water organisations which are not adequately protected. The inadequately protected organisations may be a result of a perception that the assets represent either a low probability of any incident occurring or/and low consequences of such an incident. The circumstances contributing towards this perception could be associated with a small population served and/or geographically isolated location. This is directly linked to the local water organisation risk assessment dependent on a variety of factors taken into consideration, including the financial outlay compared with the level of perceived risk.

Unfortunately, the history of risk and adverse incidents illustrates that fully adequate protective strategies are often only implemented after significant incidents, or subsequent to an enquiry or review. In consideration of the continual adaptive and evolving nature of transnational terrorism with associated post attack target hardening, it is the authors' opinion that the global water sector requires to be proactive, 'forewarned, forearmed' and prepared.

14.6.3 Protection Strategies

As part of water and wastewater CIP strategies, fundamental structural protective security principles are implemented in all States of Australia. These fundamental protection strategies contribute to a general risk reduction and mitigation which displays significant cost benefits to the water organisation as demonstrated in Fig. 14.7.

The protection strategies as indicated in Fig. 14.7 may additionally include planning for business continuity, and implementation of regular emergency training and scripted exercises, referred in this paper to as Crisis Simulation Exercises, to test systems, plans, procedures, personnel and equipment (Crowe et al. 2011).

14.6.3.1 Business Continuity Plans

Business Continuity Plans (BCP) ensure that the water organisation will likely continue to operate effectively during any future business interruption and during the recovery phase post incident. The BCP additionally supports the visible interactions and relationships with external agencies, which occur during an actual incident.

When drafting a BCP, dependencies and interdependencies of water and wastewater CI are considered forming an integral component of effective

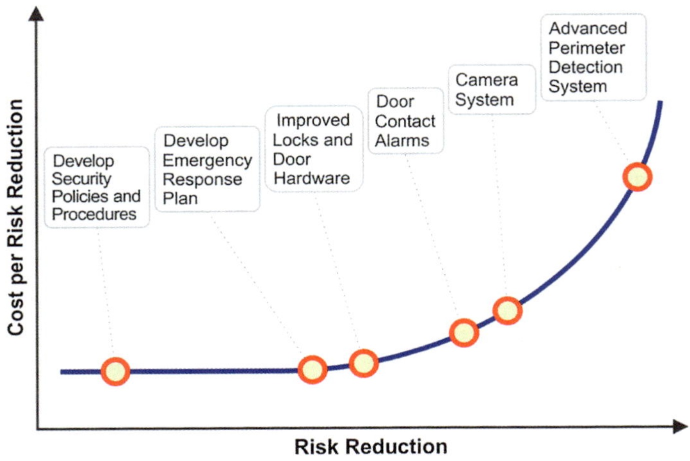

Fig. 14.7 Water and wastewater protective strategy cost analysis. *Source* AWWA (2006)

emergency planning. This is to ensure that the water and wastewater services to customers are maintained and continue as long as possible under emergency or crisis situations.

14.6.3.2 Crisis Simulation Exercises

Crisis Simulation Exercises (CSEs) were, as indicated previously, implemented as part of the Victorian Government legislation (Victorian Government 2003). This proactive initiative aims to train and rehearse water organisations to effectively mitigate the risk to key water CI. The CSEs are conducted on an annual basis and assessed by the VicPol and DSE. Interestingly, the CSEs are more broadly and eagerly implemented by the private sector, where there are significant Board assurance requirements and floating share values on the stock exchange to be considered in the performance and business continuity to the relevant organisation.

A typical model displaying the linkage between a terrorist incident as it would occur in reality and the simulated exercise is displayed in Fig. 14.8. This model identifies the commonalities of a deliberate external intervention providing the primary 'trigger' and focus relating to either a terrorist incident or all hazard event. The CSE also identifies the synergies of an external influence, providing the primary trigger of a crisis. Moreover, the defined targeted organisation can be perceived and visualised within the model with a demonstration of the various forms of external monitoring, which actually occurs in a real incident and which is simulated during a CSE. More detailed information about the CSEs can be found in (Birkett et al. 2011).

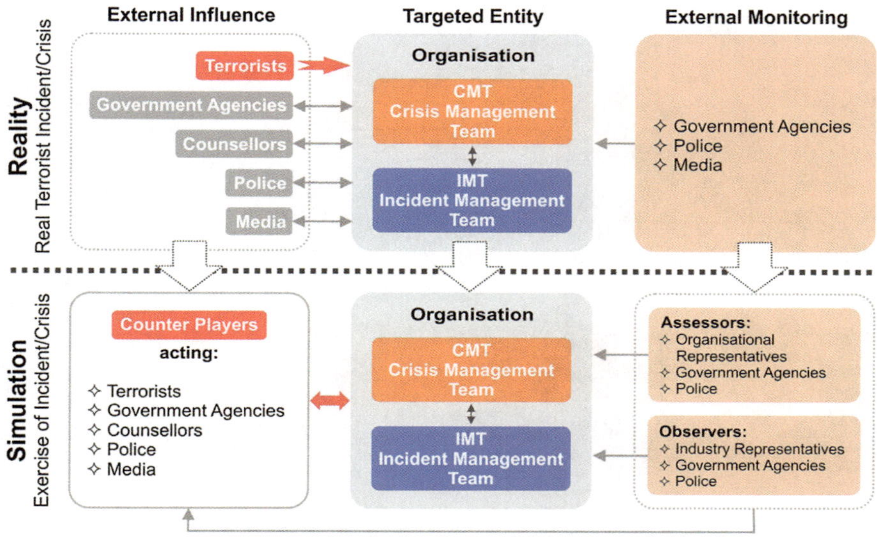

Fig. 14.8 A typical model displaying linkage between a terrorist incident as reality and simulation. *Source* Birkett and Mala-Jetmarova (2011)

The practical application of the CSE consists of exercise management and control, to 'ring-fence' the communications within the exercise loop and the active simulated element as illustrated in Fig. 14.9 (Crowe et al. 2011). The management and control elements include the exercise manager, controller, assessors and observers, whereas the active simulated component consists of counter players, the Incident Management Team (IMT) and the Crisis Management Team (CMT). The counter players are the actors representing the exercise script, with the IMT (i.e. tactical) reacting to the scripted input handling the protection of life, property and the environment and the CMT (i.e. strategic) protecting the reputation, litigation and business continuity.

14.7 Response and Recovery

The Australian response phase identifies the scale and consequence related to each specific incident and initially activates the relevant emergency plan to deal with the tactical issues associated with the emergency such as the protection of life, property and the environment. This may occur on a broad scale from an individual corporation incident response to State or Commonwealth Government intervention, dependent on the scale and size of the incident. In significant incidents, the various overlapping governmental layers respond reacting to various identified trigger points to provide response resources as required.

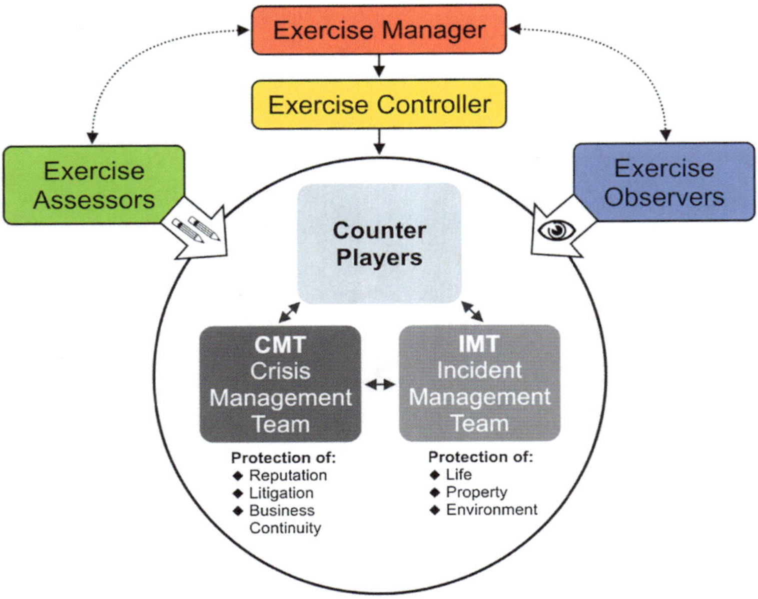

Fig. 14.9 A typical CSE operation. *Source* Crowe et al. (2011)

There is also a recovery component, which may be viewed more dramatically in relation to recent catastrophic bushfires and major flooding disasters in Australia. More specifically, over the past 2 years, there have been devastating floods in Queensland from a previous drought period, followed by cyclones again in Queensland, significant historical floods across Victoria in 2011, subsequent to the fatal bushfires of 2009, and horrific bushfires in Western Australia in 2011 and 2012.

As a consequence of those events, CI damage was massive. For example, insurance estimates of the Victorian bushfire in February 2009 indicate a total damage bill of A$1.07 Billion (Flynn and Naseby 2011). The recovery from these historical events also requires high levels of State and Commonwealth funding to recover. It is important to highlight that the recovery phase of such emergencies can last from months to years, which has been recently observed in those catastrophic disasters evident in the Eastern and Western States of Australia (Writer 2011).

14.8 Conclusion

Aspects of the mega-city phenomenon with reliance of large population clusters on extensive water and wastewater CI with associated increased consequence of its failure represents a challenge in such infrastructure planning and protection.

Although the Australian large cities are not yet categorised as global mega-cities, increased reliance and dependency on water and wastewater CI remain a significant issue. This is especially highlighted by the documented history of natural disasters and terrorist incidents.

The Australian concept for protecting water and wastewater CI illustrates a cascading and interwoven model from the Commonwealth Government through the State and Territory Government agencies and business units to the Local Government, private sector and local communities. This model, which captures both all hazard events and external human interventions, represents a shared fabric and partnership with direction and leadership provided by the Australian Government Attorney-General's Department under the umbrella of TISN and the WSIAAG. The aspect of prior preparedness and preparation leads the model's direction with the Victorian legislative system being observed with interest by other Australian States.

The Australian model for protecting water and wastewater CI is an evolving process, striving for continual improvement and effectiveness. Although Australians display an endemic national strength to rebuild and reform in the face of adversity, it is more than a reflection of preparation, planning, procedures and systems in Australia designed to act and take action in the event of any incident or emergency. This may result in a more resilient society for the effective protection of water and wastewater CI and ensure the continuity of water and wastewater services during any potential future crisis.

References

AFP (2012) Clinton Warns of Instability over Water. In: Canberra Times March 23, 2012, Canberra, ACT, Australia. http://www.canberratimes.com.au/environment/water-issues/clinton-warns-of-instability-over-water-20120323-1vnjz.html. Accessed 11 June 2012

Australian Government (2004a) Natural disasters in Australia—reforming mitigation, relief and recovery arrangements. Department of Transport and Regional Services on behalf of the Council of Australian Governments (COAG), Canberra, ACT, Australia

Australian Government (2004b) Protecting Australia against terrorism. Department of Prime Minister and Cabinet, Australian Government, Barton, ACT, Australia

Australian Government (2004c) Transnational terrorism: the threat to Australia. Department of Foreign Affairs and Trade (DFAT), Australian Government, Canberra, ACT, Australia

Australian Government (2009) Australian emergency management arrangements. Attorney-General's Department, Australian Government, Canberra, ACT, Australia

Australian Government (2010) Counter-terrorism white paper, Securing Australia - protecting our community. Department of Prime Minister & Cabinet, Australian Government, Canberra, ACT, Australia

Australian Government (2011) National guidelines for protecting critical infrastructure from terrorism. National Counter Terrorism Committee, Attorney-General's Department, Australian Government, Barton, ACT, Australia

AWWA (2006) Guidelines for the physical protection of water utilities. Standards Department, American Society of Civil Engineers (ASCE), Denver, Colorado, USA

Bahadur R, Samuels WB, Pickus J (2003) Case Study for a Distribution System Emergency Response Tool. Jointly sponsored by AWWA Research Foundation and U. S. Environmental Protection Agency (EPA), AWWA Research Foundation, Denver, Colorado, USA

Bennett BT (2007) Understanding, assessing and responding to terrorism—protecting critical infrastructure and personnel. Wiley, New Jersey

Birkett D, Mala-Jetmarova H (2011) Terrorism and simulation of terrorist incidents across critical infrastructure. In: MODSIM2011, Modelling and Simulation Society of Australia and New Zealand, Perth, Western Australia, pp 3240–3245, 12–16 Dec 2011

Birkett D, Mala-Jetmarova H (2012) Are risk mitigation strategies of water critical infrastructure adequate within a European environment of 21st Century? In: 10th international conference on hydroinformatics HIC2012, Hamburg, Germany, 14–18 July 2012

Birkett D, Truscott J, Mala-Jetmarova H, Barton AF (2011) Vulnerability of water and wastewater infrastructure and its protection from acts of terrorism: a business perspective. In: Clark RM, Hakim S, Ostfeld A (eds) Handbook of water and wastewater systems protection, series protecting critical infrastructure. Springer, New York

Brinkhoff T (2012) The principal agglomerations of the world. http://www.citypopulation.de/world/Agglomerations.html. Accessed 25 April 2012

Crowe J, Birkett D, Mala-Jetmarova H (2011) Crisis simulation exercises of terrorist incidents in the Australian water industry. In: MODSIM2011, Modelling and Simulation Society of Australia and New Zealand, Perth, Western Australia, pp 3246–3250, 12–16 Dec 2011

Crown J (1986) Australia: the terrorist connection. Sun books. The MacMillan Company of Australia, Melbourne

Evans PM, Granadino JA, Saunders J (2003) Protecting British Columbia's water and wastewater Infrastructure. Stantec Consulting Ltd. http://www.bvsde.paho.org/bvsacd/cwwa/stan.pdf. Accessed 04 June 2012

Flynn H, Naseby E (2011) Courage in crisis, A history of Australia's worst disasters. Random House Australia, North Sydney

Gillette J, Fisher R, Peerenboom J, Whitfield R (2002) Analysing water/wastewater infrastructure interdependencies. In: 6th International conference on probabilistic safety assessment and management (PSAM6), 23–28 June 2002, San Juan, Puerto Rico, USA

Gleick PH (2004) The water conflict chronology. In: Gleick PH (ed) The world's water 2004–2005: the biennial report on freshwater resources. Island Press, Covelo, pp 234–255

Gleick PH (2006) Water and terrorism. Water Policy 8(2006):481–503

Irvine D (2012) ASIO report to parliament 2011–12. Australian Security Intelligence Organisation (ASIO), Canberra, ACT, Australia

O'Sullivan P (2008) ASIO report to parliament 2007–08. Australian Security Intelligence Organisation (ASIO), Canberra, ACT, Australia

Ostfeld A, Uber JG, Salomons E, Berry JW, Hart WE, Phillips CA, Watson JP, Dorini G, Jonkergouw P, Kapelan Z et al (2008) The Battle of the Water Sensor Networks (BWSN): a design challenge for engineers and algorithms. J Water Resour Plan Manag 134(6):556–568

Rinaldi SM, Peerenboom JP, Kelly TK (2001) Identifying, understanding, and analyzing critical infrastructure interdependencies. Control Syst Mag IEEE 21(6):11–25

Standards Australia (2004) AS/NZS ISO 31000:2009 Risk management—principles and guidelines, 3rd edn. Standards Australia and Standards New Zealand, Sydney and Wellington

TISN (2004) Critical infrastructure protection national strategy, version 2.1. Trusted Information Sharing Network (TISN) for Critical Infrastructure Protection, Canberra, ACT, Australia

TISN (2007) Critical infrastructure protection modelling and analysis (CIPMA) program. Trusted Information Sharing Network (TISN), Attorney-General's Department, Australian Government, Canberra, ACT, Australia

TISN (2012) Diagram of the TISN. Trusted Information Sharing Network (TISN). http://www.tisn.gov.au/Pages/the_tisn.aspx. Accessed 16 June 2012

US Army TRADOC (2006) DCSINT handbook no. 1.02, critical infrastructure threats and terrorism. US Army Training and Doctrine Command (TRADOC), Fort Leavenworth, Kansas, USA

Van der Ploeg F, Poelhekke S (2008) Globalization and the rise of mega-cities in the developing world. Cambridge J Reg Econ Soc 1(3):477–501

Victorian Government (2003) Terrorism (community protection) Act 2003. Parliament of Victoria, Melbourne, VIC, Australia

Writer L (2011) The Australian book of disasters. Murdoch Books, Millers Point

Chapter 15
Latvian Practices for Protecting Water and Wastewater Infrastructure

Anatolijs Zabašta, Tālis Juhna, Kristina Tihomirova, Jānis Rubulis and Leonīds Ribickis

Technical Terms and Definitions

AMR	Automated meter reading
AOC	Assimilable organic carbon
BDOC	Biodegradable dissolved organic carbon
CEDM	Centre of emergency and disaster medicine
CERT	Latvian Information Technology Security Incidents Response Institution
CHA	Hydrophilic charged
CI	Cast iron
CPL	Civil Protection Law
DBP	Disinfection by-products
DOC	Dissolved organic carbon
DS	Distribution system
ESL	Epidemiological Safety Law
EU	European Union
GPRS	General packet radio service
GSM	Global system for mobile communications
HS	Humic substance
IC	Infectology Center
ICT	Information and telecommunication technology
IIS	Information systems
ISP	Internet service provider

A. Zabašta (✉) · T. Juhna · K. Tihomirova · J. Rubulis · L. Ribickis
Riga Technical University, 1 Kalku Str, Riga 1658, Latvia
e-mail: Anatolijs.Zabašta@rtu.lv

T. Juhna
e-mail: Talis.Juhna@rtu.lv

K. Tihomirova
e-mail: Kristina.Tihomirova@rtu.lv

J. Rubulis
e-mail: Janis.Rubulis@rtu.lv

L. Ribickis
e-mail: Leonids.Ribickis@rtu.lv

R. M. Clark and S. Hakim (eds.), *Securing Water and Wastewater Systems*, Protecting Critical Infrastructure 2, DOI: 10.1007/978-3-319-01092-2_15,

IT	Information technology
LD	Loose deposits
MW	Molecular weight
NEU	Hydrophilic neutral
NOM	Natural organic matter
NPOC	Non-purgeable organic carbon
NSL	National Security Law
NSSC	National State Security Council
OM	Organic material
PE	Polyethylene
PVC	Polyvinylchloride
RPM	Resuspension potential measurements
SHA	Slightly hydrophobic acids
SRD	Short range devices
TW	Talsi water
UDF	Unidirectional flushing
VHA	Very hydrophilic acids
WDN	Water distribution networks
WRT	Water retention time
WSN	Wireless sensor networks

15.1 Introduction

15.1.1 Government Policies Concerning the Protection of Critical Water and Wastewater Infrastructure

The Republic of Latvia has promulgated a number of laws that govern the national response to emergencies in water and wastewater. The Republic of Latvia has passed an Epidemiological Safety Law (ESL 1999) which describes the procedures for the evacuation and the isolation of patients during emergencies. If during routine analyses a company identifies a contamination event, it has to determine the preventive action to be taken. According to the Epidemiological Safety Law if smallpox, botulism, tularemia, plague or anthrax is identified in a person, the hospital is required to inform the Infectology Center (IC) of Latvia and isolate the patient until IC specialists take over the case. If additional infected persons are identified the Centre of Emergency and Disaster Medicine (CEDM) should be informed. CEDM is responsible for assessing the level of threat to the public (local, regional or national) and to take necessary actions.

The Civil Protection Law (CPL 2006) applies to disaster management, and the provision of legal and organizational support for the protection of public, property, and the environment in the cases of a disaster. The main tasks of the law are as follows: To carry out disaster management; to provide aid to victims of disasters;

to reduce the possible damage to property, and the environment caused by the disasters; and if a military invasion or war occurs—to support the National Armed Forces with resources.

A National Security Plan regulates the responsibility of involved authorities for carrying out disaster management plans. The State Cabinet of Ministers is responsible for risk management and remediation of the consequences from terrorist attacks.

Latvian legislation does not require every water supply company to have a plan in case of a terrorist attack. If a disaster occurs, the company would try to solve this problem using their own means. Should that prove to be impossible the company is obliged to inform the National or State Security Council (NSC), which consists of representatives of the Latvian Government and the President. The head of the NSC is the Prime Minister. If during a routine analyses a contamination event is identified a company has to decide which preventive action should be taken (e.g., closing a part of distribution system to prevent further contamination). All water supply companies must have an Action Plan for emergency situations in case of a limited water supply. No specific guidelines covering the technical measures to be taken for drinking water supply system decontamination are in existence. If the company cannot provide the necessary amount of drinking water to the customers from the water supply system, they must provide it in containers. An emergency technical council meeting is convened to identify the reasons for the emergency situation. The rapid involvement of the appropriate services to solve the accident is employed and the consequences of the incident are identified.

According to the National Security Law (2000) the Cabinet of Ministers is responsible for risk management and its aftermath. In case of a threat the crisis management council coordinates civilian-military cooperation, and directs the government institution's operational risk management measures.

From March 2001 the Law "On Pollution" has been in force. The purpose of the law is to prevent or reduce the impact of pollution on human health, to minimize property and environmental damage, and to prevent negative consequences. Based on the Pollution Law "Regulations on emissions of pollutants in water" was accepted by The Cabinet of Ministers in 2002. These Regulations prescribe effluent emission limits and prohibit pollutant emissions to water, including the order in which the operator controls the emissions of pollutants in water and so on.

15.1.2 Characteristics of the Water Utility Industry in Latvia

The Latvian Water Supply and Sewerage Enterprises Association (Krauze 2011) unites 27 enterprises, located mostly in former regional centers. An average amount of water supplied to a customer in 2010 was about 215 thou cu m (thousand cubic meters–7591.6 thou cu ft) per day, but the average amount of wastewater exceeded 260 thou cu m per day (9180.6 thou cu ft per day). The total length of water distribution pipe was 2965 km (1842.5 mi); the total length of sewer pipe was 2560 km (1590.8 mi). The members of the Association comprise 90 % of the Latvian water utility market, but the remaining 10 % is divided among small private companies. Over 71 % of water services are consumed by residents living in apartment buildings and individual houses, but the remaining 29 % are used by different industries and the public sector.

Table 15.1 indicates the large differences in water production among water utilities; therefore, the water volume processed by the Riga Water exceeds 60 % of the total association contribution, while the contribution of 14 water utilities is less than 1 %.

In 2009 Latvia moved from a two-level local governmental (regions, municipalities, and cities comprised the first level, but parish councils comprised the second-level of local government) structure to a local level—110 counties and 9 cities (Daugavpils, Jelgava, Riga, Jurmala, Liepaja, Rezekne, Riga, Valmiera, and Ventspils). Before the administrative reform (ATRL 2008) water management had been operating in each parish or town. Water infrastructure ownership has been shifted from the parish level to county and municipal ownership. Therefore,

Table 15.1 Supplied water and wastewater volumes distribution among Association's enterprises in the year 2009 (Krauze 2011)

Nr.	City	Water supply (thou cu m/day)[a]	Wastewater purification (thou cu m/day)[a]	W + WW (thou cu m/day)[a]	W + WW % of total volume
1.	Rīga	140,851	160,381	301,232	60.54
2.	Liepāja	10,654	26,760	37,414	7.52
3.	Daugavpils	15,604	13,898	29,502	5.93
4.	Jūrmala	10,160	9,832	19,992	4.02
5.	Jelgava	11,500	7,900	19,400	3.90
6.	Ventspils	7,632	10,342	17,974	3.61
7.	Rēzekne	4,520	7,297	11,817	2.38
8.	Valmiera	3,821	4,545	8,366	1.68
9.	Cēsis	3,017	3,805	6,822	1.37
10.	Jēkabpils	3,179	3,436	6,615	1.33
11.	Tukums	1,851	2,915	4,766	0.96
12.	Talsi	1,930	2,108	4,038	0.81
13.	Bauska	1,609	2,228	3,837	0.77
14.	Saldus	1,411	2,083	3,494	0.70
15.	Sigulda	1,647	1,750	3,397	0.68
16.	Aizkraukle	1,398	1,986	3,384	0.68
17.	Kuldīga	1,095	2,059	3,154	0.63
18.	Dobele	1,301	1,481	2,782	0.56
19.	Gulbene	711	1,429	2,140	0.43
20.	Līvāni	901	1,217	2,118	0.42
21.	Madona	930	791	1,721	0.35
22.	Alūksne	766	903	1,669	0.34
23.	Vangaži	500	500	1,000	0.20
24.	Limbaži	320	623	943	0.19
	Totals	227,309	270,268	497,577	100.00

[a] To convert from thou cu m/day to thou cu ft/day multiply by 35.31

counties have faced an important problem on how to organize water services throughout the county area. It has been found that when each parish administration deals with the problem itself (in rare cases getting contributions of budget and resources from the county), water quality is unacceptable, and financial and human resources are not used optimally. The majority of parishes do not have sufficient competence and lack the resources to maintain adequate water supply infrastructure even when built with the support of EU Funds. Therefore, the various county governments have started creating a single system for providing water services across the jurisdiction.

A study carried out by the authors covering three counties in the Kurzeme region revealed common problems (Zabašta 2010). Each parish independently performed water accounting and obtained payments from customers, so that the county administration did not have correct information concerning the overall situation in the county. Since each parish maintained its own customer billing and property accounting system, the county administration was not able to provide a common policy in relation to clients and debtors, due to a lack of timely information. A significant part of the municipal property was not equipped with water meters at the entrance to the building; thus, water consumption in many cases was determined by the consumption standards per person and sometimes by the number of animals owned by landlord. Different water tariffs were applied, which were not determined on the basis of actual costs. As a result of privatization formerly public water supply and sewerage infrastructure, in many parishes, ended up in private hands and the new owners were able to charge at any level they desired, because the actual cost burden was not corroborated in the Land Register.

In many cases water supply facilities and the trunk network (pumping stations, iron removal plants, water main, sewer pump stations, etc.) in parishes are not equipped with water supply monitoring and record keeping equipment; as a result leaks are detected with delay. In some counties water supply network depreciation has reached 70 % (Zabašta 2010).

There are several problems that are specific to both the county and urban water industry. Starting in 1990 both residential and industrial water consumption has been steadily decreasing. In the last 20 years water consumption has decreased significantly increasing the cost of supplied water per m^3 (Krauze 2011). Another problem is customers growing debt; however, current legislation does not permit disconnecting water, even if a subscriber does not pay their water bill. The existing legislation affecting water supply and sewer facilities regarding financing and construction is not conducive to the use of high-quality materials and advanced technologies, because only construction costs are taken into account and operating costs and facility life are neglected.

Since Latvian and Lithuanian and Estonian water utilities have encountered similar problems, the two countries have joined efforts in order to introduce information and telecommunication technology (ICT) solution for monitoring and controlling water distribution networks (WDN). For example in 2010, four Latvian and one Lithuanian counties initiated a project "Innovative e-services for water supply management" (E-Water 2010).

15.1.3 Primary Water Sources and Treatment

Due to a relatively cold climate and an abundance of soils rich in organic carbon (Broo et al. 1999), the concentration of raw water natural organic matter (NOM) in the Boreal region (Latvia) is high and its removal during conventional water treatment is complicated. The high concentration of NOM can be responsible for objectionable tastes, odors, and color in drinking water, and can be a factor in the formation of potentially carcinogenic disinfection by-products (DBPs) after reacting with disinfectants used in water treatment, and it affects biological stability and biological regrowth in distribution systems.

A water distribution system (DS) can be characterized as a biological and chemical reactor that interacts with water in the network (Gauthier et al. 2001) which results in water quality changes during its transport. These changes are dependent on many factors including the size and complexity of the DS, etc. (Servais et al. 1992). NOM's influence on water quality change is usually related to biodegradable organic carbon (BDOC) in drinking water. This may lead to the regrowth of microorganisms and coliforms, including opportunistic pathogenic bacteria in drinking water DS (Van der Kooij et al. 1982; Servais et al. 1992) and may increase the health risk for immunocompromised people. However, water with high levels of NOM contains humic substances (HS) and slowly degradable BDOC (Eikebrokk et al. 2007), which may influence water quality as well. It has been shown earlier that HSs absorb on the surface of corrosion products in iron pipes (Camper 1996), whereas slowly biodegradable compounds of NOM in water are used as substrate for bacteria. NOM constitutes a large fraction of loose deposits (LD) found in the distribution network (Gauthier et al. 2001).

15.1.4 The Problems of Monitoring Water Consumption and Identification of Leakages

Latvian water utilities are faced with growing operational and maintenance costs as a result of aging infrastructure. Leaks, ruptures in water supply pipelines, and water theft are very expensive, thus monitoring and repairing underground infrastructure presents a severe challenge.

During the Soviet period in which the economy was centrally planned, residential water consumption in Riga, Latvia, increased continuously. Water use for the population of 886,000 inhabitants was as high as 450,000 m^3 per day (15.9×10^6 ft^3 per day) at that time. The major drinking water consumption was for the vast industrial sector, which covered almost all of the expenses spent for water production and delivery of drinking water, including uncounted-for water. Household consumption was estimated according to inappropriate standards. The real cost of the water supplied for residential consumption was actually much

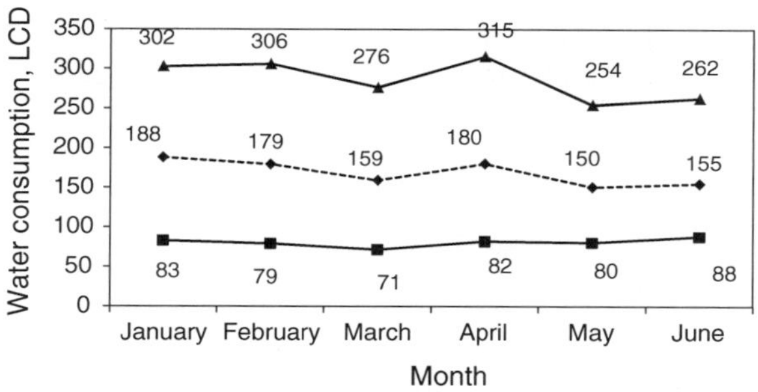

Fig. 15.1 Water consumption in liters per capita per day in apartment buildings in Riga in 1999. Water consumption is depicted: by consumers with individual meters (*solid line* with *squares*), by the municipal water supply company (*dashed line* with *diamond*) and by consumers which paid according to flat-rate tariffs (*solid line* with *triangles*)

higher than the cost charged to the household consumer, resulting in overuse by the consumer at the tap.

After Latvia achieved political independence, the market economy caused a recession in the country and industrial demand declined dramatically. As a result, the domestic sector became the major drinking water consumer. To ensure payments that covered the real cost of water, the municipal water supply enterprise of Latvia began installation of water meters in apartment buildings on the inlet pipes to buildings, while citizens were encouraged by the city council to install individual water meters in their flats. With this double accounting system (Fig. 15.1), it soon becomes clear that there was an enormous discrepancy between the volumes of water measured by the building water meter and that charged to residents with individual water meters.

Depending upon the building approximately 40–80 % of all consumers used apartment or flat meters (Rubulis et al. 2001). Even in a model apartment building in which all flats were supplied with impulse water meters together with accessories for data processing, utilizing ZENNER® interface and software products, there were discrepancies of 6–23 % between the total water amount measured by the building meter and that estimated by summing measurements from water meters in the apartments (Rubulis et al. 2001).

An inspection of water meters in apartments showed that 10 % of all the meters were installed on horizontal pipes with a horizontal clock-face, 15 % were installed on vertical pipes, and 75 % were installed on horizontal pipes with a tilted clock-face (Fig. 15.2). There are two main reasons why 75 % of water meters were installed on horizontal pipes with tilted clock-face:

1. Internal water pipes in apartment buildings which were built during the Soviet period were not suitable for the proper installation of water meters because the space between cold and hot water pipes as well as between the pipe and the wall was not sufficient for installation of water meters with a horizontal clock-face;

Fig. 15.2 Typical solution for installation of water meter with rotated clock face in internal piping

2. Easier access to the water meter (in most cases) occurs when it is located under the sink and it is also easier to record readings.

The performance of water meters on a special test-stand in the laboratory showed that the meters characteristics are significantly influenced by the type of installation (Fig. 15.3.) and all water meters met the Class A standard in cases when they were not installed on a horizontal pipe with a horizontal clock-face and minimal flow rate was 0.06 m^3/h (2.12 ft^3/h).

- The position of the meter clock-face is given as the angle (degrees) from the vertical axis of meter. Briliute et al. (2008) tested different types of water meters and concluded: similar to the results in Riga, the rotation around the pipe axis of the meter (single-jet or multi-jet vane wheel) at the 45° and 90° angle causes additional error which at the low flow range which may be as high as 30 %, whereas in the case with the volumetric meter it achieves (1.5…2.0) % in the worst case. After rotation the vane wheel meter even at the 45° angle, it is not able to meet the requirements of class A;

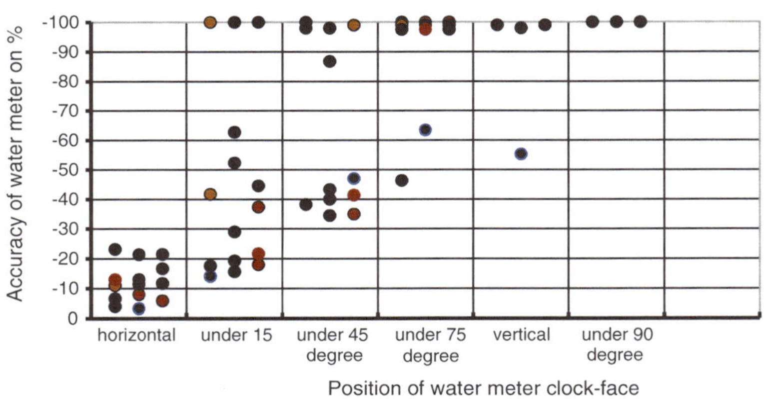

Fig. 15.3 Measurement accuracy in relation to installation style of the water meter

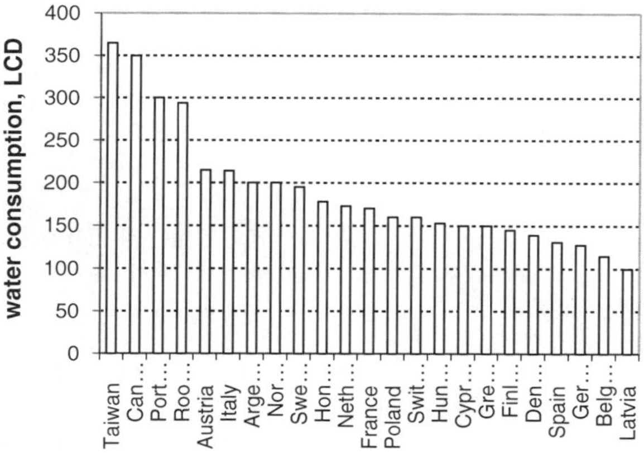

Fig. 15.4 Drinking water consumption worldwide

- when the meter is installed in the vertical pipe line, the additional error in the low flow range is almost as twice as great with the single-jet meter as compared to the multi-jet one, and may reach up to (25…27) %;
- installing the strainer upstream of the water meter has a minor influence (2 % at minimal flow), whereas when installing the strainer the bleeder plug should be directed downwards;
- installation of a strainer before the meter decreases the additional error (30…40) % of water meters due to particles which may interrupt rotation of the vane wheel;
- reducing the gasket bore inner diameter for the single-jet vane wheel meter installation always results in additional error. In the case analyzed, reducing of the gasket bore to 8.7 mm (0.03 in) led to an increase in the measurement error up to (20…21) %. The multi-jet concentric vane meters and the volumetric meters are not sensitive to reducing the gasket bore diameter (Briliute et al. 2008).

The usage of water meters with nominal capacity, 1.5 m^3/h (18.04 ft^3/h) led to water savings by consumers (Fig. 15.4) in Latvia and together with improper performance significantly decreased the total amount of measured drinking water (Fig. 15.5). As an example of the problem associated with correct meter reading is the fact that a water utility, which supplies drinking water to 16,000 householders in one Latvian city with 45,000 inhabitants, estimated its loss of revenue to be approximately 1,140,000 EUR in 2009 (Zabasta et al. 2011). Adding another 20–50 % to the consumed water volume shown on customer bills (IESM 2011) might be difficult. Currently, the absolute majority of water flow meters are still

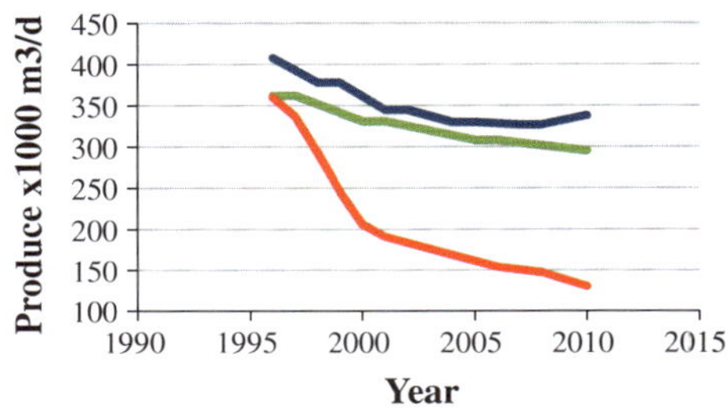

Fig. 15.5 The forecast of company "Rust VA-Projekt" (*blue line*) and "Sweden water" (*green line*) of produced drinking water amount in Riga city and actual delivered water to the city (*red line*)

monitored by visiting sites or by "walking-by"; clients frequently deliver metering data to water utilities by phone, by post, or provide them "in hand."

Therefore, due to the lack of reliable data concerning the state of WDN it is difficult to identify local problems and to minimize losses caused by water leakage, illegal connections, and customer fraud (IESM 2011). Unfortunately utilities owned by municipalities, which suffer from an economic crisis, are reluctant to invest in the development of WDN monitoring tools and methods for leaks detection.

15.1.5 Cyber Security

Nowadays, in the water sector the internet is used mostly for customers' information about tariff, emergency services, and customer services. Due to the installation of automated meter reading (AMR) systems in municipal WDN the role of ICT is growing rapidly. The internet is used not only to present the utility's home page, but as a web interface for access to metering data and to services, and therefore the number of system users are growing dramatically. The other aspect of the growing use of the Internet is the increasing dependence from telecommunication and internet services providers (ISP) due to a growing volume of transmitted data. Therefore, data security and integrity has become a crucial aspect for Latvian utilities.

According to the Latvian IT Security Incidents Response Institution (CERT) about 3,300 IP addresses were infected (by bots) in the year 2011. In the year 2011 CERT dealt with more than 15,000 security incidents (Kaškina 2012). Websites of large state institution are compromised on a regular basis, for example: Latvian Health Emergency Service and Riga Municipal Police in October 2011, and energy monopolist Latvenergo in April 2011.

Internet and telecommunication security issues are regulated by the "Information technology security law" (promulgated in 2010) and by the government regulation "Information technology critical infrastructure security measures planning and implementing" which was promulgated in 2011. One of the security measures was creation of the state funded CERT in February 2011. The CERT responsibility is delegated to the Institute of Mathematics and Computer Science,

University of Latvia. The scope of CERT's responsibilities covers public sector institutions, ISPs, and Critical IT infrastructure owners. These institutions are obliged to submit an "Action plan for continuous operations" and to report the major security incidents. CERT has the right to request IT security documentation, to initiate IT security audits and even to disconnect an end user (Andžans et al. 2012).

The National IT Security Council coordinations tasks and performance measures set by the IT Security Law. The Council is chaired by a representative of the Ministry of Transport and Communications. The members of the Council represent Latvian ministries, Bank of Latvia, CERT, and state security agencies. The application of security monitoring systems appears to be, that because of the quick acquisition of ICT tools water utilities will likely be able to meet new security challenges in the future.

15.1.6 Literature Review

NOM is one of the components connected with biofilm growth and heterotrophic bacteria proliferation, because it is a substrate for microorganism growth and regrowth. NOM is closely related to DOC and part of DOC supports exoenzymatic hydrolysis reactions and includes the biodegradable fraction (BDOC) which includes rapidly and slowly hydrolyzed classes of BDOC (Servais et al. 1992, 1995).

Many researchers have shown a correlation between assimilable organic carbon (AOC) and bacterial regrowth in distribution networks (see e.g., Van der Kooij and Hijnen 1991). AOC is related to coliforms and heterotroph regrowth in well-mixed system (Camper 1996). Fixed bacteria are linearly correlated to suspended bacteria in water (Servais et al. 1992) and regrowth cells and water residence time in the distribution network (Hammes et al. 2010).

The NOM fraction as humic substance (HS) can act as a substrate and an energy source for biofilm formation and the number of cultivable bacteria in biofilm have shown to be similar to that which was produced on an organic substrate as amino acids or carbohydrate (Camper 2004) and its concentration can be limiting for biofilm formation (Servais et al. 1995). Some researchers have observed a loss of cultivability at low DOC concentrations (Boualam et al. 2002). The BDOC fraction is a complex mixture of organic carbon and is available as substrate for bacteria. After the reaction with disinfectant it forms DBPs (Butterfield et al. 2002). Thus, BDOC is a major factor for control of bacterial growth (Servais et al. 1992, 1995) and for the biological stability of drinking water, especially in distribution networks with long WRT (Raczyk-Stanisławiak et al. 2004).

Drinking water contains organic and inorganic matter, which can accumulate at pipe surfaces in drinking water pipelines (Gauthier et al. 1999; Zacheus et al. 2001). After some years of use of distribution network, pipes are covered with deposits of corrosion products (Sarin et al. 2001) or soft and LD (Zacheus et al. 2001; Prevost et al. 2005). Changes of water flow or pressure in network can detach deposits which, in turn, increase turbidity, electrical conductivity, bacterial numbers in water; thus, deteriorating drinking water quality (Mustonen et al. 2008).

Some studies (Sarin et al. 2001, 2004) have shown that the structure of iron pipes deposits can be divided into four layers: Corroded floor, porous core, shell-like layer, and surface layer. A corroded “floor” forms on the metal surface during corrosion. A porous core layer is composed of small particles of different iron solids as iron (II) hydroxide, oxide, carbonate, or dissolved iron. A shell-like layer contains iron-oxidizing bacteria, and finally, the top surface layer is in contact with water and is largely influenced by the water quality. The top surface layer contains amorphous iron (III) hydroxide, silicate, phosphates, carbonates, and NOM molecules; thus, it can absorb bacteria from drinking water and form biofilms (Herro and Port 1993; Sarin et al. 2001, 2004; Prevost et al. 2005). The accumulation of bacteria and NOM on pipeline surfaces and the formation of deposits depends on the microbiological and chemical quality of drinking water in the distribution network: surface material, application of corrosion inhibitors at the water treatment plant, pH, alkalinity, mineral concentration, temperature, microbiological activity, as well as the type and concentration of NOM (Tuovinen et al. 1980; Zacheus et al. 2001; Sarin et al. 2001, 2004; Ndiongue et al. 2005; Prevost et al. 2005). In addition, biofilm may promote the deterioration of metallic pipes (bio-corrosion) and the dissolved organic compounds and humic colloids present in water act as catalysts in cast iron corrosion (Tuovinen et al. 1980; LeChevallier et al. 1993; Korshin et al. 2005). NOM interaction with pipe surfaces includes such processes as sorption, precipitation, dissolution, aggregation, comlexation, and degradation, which depend on different parameters, such as water pH, nutrient with electrolyte properties concentration (e.g., KNO_3), reaction time or WRT, carbonate, and calcium concentration (Ca^{2+} forms complexes with negative groups of NOM) (Day et al. 1994; Kaplan and Newbold 2000; Korshin et al. 2005). The reactivity of NOM depends on the physicochemical properties of organic matter such as molecular weight (MW), aromaticity, functional groups content, and hydrophobicity/hydrophilicity (Larson 1966; Sontheimer et al. 1981; Broo et al. 1999; Swietlik et al. 2004; Korshin et al. 2005). Many studies (Gruškeviča et al. 2008) have shown that non-humic hydrophilic NOM fractions are more corrosive than HS in copper and iron pipes at stagnant conditions and this effect is stronger in iron pipes.

The deployment of wireless sensor networks (WSNs) for monitoring water flow, pressure, and vibration at a large number of locations was proposed by (Lin et al. 2008). WSN can address the challenge of near real-time monitoring and eventually system control. The ability of a cross correlation algorithm to locate a leak in a pipeline when two sensors detect a leak was tested by Stoianov et al. (2007). Several leak detection methods are described by Hunaidi and Wang (2006), for example, water flow and pressure monitoring by measuring the minimum night flow rate on a continual basis and a system for locating pipe leaks based on the cross-correlation method. A pipe detection method based on the observation of abnormalities in water flow and pressure is presented by Bicik et al. (2011).

15.2 Research and Practices

15.2.1 Research on Control of Bacterial Regrowth

Drinking WDN consists of many different types of material (polyethylene—PE, polyvinylchloride—PVC, cast iron—CI) but water residence time depends on constant factors as pipe length and diameter, and also a variable factor—the water consumption. Networks contain different types of LD and biofilm depending on the age of the water supply system and its water quality; therefore, it can be defined as a chemical and biological system (Bose and Reckhow 2007). It is known that LD in a network also contains different concentrations of NOM (Amy and Her 2004), where bacterial regrowth in the distribution network takes place (Brandt et al. 2004). Therefore, the goals of this research were (i) to evaluate NOMs ability to accumulate in LD in different points in the distribution network and its influence on water quality in the distribution network; (ii) to evaluate the influence of NOM on water quality in the network.

The concentration of TOC and BDOC depending on WRT and the biodegradation kinetics of organic carbon in water samples from different points of the 1374 km (853.8 mi) distribution network in Riga city, Latvia (Fig. 15.6) was determined. The concentration of TOC ranged from 2.2 to 8.1 mg/l measured at different places ($n = 30$) in the network in Riga using a submersible, two beam spectrophotometer spectro::lyser™ (s::can Meßtechnik GmbH, Vienna, Austria) with on-line UV/Vis measurement. Results showed strong positive correlation between TOC and WRT in the distribution network (Moment Correlation coefficient, $r = 0.81$; Fig. 15.7a).

To assess BDOC changes in the drinking water distribution network data were collected during two sampling periods (Fig. 15.7b). The concentrations of BDOC ranged from 0.33 to 2.38 mg/l ($n = 25$) in all samples. Curve A displays BDOC changes in sample values for the April test series, when the water temperature ranged from 10 to 16 °C (50–61 °F). The exponential decay rate for BDOC in the network was equal to 0.05 h^{-1} in this period. Curve B shows the BDOC data from samples collected from different periods (February–July). Temperature changes during this period were greater from 9 to 21 °C (48.2–69.8 °F), and the biodegradation rate was higher also (0.31 h^{-1}). BDOC changes indicated that the biodegradation process in water during colder periods was slower than during the warmer periods. BDOC ranged from 0.85 to 1.79 and the degradation rate in the network was 6 times higher in seasons with high temperatures. Statistical data indicated that WRT determined by using hydraulic models can be divided in three phases, which display strong negative correlations with the concentration of BDOC in the distribution network for periods 2.82–13.72 h and 14.27–17.62 h ($r = -0.72$, $n = 8$ and -0.74, $n = 6$, respectively) and no correlation between BDOC values and WRT > 18.03 h ($n = 11$).

The biodegradation rates determined in samples taken at different places in the distribution network in Riga ranged from 0.24×10^{-2} to 3.15×10^{-2} min^{-1}

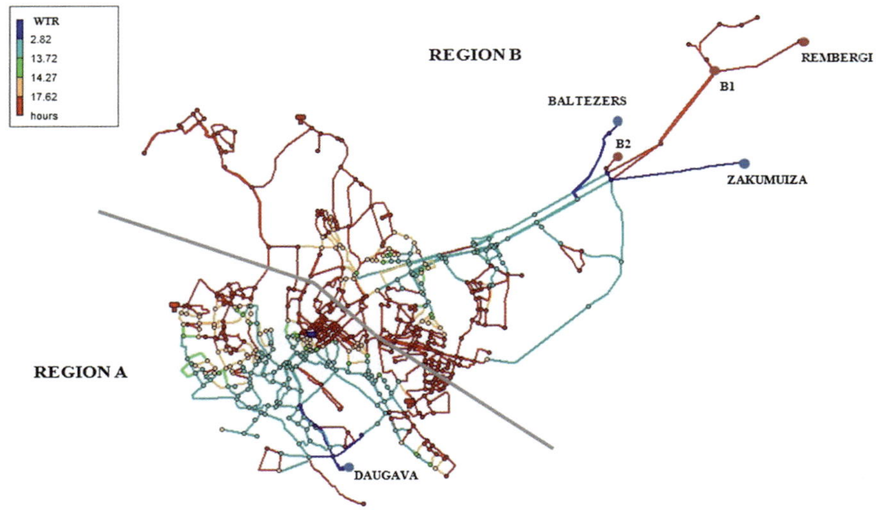

Fig. 15.6 Service areas of Riga water treatment plants in EPANET 2.0 with three reservoirs and WTP viz. Baltezers, Zakumuiza, Baltezers-1 (*B1*), Baltezers-2 (*B2*), Rembergi, and Daugava. *Region A* supplied by Daugava WTP, *region B* supplied by Baltezers, Zakumuiza, *B1*, *B2*, and Rembergi

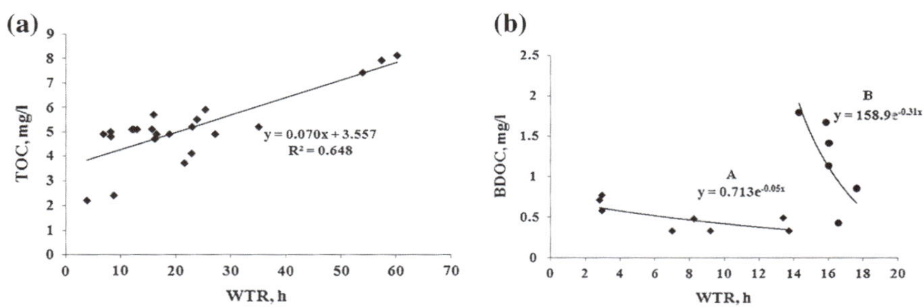

Fig. 15.7 TOC concentration in water samples correlation with WRT in distribution network (**a**); BDOC concentration changes in water samples depended on WRT (**b**)

(n = 13) and there was a strong correlation with concentration of BDOC in these samples (r = 0.78).

The decrease in BDOC concentration for the first hours (<24 h) within distribution network can be explained by consumption of substrate by bacteria. For more distant areas in network (>24 h) the broad range of BDOC values and increases in TOC indicate the leaching of NOM from deposits within the distribution network. Results showed that DOM is an important parameter and can be used as indicator to evaluate drinking water quality depending on the WRT in distribution network.

The DOC concentration ranged from 2.5 to 7.7 mg/l in drinking water samples from the DS during the study period. To determine the NOM concentrations in LD, 38 samples were collected and the average amount of NOM measured as NPOC was determined. The NPOC in deposit samples collected from the DS supplied with groundwater varied from 0.18 to 21.01 mg/g ($n = 21$). In the samples collected from the DS supplied with surface water (River Daugava) NPOC ranged from 1.21 to 18.99 mg/g ($n = 12$). In the samples collected from different drinking water reservoirs in Latvia, NPOC values ranged from 0.20 to 3.11 mg/g ($n = 5$).

Results obtained from the study (Gruškeviča et al. 2009) did not show any significant relationship between NPOC and any other parameters (WRT in DS, diameter of pipe, source of raw water) studied, except for the pipe materials. PVC pipes showed an average 1.48 ± 1.77 mg/g (ranged from 0.18 to 3.50 mg/g); CI pipes—4.36 ± 3.13 mg/g (ranged from 1.83 to 14.33 mg/g); but PE pipes—14.29 mg/g (ranged from 9.99 to 21.01 mg/g). It appears that organic matter (OM) originates not so much from the source as from the pipe material, which leaches organic carbon substances. CI pipes are the oldest pipes (about 50 years) in the networks, whereas PE and PVC pipes are not older than 15 years. Thus, the NOM concentration in deposits depends on the pipe material itself, rather than the length of exposure to organics (PE and PVC).

Rapid fractionation results (Fig. 15.8) showed that the LD samples obtained from the DS which was supplied from groundwater OM were predominantly composed of VHA (44 %) and other fractions were distributed as follows: SHA—12 %, CHA—13 %, and NEU—31 %. In the samples obtained from surface water supply DS dominant NEU was the main fraction (41 %) and other fractions: VHA—24 %; SHA—22 %; and CHA—13 %. This can be explained by the fact that the VHA and SHA fractions were separated by 70 % (Fig. 15.8) during the water treatment process.

The NOM balance in the drinking water distribution network is described in Fig. 15.9, based on results presented in this research and published earlier (Servais et al. 1992; Frimmel 1998; Gauthier et al. 1999; Rubulis et al. 2007; Vreeburg 2007).

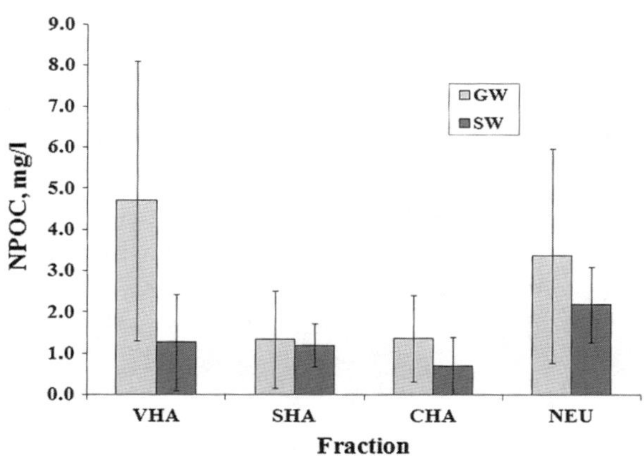

Fig. 15.8 RF results for isolated organic matter samples from the LD obtained from systems supplied by groundwater (GW, $n = 6$) and surface water (SW, $n = 5$)

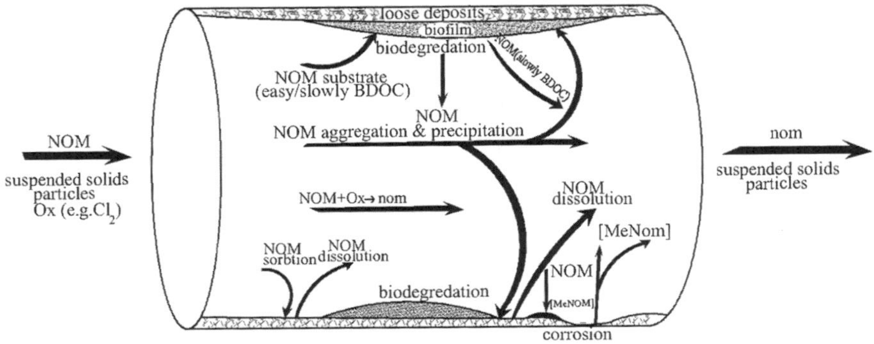

Fig. 15.9 NOM balance in drinking water distribution network

The network is the reactor where biological and chemical processes take place and pipes are covered with LD and biofilm in the distribution network. NOM molecules can be transported through the network. They have been involved in oxidation reactions (with disinfectants used in water treatment), corrosion processes (form complexes with metals) and can be used as substrate for microorganisms (biodegradation). Results showed that the TOC concentration increased in distribution network with increased WRT (caused by NOM dissolution, deterioration of the LD, and leaching from the pipe walls) and that BDOC concentration decreased with increased WRT (consumption of substrate by bacteria).

Hammes et al. (2010) showed that cell concentrations increased with increased WRT in the distribution network in Riga. Another study (Boualam et al. 2002) has shown the loss of cultivability of heterotrophic and coliform bacteria in water at low DOC concentrations (0.5 mg/l), suggesting that DOC could be the limiting factor for bacterial occurrence in drinking water. Thus, TOC concentrations (from 2.2 to 8.1 mg/l) determined at different places ($n = 30$) in the distribution network in this Chapter are high. There is a strong correlation between TOC and WRT which showed that substrate available for bacteria in the network increased with a decrease in BDOC suggesting that substrate was partially consumed.

A previous study (Gruškeviča et al. 2009) showed that LD samples collected from the DS contained goethite and quartz. NOM sorption isotherms for all media were not linear and the sorption rate was faster than desorption (Chi and Amy 2004), indicating that NOM forms relative strong bonds with iron surfaces. Thus, different amounts of nonpurgable organic carbon (NPOC) (from 0.18 to 21.01 mg/g, $n = 38$) in samples confirmed that NOM was accumulated in the LD. As the result of changes in hydraulic conditions or pressure shocks (Mustonen et al. 2008), NOM together with LD may be detached from the pipes leading to aesthetic complains of consumers.

15.2.2 Practical Examples of Flushing a Water Distribution Network

Drinking water discoloration is the most reported complaint by consumers of drinking water supply companies (Vreeburg and Boxall 2007; Polychronopolous et al. 2003; Matsui et al. 2007) in regions where scarcity of drinking water is not so evident. In addition even in the centralized water supply systems of Latvia discoloration is quite often referred as a problem (Juhna 2007; Neilands et al. 2009). Unpleasant tastes and color of water significantly influences the confidence of consumer regarding the quality of tap water and the water company's service quality and causes consumers to want to pay less for their water (Kelay et al. 2008). Nevertheless, consumers in Latvia have fewer complaints with water suppliers (0.07 contacts/1000 pop/year) as compared with countries such as the Netherlands (0.5–1 contacts/1000 pop/year) and United Kingdom (4 contacts/1000 pop/year).

The accumulation and erosion of particles in drinking water system have been broadly described in the last years and there have been models developed models which describe the discoloration process (Ryan et al. 2008; Boxall et al. 2001). The main cause of particle accumulation in the drinking water network in Latvia is due to high iron concentrations in the source (HIL 2011) (background pollution) and inefficient treatment. Within the EU 6FP Techneua project the application of the resuspension potential method (RPM) developed in Netherlands (Vreeburg et al. 2004) for online turbidity monitoring in treatment plants and the unidirectional flushing (UDF) technique (Antoun et al. 1999; Friedman et al. 2002) for water quality control/improvement were carried out. These techniques were applied in several municipalities (population from >1,000–700,000; network length: 2–1,374 km; DN 100–DN 400). A mobile cart with installed online flow and turbidity meters and manometer, as well as a tap for grab samples (Fig. 15.10) was developed for this purpose.

Fig. 15.10 The mobile hand cart with equipment for UDF and RPM procedure in Latvia

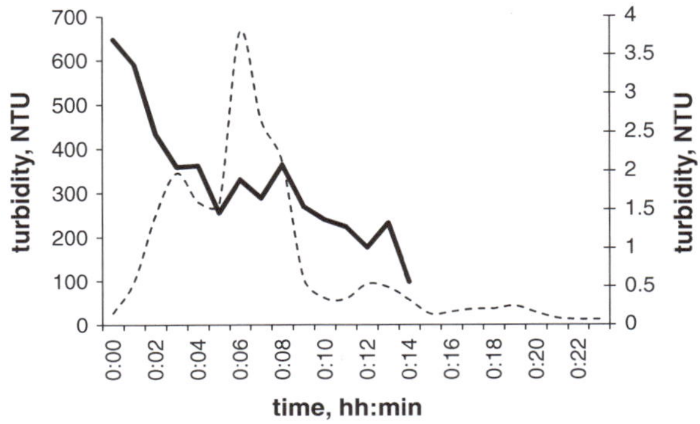

Fig. 15.11 Application of RPM before (*solid line* with values on primer *y-axis*) and after (*dashed line* with values on secondary *y-axis*) the network flushing

Innovative approaches like RPM, UDF (Fig. 15.11), and online monitoring in water treatment plants could be used effectively to better understand the functioning of a water supply system and to help water utilities control water quality deterioration in WDN.

The case study showed that effective flushing is possible even in a big city such as the capital of Latvia—Riga (population 700,000) to eliminate discoloration problems. While comparing the UDF in a city with a population of 700,000 with the municipality with a population less than 10,000 it can be concluded that flushing is much more difficult in the larger city because of the intense traffic, limited space, and minimal capability to release the flushed water and because of the very complicated distribution network.

The pictures showing observed technical problems during the flushing program are presented (see Figs. 15.12, 15.13, 15.14).

Fig. 15.12 Flooded and filed with sand *Riga* type hydrant (2 m in deep) due to raised groundwater level

Fig. 15.13 By third person damaged manhole of hydrant in Riga case study

Fig. 15.14 By third person buried hydrants with clay cover

15.2.3 Wireless Sensor Networks Application at the WDN in Talsi

The system piloted in the municipality of Talsi consisted of 347 water flow meters 19 water pressure meters, metering data reading and transmitting equipment, server, software for network monitoring, and metering data analysis and reporting equipment. One of the project's tasks was to work out an interface between existing Talsi Water (TW) bookkeeping and billing information systems (IS) and a new metering data base. The pilot project was implemented in cooperation with the Talsi municipality and TW. After project completion incremental improvements have being made in order to introduce new elements of WDN management (CWDN 2010).

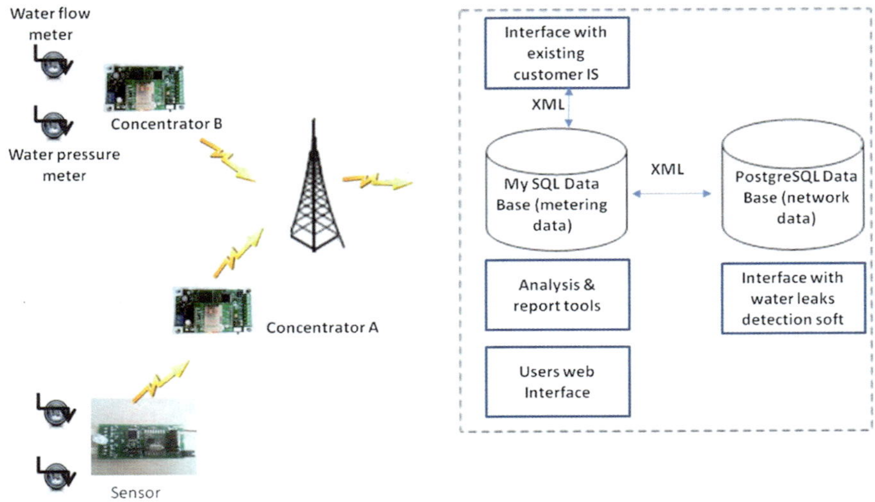

Fig. 15.15 A block scheme of a system for integrated water resources management (CWDN 2010)

A block scheme in Fig. 15.15 describes a system for integrated water resources management in a municipality.

Sensors—transmitters convert analog signals generated from water flow and water pressure meters into digital signals, process them into messages and send to gateways—concentrators using Short Range Devices (SRD) unlicensed telemetry band 868 MHz. Furthermore, a global system for mobile communication (GSM) network and a general packet radio system (GPRS) is used for data transmission between concentrators and a central server. MySQL database is used for metering data storage and processing, but the PostgreSQL database is used for WDN data storage and processing. The system's users: water utility staff, housing services providers, residents, and other clients accesses their data via a web interface.

A concentrator (gateway) shown in Fig. 15.16 periodically reads messages from meters and delivers data to the server via the GSM network using an IP protocol. The concentrator consists of: a microcontroller (Atmel), which controls the receiver and all concentrator operations, and stores metering data obtained from dedicated sensors. Receivers also use a short range devices (SRD) unlicensed telemetry band, 868 MHz. A TELIT GM864/865 is used as GSM modems for data exchange with the server. It has an internal processor and memory for programming in Python language.

In order to make the system more robust and to reduce power consumption the sensors—transmitters have only two regimes: registration of impulses from water meters and transmission of messages several times per hour; therefore, synchronization is not needed for the network's sensors. The approximate probability of message collision is about 1:10,000, so up to this point only a few cases have been

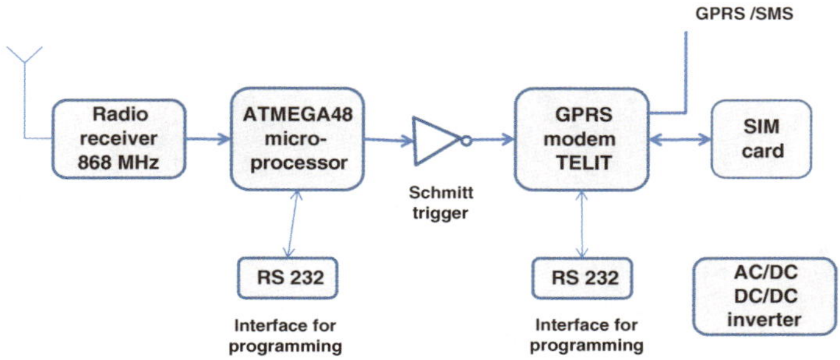

Fig. 15.16 A block scheme of A-type concentrator

noted. Such an approach has also helped to keep the cost of the system within the required boundaries, i.e., less than 35 euro per unit.

The power consumption of the sensors—transmitters was estimated to be capability to operate at least 10 years without battery replacement. Such an operational lifetime is achieved thanks to a 7,500 mAh battery (at the end of the project a new version of sensors—transmitters with 2500 mAh was created) and efficient operation of the transmitters. Depending on the settings a microprocessor switches on a transmitter two or more times per hour with a transmitting time of 20–50 ms. Between sessions the transmitter is switched off.

In order to diminish the risk of tampering with meters, which could be caused by magnets or other methods, a soft alert that occurs when hercon relay contacts show unusual behavior, was tested just after project completion.

Exploitation of GSM operator's networks helped to save capital investment and to achieve the best price. An agreement between the mobile operator and TW restricts a discounted volume of GPRS traffic for one concentrator by 10 Mbit per month. In order to reduce the traffic between concentrators and server text abbreviations and text cut are applied to data files, moreover files are split into short messages applying Md5 sum aiming to check integrity of files (MASOM 2010). Moreover, the initial TW requirement to transmit water pressure metering at 10 min intervals was revised; thus, data from the meters are now collected and transmitted each 30 min.

The system uses a Linux operational system, Ruby on Rails an open source full-stack web application framework and MyXQL database for storage and processing metering data. Free Open street map software is used for visualization of WDN topology. For the data integration with the third part IS systems (bookkeeping, billing, real estate management, etc.,) were used web-based XML requests.

The pilot project also had a target to create a framework for pinpointing leaks. This target was achieved by using hydraulic modeling tools that approximate defective sections of the water network. For this purpose a hydrodynamic model for network diagnostic application was created using a restricted number of nodes, pipes, and water flow and pressure meters. The model developed was linear, because there are no active elements such as pumps or tanks, which are depend on

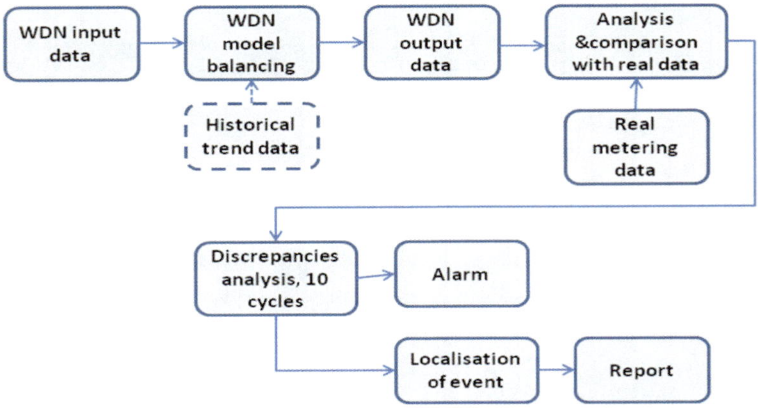

Fig. 15.17 An algorithm of network diagnostic

time or are caused by other nonlinearities; therefore, water flows freely from two reservoirs located at the highest points of the water network. The network structure was stored in a relational database as a directed graph. Using directed graphs allows for hydrodynamic model calculations and compares calculated and real metering data with the objective of identifying abnormities (for example, leakage events).

The diagnostic application calls on the WDN main application, which sends metering data as an XML file. The diagnostic application performs network modeling calculations and appropriate diagnostics and returns computational results as XML files, which is used by the network main program for visualization of results or for generation of alarms (for example SMS).

Figure 15.17 shows the WDN diagnostic algorithm, which comprises: metering data input, WDN model balancing, analysis and comparison with real data, and performs a discrepancies analysis providing for cycles, generation of alarms, identifying location of events, and reporting.

Due to the restricted scope and time of the project, the lack of historical records and due to the lack reliable data about WDN parameters, the diagnostic application was developed and tested as a trial model of the water distribution network. Although only approximate WDN parameters and conditional data were used for the simulation, the concept of diagnostic application proved its viability.

15.3 Discussions and Proposed Actions

15.3.1 Discussions About Results of Implemented Projects

The influence of high concentrations of NOM on the drinking water quality in the distribution network was examined in a project where the WRT in the supply system was determined using a hydraulic model, changes in the biological stability of water and NOM concentration changes in network, and in LD in the drinking water pipelines were taken into account.

Results suggest that NOM molecules are not simply transported through the network. Pipes are covered with deposits of corrosion products (Sarin et al. 2004) and NOM concentration in the deposits varies widely (Gauthier et al. 1999). Measurements of NOM concentration in the LD showed that it varies significantly, and its concentration level depends on the pipe material (e.g., PVC, PE, or CI) in the network. TOC concentrations increased in the distribution network with increased WRT indicating NOM dissolution, deterioration from the LD, and leaching process from the pipe walls. The BDOC concentration decreased with increased WRT indicating the consumption of substrate by bacteria. The BDOC reduction can be confirmed by previously published results (Van der Kooij and Hijnen 1991; Hammes et al. 2010) which have shown that the bacterial quantity in the distribution network increased with WRT as did the consumption of the AOC fraction. Simulated changes of water flow or pressure in network can detach deposits; therefore, result in increased turbidity, electrical conductivity, bacteria numbers in water (Mustonen et al. 2008) as well as TOC, thus deteriorating the drinking water quality.

The study has shown that grab sampling which is performed on a regular basis by water suppliers is not an accurate and effective method for determination of the risk of discoloration in the distribution networks. The resuspension of particles in distribution networks cannot be induced with current grab sampling procedures which use sampling from taps. Therefore, methodology of resuspension of particles like in RPM should be incorporated in the legislation.

The registration of complaints should be improved and consumers must always be informed about UDF.

The installation of telemetry at the Talsi WDN encouraged TW dispatchers to apply water flow and water pressure data monitoring in order to find abnormalities. They do it mostly at night-time, when the flow rate is minimal and any unusual event becomes more visible. TW dispatchers reported several cases, when water pressure meters incorporating remote monitoring helped them to identify pipe bursts and to identify local problem areas before leaks became visible. Water flow meter records also helped to explain to clients, why a particular monthly bill was larger than usually.

One of the problem issues is the reliability of the GSM network used for transmitting data between concentrators and a central server. Although the GSM network appears to have 100 % coverage and is transmitting almost 100 % of the time, delays in data transmission and unavailability of connections cause the system problems more frequently than other factors. A possible reason is the overload of the GSM network and lower priority of GPRS in comparison with voice and SMS traffic.

15.3.2 Suggestion for Standards and Legislations

The results showed that grab sampling alone in the distribution networks is not an adequate method for understanding discoloration problems, and tools such a RPM and UDF offers the possibility of evaluating and mitigating the turbidity risks in small and large cities.

The author of this chapter asked an officer of the Ministry of Environment Protection and Regional Development, which is the leading state administrative institution in the field of environment protection, about penalization utilities responsible for water loss in WDN. Since discussions regarding an emerging "Law on water services" have been started the issue of accountability for water loss could be one of the Law's topics. However, the Ministry still is reluctant concerning establishment of penalties due to monopolistic market for water supply services. Therefore, because customers are not able to change suppliers, losses, and penalties possibly would be assessed against clients as increasing tariffs for water supply services.

The following suggestions regarding the monitoring of drinking water consumption by residents should be considered:

- The application of impulse water meters or wireless sensor network together with data processing system;
- To overcome water theft due to usage of magnets the volumetric type of water meters should be used or alternatively vane wheel water meters with magnetic protection. Installing the water meters outside the flats could combat this problem as well as decrease the number of inaccurate readings;
- A solution against water theft could be the use of software to analyze the unusual behavior of meters;
- More sensitive water meters should be used (Class C).

15.4 Summary and Conclusions

NOM molecules are not simply transported through the network. The concentration of OC in the LD varies significantly, and its level depends on pipe material (e.g., PVC, PE, or CI) in the water distribution network. TOC concentrations in distribution networks increased with increased WRT indicating NOM dissolution, deterioration from the LD and leaching from the pipe walls. BDOC concentrations decreased with increased WRT indicating consumption of the substrate by bacteria.

The overall conclusion is that UDF is very effective method for reduction of turbidity and is more applicable in small communities then in large. In contrast to grab sampling, the RPM measurements showed the highest risk of discoloration risk in all municipalities.

The automatic wireless water meters reading system in Talsi town applies a two stage data collection and processing system; the sensors transmit data to the concentrators using SRD band 868 MHz, furthermore the traffic between concentrators and central server is ensured by GPRS. The system provides opportunities to collect data from difficult to access meters, to utilize existing mobile operator networks and to connect new customers without major investment. The increasing dependence on telecommunication and internet service providers (ISP), due to the growing volume of transmitted data, will force Latvian utilities to pay attention to data security and integrity aspects.

A hydrodynamic model for network diagnostic application has been created and simulated using an example model of a water distribution network. However, due to the limited scope of the project the network diagnostic application was tested using only a conditioned network segment, i.e., without application of historic and real data. Therefore, a new research project should be initiated in order to strengthen the initial research results and to make help make initial investments more effective.

References

Amy G, Her N (2004) Size exclusion chromatography (SEC) with multiple detectors: a powerful tool in treatment process selection and performance monitoring. Water Sci Technol: Water Supply 4(4):19–24

Andžāns M, Kaškina B, Teivāns V (2012) Situation regarding IT security in Latvia. The meeting with the State President in Riga Castle, 23 April 2012, pp 1–33

Antoun EN, Dyksen JE, Hiltebrand DJ (1999) Unidirectional flushing: a powerful tool. J AWWA 91(7):62–71

ATRL (2008) Administrative-Territorial Reform Law. “Administratīvi teritoriālās reformas likums”. http://www.likumi.lv/doc.php?id=51528/

Bicik J, Kapelan Z, Savić DA (2011) Challenges in the implementation of a DSS for real time WDS management. In: Proceedings of the CCWI 2011 conference, Exeter, UK, pp 1–7

Bose P, Reckhow DA (2007) The effect of ozonation on natural matter removal by alum coagulation. Water Res 41(7):1516–1524

Boualam M, Mathieu L, Fass S, Cavard J, Gatel D (2002) Relationship between coliform culturability and organic matter in low nutritive waters. Water Res 36(10):2618–2626

Boxall JB, Skipworth PJ, Saul AJ (2001) A novel approach to describing sediment movement in distribution mains, based on measured particle characteristics. Water Softw Syst 1:263–273

Brandt BW, Kelpin FDL, Van Leeuwen IMM, Kooijman SALM (2004) Modeling microbial adaptation to changing availability of substrates. Water Res 38(4):1003–1013

Briliute I, Bonckus A, Zygmantas G (2008) Influence of installation effects on the performance of residential water-meters. In: The 7th international conference environmental engineering, vol 2, pp 785–792

Broo AE, Berghult B, Hedberg T (1999) Drinking water distribution—the effect of natural organic matter (NOM) on the corrosion of iron and copper. Water Sci Technol 40(9):17–24

Butterfield PW, Camper AK, Ellis BD, Jones WL (2002) Chlorination of model drinking water biofilm: implifications for growth and organic carbon removal. Water Res 36(17):4391–4405

Camper AK (1996) Factors influencing microbial growth in the distribution system: laboratory and pilot experiments. American Water Works Association Research Foundation, Denver, p 60
Camper AK (2004) Involvement of humic substances in regrowth. Int J Food Microbiol 92 (3):355–364
Chi FH, Amy GL (2004) Kinetic study on the sorption of dissolved natural organic matter onto different aquifer materials: the effects of hydrophobicity and functional groups. J Colloid Interface Sci 274(2):380–391
CPL (2006) Civil Protection Law, Cabinet of Ministers Regulation No. 171, "Civilās aizsardzības likums". http://www.likumi.lv/doc.php?id=146474
CWDN (2010) Project computerized water distribution network control system for Talsi District. (Datorizētas ūdenssaimniecības vadības un kontroles sistēmas ieviešana Talsu novadā 2009–2010), EEA Grants program, Identification Nr. TNP 40/10. http://www.talsuudens.lv/arhivs_2009 (unpublished)
Day GMcD, Hart BT, Hart BT, McKelvie ID, Beckett R (1994) Adsorption of natural organic matter onto goethite. Colloids Surfaces A: Physicochem Eng Aspects 89(1):1–13
Eikebrokk B, Juhna T, Melin E, Osterhus SW (2007). Deliverable 5.3.2a. Water Treatment by Enhanced Coagulation and Ozonation-Biofiltration. Intermediate report on operation optimization procedures and trials. Report of Integrated Project TECHNEAU Funded by the European Commission under the Sustainable Development, Global Change and Ecosystems Thematic Priority Area. http://www.techneau.org/index.php?id=120
ESL (1999) Epidemiological Safety Law Republic of Latvia, "Epidemiologiskās drošības likums", http://www.likumi.lv/doc.php?id=52951
E-Water (2010) Innovative e-services for water supply management. Latvia—Lithuania Program, http://www.latlit.eu/eng/running_projects
Friedman MK, Gregory J, Antoun E (2002) Developing and implementing a distribution system flushing program. J AWWA 94(7):48–56
Frimmel FH (1998) Characterization of natural organic matter as major constituents in aquatic systems. J Contam Hydrol 35(1–3):201–216
Gauthier V, Gerard B, Portal JM, Block JC, Gatel D (1999) Organic matter as loose deposits in a drinking water distribution system. Water Res 33(4):1014–1026
Gauthier V, Barbeau B, Millette R, Block JC, Prevost M (2001) Suspended particles in the drinking water of two distribution systems. Water Sci Technol: Water Supply 1(4):237–245
Gruškeviča K, Tihomirova K, Osterhus S, Juhna T (2008) Influence of water velocity and NOM composition on corrosion of iron pipes. Programme and abstracts book, cost action 637 METEAU materials and related substances in drinking water. COST, Lisbona, pp 24–25
Gruškeviča K, Tihomirova K, Rubulis J, Juhna T (2009).Organic matter isolation from loose deposits in drinking water distribution system. In: Proceedings: 1st IWA East European regional young water professionals conference, Belarusian National Technical University, Minsk, pp 278–284
Hammes F, Vital M, Briedis A, Rubulis J, Juhna T (2010) Tracking bacterial growth in a full scale drinking water distribution network with a flow cytometry, ATP analysis and hydraulic modeling. In: Proceedings: The water research conference. COST, Lisbon
Herro HM, Port RD (1993) The Nalco guide to cooling water system failure analysis. McGraw-Hill, New York, 459 p
HIL (2011) Health Inspectorate of Latvia. An overview of drinking water quality in year 2011, http://www.vi.gov.lv/uploads/files/parskats_2011_2.pdf
Hunaidi O, Wang AA (2006) New system for locating leaks in urban water distribution pipes. Int J Manag Environ Quality 17(4):450–466
IESM (2011) Research on Justification of Development of the System for Monitoring and Control Water Resources, Latvia—Lithuania Cross-border cooperation program, project Innovative e-services for water supply management, LLIII-127 E-Water, Ventspils High Technology Park, 2011, 94 p (unpublished)

Juhna T (2007) Atdzelžošanas principi un to pielietojums dzeramā ūdens sagatavošana. Rīga, SIA Fotomart, p 40

Kaplan LA, Newbold JD (2000) Surface and subsurface dissolved organic carbon. In: Jones JB, Mulholland PJ (eds) Streams and ground waters. Elsevier, Academic Press, San Diego, 425 p

Kaškina B (2012) Dealing with the whole country: creating a National CSIRT. In: 35th TF-CSIRT and FIRST meeting, 31 January 2012, Rome, Italy, pp. 1–30. www.cert.lv

Kelay T, Vloerbergh I, Hagegård K (2008) Deliverable 6.1.3./6.2.3.) Interim Report: Consumer Focus Groups. Report of Integrated Project TECHNEAU Funded by the European Commission under the Sustainable Development, Global Change and Ecosystems Thematic Priority Area, http://www.techneau.org/index.php?id=120

Korshin GV, Ferguson JF, Lancaster AN (2005) Influence of natural organic matter on the morphology of corroding lead surfaces and behavior of lead-containing particles. Water Res 39(5):811–818

Krauze G (2011) Latvian Water Supply and Sewerage Enterprises Association. Presentation at international conference Baltic States in transition, Jurmala, Latvia, 16 September 2011, p 13 (unpublished)

Larson TE (1966) Chemical Control of Corrosion. J Am Water Works Assoc (AWWA) 58 (3):354–359

LeChevallier MV, Lowry CD, Lee RG, Gibbon DL (1993) Examining the relationship between iron corrosion and the disinfection of biofilm bacteria. J Am Water Works Assoc (AWWA) 85 (7):111–123

Lin M, Wu Y, Wassell I (2008) Wireless sensor network: water distribution monitoring system. In: Radio and wireless symposium 2008, IEEE 2008, pp 775–778

MASOM (2010) Mazzy Aquamet software operation manual, https://mazzy.basecamphq.com/login. Micro Dators ltd., IT Hous/ITH Group 2010. 52 p

Matsui Y, Yamagishi T, Terada Y, Matsushita T, Inoue T (2007) Suspended particles and their characteristics in water mains: development of sampling methods. J Water Supply: Res Technology—AQUA 56(1):13–24

Mustonen SM, Tissari S, Huiko L, Kolehmainen M, Lethola MJ, Hirvonen A (2008) Evaluating online data of water quality changes in a pilot drinking water distribution system with multivariate data exploration methods. Water Res 42(10–11):2421–2430

National Security Law (2000) "Nacionālās drošības likums". http://www.likumi.lv/doc.php?id=14011

Ndiongue S, Huck PM, Slawson RM (2005) Effect of temperature and biodegradable organic matter on control biofilms by free chlorine in model drinking water distribution system. Water Res 39(6):953–964

Neilands K, Rubulis J, Juhna T (2009) Application of online monitoring, resuspension potential method and unidirectional flushing for control of drinking water quality in a distribution networks, Latvia. In: Proceedings 1st Eastern European regional young water professionals conference, Minsk, Belarus, p 6

On Pollution Law (2001) "Par piesārņojumu likums". http://www.likumi.lv/doc.php?id=6075

Polychronopolous M, Dudley K, Ryan G, Hearn J (2003) Investigation of factors contributing to dirty water events in reticulation systems and evaluation of flushing methods to remove deposited particles: a review. Water Sci Technol: Water Supply 3(1–2):295–306

Prevost M, Joret JC, Laurent P, Servais P (eds) (2005) Biodegradable organic matter in drinking water treatment and distribution. American Water Works Association, Washington, 300 p

Raczyk-Stanisławiak U, Świetlic J, Dabrowska A, Nawrocki J (2004) Biodegradability of organic by-products after natural organic matter oxidation with ClO_2-case study. Water Res 38 (4):1044–1054

Rubulis J, Šnīdere L, Briedis V (2001) Problems with drinking water metering in apartment buildings and flats in Riga city, Latvia. In: Ulanicki B, Coulbeck B, Rance J (eds). Research Studies Press, UK, pp 349–356

Rubulis J, Lukjaņeca I, Tihomirova K, Koliškins A, Nazarovs S, Vital M, Hammes F, Juhna T (2007) Simulation of bacterial regrowth in water networks by modeling of pseudomonas fluorescents growth in a lab biofilm reactor. In: CCWI2007 and SUWM2007 conference, water management challenges in global change, Leicester, pp 229–236

Ryan G, Mathes M, Haylock G, Jayaratne A, Wu J, Noui-Mehidi N, Grainger C, Nguyen BV (2008). Particles in water distribution systems, Tech. Rep. 33, Cooperative Research Centre for Water Quality and Treatment

Sarin P, Snoeyink VL, Bebee J, Kriven WM, Clement JA (2001) Physico-chemical characteristics of corrosion scales in old pipes. Water Res 35(12):2961–2969

Sarin P, Snoeyink VL, Lytle DA, Kriven WM (2004) Iron corrosion scales: model for scale growth, iron release, and colored water formation. J Environ Eng ASCE 130(4):364–373

Servais P, Billen G, Laurent P, Levi Y, Randon G (1992) Studies of BDOC and bacterial dynamics in the drinking water distribution system of the Northern Parisian suburbs. J Water Sci, Revue des science de L'EAU, 5(Special):69–89

Servais P, Laurent P, Billen G, Gatel D (1995) Development of a model of BDOC and bacterial biomass fluctuations in distribution systems. J Water Sci, Revue des science de L'EAU 8:427–462

Sontheimer H, Kolle W, Snoeyink VL (1981) The siderite model of the formation of corrosion-resistant scales. J Am Water Works Assoc (AWWA) 71(11):572–579

Stoianov I, Nachman L, Madden S (2007) PIPENET: a wireless sensor network for pipeline monitoring. In: Proceeding of IPSN '07; Proceedings of the 6th international conference on Information processing in sensor networks. ACM, New York, pp 264–273

Swietlik J, Dabrowska A, Raczyk-Staniskawiak U, Nawrocki J (2004) Reactvity of natural organic matter fractions with chlorine dioxide and ozone. Water Res 38(3):547–558

Tuovinen OH, Button KS, Vourinen A, Carlson A, Mair DM, Yut LA (1980) Bacterial, chemical, and mineralogical characteristics of tubercles in distribution pipelines. J Am Water Works Assoc (AWWA) 72(11):626–635

Van der Kooij D, Hijnen WAM (1991) Criteria for defining the biological stability of drinking water as determined with AOC measurements. In: Proceedings: water quality technology conference San Diego. American Water Works Association, Denver, pp 1281–1333

Van der Kooij D, Visser A, Hijnen WAM (1982) Determination the concentration of easily assimilable organic carbon in drinking water. J Am Water Works Assoc (AWWA) 74:540–545

Vreeburg JHG (2007) Discoloration in drinking water systems: a particular approach. Ph.D. thesis report

Vreeburg JHG, Boxall JB (2007) Discolouration in potable water distribution systems: a review. Water Res 41:519–529

Vreeburg JHG, Schaap P, van Dijk JC (2004) Measuring discoloration risk: resuspention potential method. In: IWA Leading Edge Conference, Prague, p 8

Zabašta A (2010) Ventspils region drinking water and sewage holding research and model development using information and communication technologies (ICT) solutions. Ventspils High Technology park, p 28 (unpublished)

Zabasta A, Kunicina N, Chaiko Y, Ribickis L (2011) Automatic meters reading for water distribution network in Talsi City. In: Proceeding of EUROCON 2011, April 2011, Lisbon, Portugal, IEEE 2011, pp 1–6

Zacheus OM, Lehtola MJ, Korhonen LK, Martikainen PJ (2001) Soft deposits, the key site for microbiological growth in drinking water distribution networks. Water Res 35(7):1757–1765

Chapter 16
Austrian Activities in Protecting Critical Water Infrastructure

M. Möderl, W. Rauch, S. Achleitner, A. Lukas, E. Mayr, R. Neunteufel, R. Perfler, C. Neuhold, R. Godina, H. Wiesenegger, F. Friedl, D. Fuchs-Hanusch, J. Lammel, P. Hohenblum, F. Skopik, T. Bleier, K. Weber, F. Eder and M. Brugger

Technical Terms and Abbreviations

AC	Asbestos cement
ACHILLES	Spatial vulnerability identification tool
AIT	Austrian Institute of Technology
alpS	Centre for Climate Change Adaptation Technologies
APCIP	Austrian program for critical infrastructure protection
AQUASEC-AUT	Austrian Crisis Management Laboratory

M. Möderl (✉) · W. Rauch · S. Achleitner
Institute of Infrastructure, University of Innsbruck, Technikerstr. 13, 6020 Innsbruck, Austria
e-mail: michael.moederl@gmail.com

W. Rauch
e-mail: wolfgang.rauch@uibk.ac.at

A. Lukas · E. Mayr · R.Neunteufel · R. Perfler
Institute of Sanitary Engineering and Water Pollution Control, University of Natural Resources and Life Sciences, Muthgasse 18, 1190 Vienna, Austria
e-mail: aditya.lukas@boku.ac.at

E. Mayr
e-mail: ernest.mayr@boku.ac.at

R. Perfler
e-mail: reinhard.perfler@boku.ac.at

C. Neuhold
Institute of Water Management, Hydrology and Hydraulic Engineering, University of Natural Resources and Life Sciences, Muthgasse 18, 1190 Vienna, Austria
e-mail: clemens.neuhold@boku.ac.at; clemens.neuhold@lebensministerium.at

C. Neuhold · R. Godina
Austrian Federal Ministry for Agriculture, Forestry, Environment and Water Management, Federal Water Engineering Administration, Stubenring 1, 1010 Vienna, Austria
e-mail: reinhold.godina@lebensministerium.at

R. M. Clark and S. Hakim (eds.), *Securing Water and Wastewater Systems*, Protecting Critical Infrastructure 2, DOI: 10.1007/978-3-319-01092-2_16,

BMLFUW	Austrian Federal Ministry of Agriculture, Forestry, Environment and Water Management
BMVIT	Austrian Federal Ministry for Transport, Innovation and Technology
CAIS	Cyber attack information system
CERT.AT	Computer emergency response team Austria
DI	Ductile iron

F. Friedl · D. Fuchs-Hanusch
Institute of Urban Water Management and Landscape Water Engineering, Graz University of Technology, Stremayrgasse 10/I, 8010 Graz, Austria
e-mail: friedl@sww.tugraz.at

D. Fuchs-Hanusch
e-mail: fuchs@sww.tugraz.at

H. Wiesenegger · F. Friedl · D. Fuchs-Hanusch
Hydrographical Service Salzburg, Salzburg, Austria
e-mail: hans.wiesenegger@salzburg.gv.at

J. Lammel
AlpS GmbH, Grabenweg 68, 6020 Innsbruck, Austria
e-mail: lammel@alps-gmbh.com

P. Hohenblum
Environmental Agency Austria, Substances and Analysis,
Spittelauer Lände 5, 1090 Vienna, Austria
e-mail: philipp.hohenblum@umweltbundesamt.at

F. Skopik · T. Bleier
Safety and Security Department, AIT Austrian Institute of Technology,
2444 Seibersdorf, Austria
e-mail: florian.skopik@ait.ac.at

T. Bleier
e-mail: thomas.bleier@ait.ac.at

K. Weber
Department of Public Law, State and Administrative Theory, University of Innsbruck,
Innrain 52d, 6020 Innsbruck, Austria
e-mail: karl.weber@uibk.ac.at

F. Eder
Department of Political Science, University of Innsbruck, Universitätsstraße 15,
6020 Innsbruck, Austira
e-mail: franz.eder@uibk.ac.at

M. Brugger
Section III - Telecommunication and Innovation, Staff Group for Technology Transfer and Security Research, Federal Ministry for Transport, Innovation and Technology, Renngasse 5, 1010 Vienna, Austria
e-mail: michael.brugger@bmvit.gv.at

EPCIP	European programme for critical infrastructure protection
EU	European union
FEIS	Failure experience improvement system
FP7	Seventh framework programme of the European community
GEDES	Risk of flood and landfill hazard tool
GIS	Geographic information system
ICS	Industrial control systems
ICT	Information and communication technology
INFOSAN	Strategic data acquisition for sewer rehabilitation planning in Austria
INSPIRE	Infrastructure for spatial information in the European community
IRA-WDS	GIS-based risk analysis tool for water distribution systems
IT	Information technology
I-VAM	Infrastructure vulnerability assessment model
KIRAS	Austrian security research programme
LDTWM	Large-diameter transmission water mains
OIIP	Austrian institute for international affairs
ORTIS	Operational risk management tool and information system
P	Phases
PE	Polyethylene
PiReM	Pipe rehabilitation management
RAMCAP	Risk analysis and management for critical asset protection
RAM-W	Risk assessment methodology for water
SAGA-GIS	System for automated geoscientific analyses—Geographic information system
SNA	Social network analysis
VSAT	Vulnerability self-assessment tool
WS	Work streams
WS1-P1	Identification of priority sectors for action
WS2-P2	Identification of vulnerabilities, threats and risks
WS2-P3	Implementation of minimum protection measures
WS3-P3	Development of specific protection measures for each national critical infrastructure
WSP	Water safety plan
ZuHaZu	Condition assessment of LDTWM

16.1 Background Information About The Austrian Water Supply Sector

The first section of the chapter gives an overview of the legal rules and standards governing the Austrian water supply sector, but it includes also interdependencies to other sectors. Further, best practices in management and planning established by Austrian utilities and results from an Austrian benchmarking survey for water suppliers are discussed.

16.1.1 Physical and Organisational Structure of the Austrian Water Supply Sector

Following the profile data given by the Austrian Association for Gas and Water (ÖVGW 2009) and the Federal Ministry of Agriculture, Forestry, Environment and Water Management (BMLFUW 2007) 7.5 million of the Austrian population—out of a total of 8.3 million—live in areas covered by central water supply. Another 0.9 million get their drinking water from private springs and wells. Austrian drinking water supply comes from groundwater and spring water sources, in many cases without any water treatment.

There exist about 5,500 water supply utilities, approximately 1,900 community based utilities, approximately 165 water supply associations and approximately 3,400 water supply cooperatives which differ in organisational structure but are under public ownership and operation, either directly or indirectly:

- Water utility as part of community administration
- Water supply association as regulated by the Austrian Water Act or the Municipalities Association Act
- Water supply cooperatives as regulated by the Austrian Water Act

The surprisingly high number of drinking water supply utilities results from the easy availability of sufficient water resources with high water quality. Regions with dispersed settlement show a large number of small and very small utilities which only accounts for a rather small proportion of total supplied population. Closed housing areas and urban areas are dominated by community based utilities or regional providers.

The 14 biggest member utilities of the Austrian Association for Gas and Water (number of supplied population bigger than 50,000 inhabitants each) supply 4 million inhabitants. There are 6.91 million out of 7.44 million of centrally supplied population supplied by utilities with more than 10,000 inhabitants.

Basic data on physical infrastructure:

- Length of distribution network: 77,000 km (47,740 mi) (including about 14,500 km (8,990 mi) transmission lines)

- Number of storage tanks: about 2,900 with a total volume of about 4 million m^3 (141.24 mil cu ft)
- Number of service connections/customer meters: about 1,39 million
- Number of spring taps: about 2,600
- Number of wells: about 1,000

The total amount of centrally supplied drinking water is about 757 million m^3 per year (2,653 mil cu ft). The average household consumption (without trade units, industry) is about 130 l (34.3 gal) per person and day. An average household with 4 persons uses about 200 m^3 (52, 840 gal) per year.

The average costs for domestic drinking water supply in Austria amounted to 1.09 €/m^3 (0.0041 €/gal) in 2008 (ÖVGW 2007). The usually applied tariff systems are based on uniform rates, where the fixed tariff components (such as metering and service charges), cover less than 10 % of the total tariff compound (Hernandez-Sancho et al. 2007). Data derived from benchmarking studies shows that 70–80 % of the total costs are represented by fixed costs (such as asset depreciation and interest) or quasi-fixed costs such as personnel, administration and so on (Theuretzbacher-Fritz et al. 2006). In addition, around 60 % of total costs are ascribed to transmission and distribution (Neunteufel et al. 2004). For this reason, the variable network length is typically the main cost driver.

In West Austria, a region with Alpine topography and climate, small and medium sized municipalities with up to approximately 100,000 inhabitants are located along the main rivers and several anabranches on the floor of the valleys. The main water sources are springs that provide potable water often from more than 100 m (328.1 ft) above the populated areas. The hydraulic residence time of the spring water in the soil ranges between 60 days and 40 years. Thus, the water quality of spring water is usually very high and no disinfection is required during normal operation. Ground water wells are installed too, but provide only about 10 % of the total volume. As the possible delivery of the springs is mostly higher than required for the municipal demand, the systems along the river are usually not connected with each other with some exceptions for emergency response. This fact is based on public opinion, which wishes an independent supply for each municipality, although integrated pipes used for balancing water between the systems would increase supply security significantly. Further, the pipes of the water distribution networks are sometimes oversized because of overestimated population growth in older technical guidelines, increasing water efficiency in usage and water saving measures.

Supply systems are often operated with several pressure zones due to high elevation differences. If not, systems constitute high supply pressures, because of the altitude of spring water from the mountains. For instance, the largest supply system in Tyrol (Austria) delivers water for more than 100,000 inhabitants with a hydraulic pressure of over 100 m (328.1 ft). At present, the operators do not intend a reduction of this high value. In other cases, pressure reducing valves (PRV) are implemented to get a sufficient pressure level in the supply area. For instance, the tanks of the supply system of the municipality Hall have the same pressure head as in Innsbruck, but the supply pressure for 10,000 inhabitants ranges between 50 and 70 m (164 and 230 ft), because PRVs are used to control the excess energy. In both cases, the pressure head is high enough to prevent intrusion of toxic substances under regular conditions. The average water demand of the consumers in Hall is 60 l/s (15.9 gal/s), while the minimum delivery of the springs is approximately 330 l/s (87.2 gal/s). Therefore, two hydropower plants are installed at the tank locations which are designed for a flow rate of in total 270 l/s (71.3 gal/s). This water for energy production can be used for supply in emergency cases.

16.1.2 Results from an Austrian Benchmarking Survey for Water Suppliers

Every four years, the Austrian Gas and Water Association (ÖVGW) conducts a benchmarking survey for water suppliers. All domestic water suppliers are eligible to participate in this benchmarking survey, regardless of whether or not they are members of ÖVGW. The aim of this survey is to:

- compare water suppliers based on standardized methods, using clear definitions and nationally as well as internationally comparable parameters;
- compare as many water suppliers as possible to get a representative result.
- compare domestic companies with foreign ones, thus presenting Austrian companies and the quality of their services within a wider context. This comparison has already revealed that most Austrian water suppliers have very high performance standards.

Neunteufel (2011) gives an overview of parameters which are covered by benchmarking: resource situation, technical safety, water quality and other related services, inspection and maintenance, damage quotas and water losses, service quality and customer service, technical and economic sustainability, cost, staff and energy efficiency as well as other factors such as the age of the distribution system, the organisational structure of the company etc.

One example for the increase of supply safety is displayed in Fig. 16.1 (Neunteufel et al. 2010). The trend analysis regarding the remote control degree indicates an increasing standard among those eleven participants (data displayed in box-plots) which took part in each of the three Austrian utility benchmarking projects (2002, 2004, 2007).

Of course the increase of the remote control degree is not necessarily interdependent with the participation in benchmarking exercises. It is rather related to the normal development of modernising supply utilities. Still several utilities reported that the realisation of improvements was triggered by the benchmarking even if the measures were already planned for the future. Undoubted, benchmarking is an appropriate tool to get changes, modernization and improvements on display.

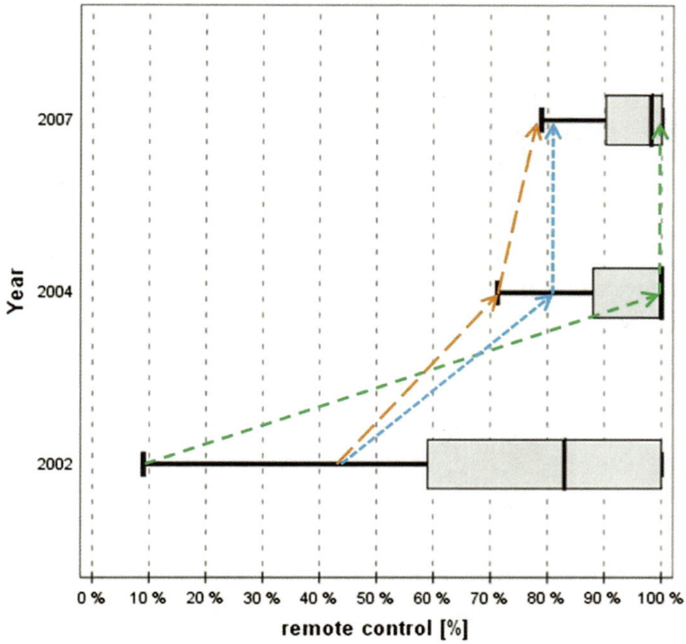

Fig. 16.1 Development of the remote control degree of participants

16.1.3 Legal Remarks on Liability Regarding Urban Water Supply

Legal framework provisions for urban water supply are spread in a number of different acts falling in the field of public and civil law. Public law requires operators of water supply establishments to seek permission according to water law and special regulations of the provinces (such as water supply acts, canalisation acts). The duties of administrative authorities, generally from the district and the community, are to supervise establishments and issue respective orders to prevent or solve disturbances and dangers. Once operators do not fulfil these obligations, penalties can be imposed and, in the worst case, the establishment can be shut down.

Civil liability applies, if third parties suffer damages from construction or operation of urban water supply establishments. Legal entities of water supply are mainly communities and communal companies. Although private operators of services in the field of urban water supply are legally permitted, they hardly ever appear in Austria. While in Tyrol and Vorarlberg the water supply falls within the responsibility of communal authorities, communities in the remaining provinces operate legally as private companies. This differentiation leads to consequences regarding civil procedure: Damages in Tyrol and Vorarlberg have to be claimed in special procedures according to the official liability act. Differently, in the remaining provinces general tort provisions of the general civil act apply. However, requirements to enforce tort claims are similar: A material damage of the aggrieved party which has to be proofed must be caused by an unlawful and culpable act (or omission). Only damage, really occurring will be covered, excluding non-material damage (such as value of personal preferences etc.). The water act includes additional special provisions regarding liability.

Apart from public law responsibility and civil law liability criminal responsibility has to be considered. If a person is injured or dies during the construction or operation of a water supply establishment, criminal law penalties due to negligent assault or killing may be imposed. Moreover, once the operation of such establishment causes damage of the environment, respective criminal prosecution may apply. However, such offences have never been imposed in the context of urban water supply in Austria.

16.1.4 Rehabilitation Management for Austrian Water Infrastructure

This part of the chapter focuses on efforts made in Austria to reduce the intrinsic risk of buried infrastructure systems caused by unsustainable maintenance and rehabilitation management. Between 2007 and 2012 two Austrian research projects funded by the Austrian government dealt with the topic of rehabilitation management and the need to achieve and maintain a high quality of the urban water infrastructure. One project focused on water supply systems (PiReM), another on sewer systems (INFOSAN) which is still on-going.

In general, the guideline ÖVGW (2007) describes strategies and methods to be followed in maintenance and rehabilitation. Further the guideline describes service levels for system performance which were derived from analyses within the Austrian Benchmarking (Table 16.1). The values given in Table 16.1 represent service levels for the entire system. For specific pipe clusters (groups of pipes with similar life expectancy) the failure rates can be significantly higher than these levels.

To get an overview about the condition of different pipe types which are used in Austrian supply systems a lifetime analyses based on failure data of 6 Austrian utilities was made within the PiReM project. By clustering pipes according to their

Table 16.1 Service levels for Austrian water systems

Failure rates	Supply pipes (failures per 100 km per year)	Service connections (SC) (failures per 1,000 SC per year)	Suggestions
Low	<7	<3	Hold standard
Moderate	7–20	3–10	Improve
High	>20	>10	Urgent need for activities

deterioration behaviour, groupings and characteristic survival functions were derived (Fuchs-Hanusch et al. 2008; Fuchs-Hanusch and Kasess 2012). Results of these analyses show that the critical pipe groupings in Austrian systems are not generally the oldest pipes, which are cast iron pipes going back to the late nineteenth century. The pipes buried between World War II and the late 1970s show higher degradation resulting in increasing failure. These groups are critical as in these periods a high activity of construction took place due to reconstruction after World War II and strong economic development in the 1960s and 1970s. Hence many utilities are faced with a large number of pipes belonging to these groups. According to the condition of these pipes the criticality results from two facts, one is low material quality, especially after world war two, the other is that new material types, like polyvinylchloride (PVC) and ductile iron (DI) were improperly handled during storage and installation. Hence utilities with long pipe length belonging to these pipe groupings show a high intrinsic pipe breakage risk and are going to face high rehabilitation needs in the next decades. This is especially true in the bigger cities. In rural areas where the systems were mainly constructed from the late 1970s up to the 1990s most of the systems currently are in a stable condition.

In the Austrian research project ZuHaZu, funded by the City of Vienna a risk assessment method to prioritize large-diameter transmission water mains (LDTWM) for inspection, maintenance and rehabilitation was developed. LDTWM are essential elements of the water supply infrastructure. Failure consequences can be disastrous and especially affect supply security and pressure conditions. Consequences are mainly governed by the characteristics of failure mode depending on orifice area, internal pressure and amount of water loss in combination with environmental conditions such as soil type, paved surface area, adjacent subsurface infrastructure and potentially endangered objects. Hence besides failure frequency the failure mode probability is of further interest. Table 16.2 gives an overview about the most common types of failure modes in LDTWM specifically for materials, environmental and operating conditions in Austria. For instance, asbestos cement (AC) is fragile for most of the failure types. In contrast, polyethylene (PE) is more resistant against the investigated failure types.

Table 16.2 Most common types of failure modes in LDTWM specifically for materials, environmental and operating conditions in Austria

Failure type	Cast iron	DI	Steel	AB	PE	PVC
Circumferential break	Typically	Rarely	Rarely	Typically	Rarely	Rarely
Longitudinal break	Typically	Possible for larger diameter	Possible for larger diameter	Typically	Rarely	Typically
Leaking joint	Typically	Rarely	Rarely	Typically	Rarely	Typically
Split bell	Typically	Rarely	Not possible	Typically	Rarely	Rarely
Holes	Typically	Typically	Typically	Rarely	Rarely	Rarely

Within the Project ZuHaZu a binary logistic regression analyses was applied to model the structural deterioration and subsequent failure mode probability of LDTWM. A validated failure-mode-probability prediction performance of 85 % was achieved by using this statistical model applied on the transmission pipe data set of 2 large water utilities (Friedl et al. 2012).

16.2 Strategies for Critical Infrastructure Protection

In this section, strategies and action plans for critical infrastructure protection are discussed.

16.2.1 Defining the Term "Critical Infrastructure"

Not only in the United States of America, but also elsewhere water supply systems are categorized e.g. according to Reid (2009) as critical infrastructure, because of their vital role for public services. Prolonged disruption of critical infrastructure for a city or urbanized region would inevitably lead to major economic losses, deteriorated public health, and eventually to population migration. There are also other definitions for such systems. For instance, in earthquake science the term lifeline is used for critical network infrastructure. In Austria and in the European Union (EU), it is not standardized yet which infrastructure sectors are seen as critical. The strategy to protect critical infrastructure in Europe is defined in an EU directive (Council of the European Union 2008). Its aim is the identification and designation of European critical infrastructures and the assessment of the need to improve their protection. As a first step, the directive focuses on the transport (road, rail, aircraft, inland and maritime navigation) and energy (power, oil and gas) sectors. Other sectors, especially telecommunication, may be analyzed in the future. Water and waste water infrastructure are not mentioned in this directive.

16.2.2 Programs for Critical Infrastructure Protection

The European programme for critical infrastructure protection (Commission of the European Union 2006) gives advice on the implementation of the Council directive and an action plan structured in three work streams (WS):

- strategic platform for EU-wide coordination
- protection of European critical infrastructure
- support for member states in national critical infrastructure

In addition three phases (P) represent a time line for the actions. Actions are e.g. identification of priority sectors for action (WS1-P1), identification of vulnerabilities, threats and risks (WS2-P2), implementation of minimum protection measures (WS2-P3) and development of specific protection measures for each national critical infrastructure (WS3-P3).

The Austrian program for critical infrastructure protection (APCIP, Bundeskanzleramt 2008) released by the Austrian government is the answer to the EU activities and follows similar objectives.

16.2.3 The Austrian Security Research Programme KIRAS

Established in 2005 by the Austrian Federal Ministry for Transport, Innovation and Technology (bmvit[1]), KIRAS was the first national security research programme within the European Union. The name originates from ancient Greek and combines the words "kirkos" (circle—being a metaphor for the holistic concept behind KIRAS) and "asphaleia" (security). Since its introduction, the initiative has been based on a broad and circumspect notion of (civil) security, giving due consideration to socio-political, economic and cultural developments not only on a national level but also Europe wide.

In line with this idea, KIRAS promotes national research projects which correspond to the strategic objectives of the programme, namely

- to improve the subjective perception & objective level of security of Austrian citizens
- to support the generation of knowledge needed for security policy
- to promote security related technology leaps
- to support the growth of the Austrian security industry
- to achieve excellence in security research and
- to integrate relevant societal questions in every project

In this context, the thematic priority addressed by KIRAS is the protection of critical infrastructures in sectors such as: energy, communication & information, financial- and health system, food, water, transport as well as public security and administration. KIRAS is embedded in several national strategies, e.g. the above mentioned APCIP and is the only research theme explicitly mentioned in the current coalition agreement.[2]

The programme is governed by a steering board which is chaired by bmvit and comprises relevant ministries (e.g. Federal Chancellery, Ministry of the Interior, Ministry for Science and Research, Ministry of Economy, Family and Youth,

---

[1] Bundesministerium für Verkehr, Innovation und Technologie (http://www.bmvit.gv.at/en/index.html).

[2] Regierungsprogramm für die XXIV. Gesetzgebungsperiode (2008–2013).

Ministry of Justice, Ministry of Defence, Ministry of Agriculture, Forestry, Environment and Water Management) as well as special interest groups (e.g. Austrian Economic Chambers, Federal Chamber of Labour, Federation of Austrian Industries) and research institutions (e.g. Austrian research Promotion Agency, Council for Research and Technology Development, Austrian Science Fund). KIRAS is furthermore characterized by a mandatory inclusion of end users as well as an obligatory integration of humanities, social sciences and cultural studies. With regard to the water sector several (municipal) waterworks, but also the Ministry of Agriculture, Forestry, Environment and Water Management, the Ministry of Defence, the Austrian Agency for Health and Food Safety and the Environment Agency Austria have been participating as end-users within KIRAS-projects.

The permanent accompanying evaluation of KIRAS confirms a strong application focus and end-user emphasis within supported projects. This integrative approach ensures demand-driven research and therewith fosters the development of marketable technological products and services. The evaluation likewise confirms that societal aspects are deeply embedded in KIRAS projects. Questions on societal issues, like the acceptability of newly developed security solutions, are taken into consideration during the whole project lifecycle and therewith represent an integral part of the research work. In addition, Austrian institutions are very successful in European security research, namely in the FP7[3] security theme, and thereby create an over-average return of research funds.

In order to further enhance networking and communication among all relevant stakeholders a new initiative, the "Innovation Platform", was started in autumn 2011. This stakeholder-driven platform aims at fostering and structuring the dialogue between the supply (researchers, industry) and demand side (end-users, procurers) in order to commonly identify promising R&D[4] priorities.

16.2.4 Safety Plans as Risk Analysis Concept for Critical Infrastructure Protection

For the assessment of threats and risks to the critical infrastructure, which is as mentioned above an action of the APCIP, a risk analysis concept is required. For this task, the concept of safety plans seems appropriate, because it is pro-active, sustainable and integrated. Further, operation and maintenance solutions are provided. This concept is applicable for any system configuration and it is also easy to integrate them into existing management systems. Safety panning is particularly powerful in a combination with Geographic Information Systems (GIS) and spatial risk assessment methods as presented in the following by

[3] Seventh Framework Programme of the European Community for research, technological development and demonstration activities (2007–2013).

[4] Research and technological development.

applying failure forecast and hydraulic models as well as functional cluster analysis and other approaches.

Minimum requirements for safety plans are seen as follows (Council of the European Union 2008):

- Identification of relevant infrastructure components
- Analysis of relevant threat scenarios and their associated consequences to assess vulnerability of infrastructure
- Evaluation, collection and ranking of pro-active actions distinguishing between permanent provisions and stepwise actions, which are executed depending on the level of risk.

The implementation of a water safety plan is standardized in Austria by ÖVGW (2008) and servers as a frame for safety management. The safety plan concept suggested in this chapter, contains the evaluation of vulnerabilities (spatial description of consequence) and the effects of potential hazards (spatial description of likelihood) on the entire system performance.

It is also suggested to enhance this concept with the development of emergency plans which is another form of proactive-action. Mayr et al. (2012) present a software-supported Water Safety Plan to implement the approach by small and medium-sized water suppliers in Austria.

16.2.5 Potential Hazards Typical for Austria

In alpine regions, floods, landslides, debris flows and avalanches are the natural hazards that are most prevalent in the Alpine region of West Austria (BMLFUW 2007). Global warming and changes in precipitation will further influence these hazard types in the future (Rauch and De Toffol 2006; Soldati et al. 2004; Höller 2007; Stoffel and Beniston 2006). All natural hazards typical for Austria impact the water sources (springs and ground water wells). Furthermore, floods impact critical water infrastructure if the pressure outside is higher than inside the systems. Floods also impact the water sources by polluting them and disabling electrical pumping equipment. Landslides and debris flows impact infrastructure by sweeping pipes along with soil material and the water resources by submerging the latter. Avalanches impact primary the water resource, because only in very little cases they whirl up soils. The natural hazard zones of Austria are categorized based on existing hazard zone maps regulated by BMLFUW (1976). In this data, seasonal temperature change is unconsidered, but also of relevance. In cold winters main trunks tend to break more often.

Also human factors and deficient integrity of infrastructures lead to hazardous events in water supply systems. For instance, in Upper Austria a helicopter accidentally dropped oil on the catchment of a spring. In another case, a farmer fertilised in error a water protection area in Vorarlberg. Most often, critical conditions are caused by neglecting rehabilitation of defect pipes. However, many

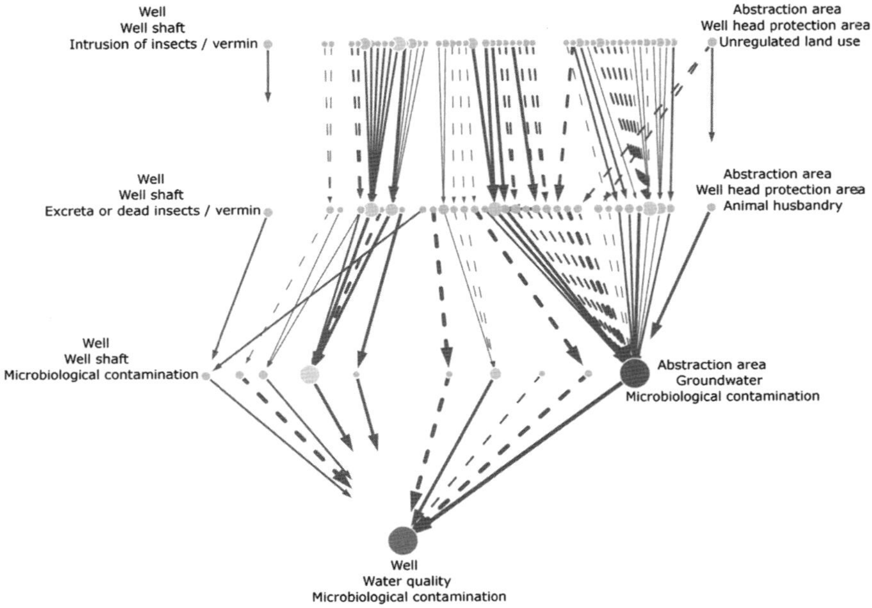

Fig. 16.2 Visualisation of the cause-effect chains upstream of the failure event "microbiological contamination at the well"

hazardous events can be prevented by appropriate management, planning and design. The hazardous event "intrusion of insects/vermin" in Fig. 16.2, for example, can be prevented by regular inspection of the integrity of vermin proofing screens.

In the following, three threats to the water supply sector are discussed in detail, because of their vital role in protecting critical infrastructure.

16.2.5.1 Flood Hazard and Landfills

Landfills are mainly located in lowland areas close to settlements inducing potential environmental flood risk. The exposure of groundwater aquifers, wells and water supply systems in general may immediately result in contamination and consequently to adverse health effects.

During recent flood events numerous landfill sites were reportedly inundated leading to erosion of landfilled material and release of pollutants threatening humans and the environment (Habersack and Moser 2003; Neuhold and Nachtnebel 2011; Young et al. 2004). Although emissions from landfills under regular operating conditions are well investigated, the behaviour and associated emissions in case of flooding are widely unknown. Geller et al. (2004) observed increased concentrations of hazardous substances in floodplain soils and river sediments caused by the Elbe River flood in 2002. Heavy metal and arsenic contamination within the Elbe catchment are to some extent associated with flooded landfills (Rank et al. 2003).

As maintenance and decomposition for landfill life cycles are assessed for 200–500 years depending on waste composition, climatic conditions and applied assessment methodologies (Belevi and Baccini 1989) even sites with flood protection schemes, generally designed to resist a 100-years flood event, are highly likely to be inundated before hazardous materials are decomposed. It has to be assumed that inundated landfills become water saturated which leads to a substantial mobilisation of pollutants, as the presence of water enhances decomposition and transport processes (Klink and Ham 1982).

16.2.5.2 Austria and the Terrorist Threat

The protection of critical infrastructure has become one of the central aims in the European Union's fight against transnational terrorism. Classified under the strand "protect", the Union seeks to reduce its vulnerability of an attack on key facilities and networks (Council of the European Union 2005). In a communication of the EU Commission to the Council and the EU Parliament in 2004, the threat of a terrorist attack on critical infrastructure was seen as an increasing one and it is stated that the supply networks of food and water (storage, treatment, etc.) were essential for major urban areas (Commission of the European Communities 2004).

This threat perception has to do with a possible change of tactics by terrorist actors, called "substitution" (Brandt and Sandler 2010; Enders and Sandler 2004; Frisch 2009). Whenever terrorists face obstacles, which deny or impede the success of an attack, they switch, after a cost-benefit-calculation, to other possible targets. In other words, the harder it is for terrorists to successfully attack government buildings and military facilities, the more likely it is that they are looking for alternative aims, which generate public attention. Critical infrastructure, and especially the supply networks of food and water, could be such tempting aims.

In the Council Directive on the identification of critical infrastructure (Council of the European Union 2008), however, food and water supply networks and their safety was not mentioned once. Only the protection of inland waterways transport was seen as a critical infrastructure sector. Similar to the Council Directive, the masterplan of the APCIP (Bundeskanzleramt 2008) ignored the possible threat of an attack on the water supply infrastructure.

This ignorance seems to be absurd, taking into consideration the significance of water for human mankind and the fact that an attack on the water supply "could lead to serious medical, public health, and economic consequences" (Meinhardt 2005). However, by taking a closer look at the chances of success of a terrorist attack on the water supply systems of modern societies, it becomes clear why this threat is more a virtual than a real one. As Gleick (2006) shows there are two main scenarios for a terrorist attack on water supplies: the use of conventional explosives to disrupt the infrastructure as such or to add chemical or biological agents into the local water.

Both scenarios have massive disadvantages in the eyes of terrorists. First, the use of conventional explosives cannot inflict enough damage on the water infrastructure, which is conceptualized as redundant systems. Other parts can compensate the breakdown of one part of the system quickly. Second, adding chemical or biological agents is more difficult than generally assumed. Such agents must be weaponized and be appropriate for water, they should be enough infectious, virulent or toxic in order to inflict mass damage and they must be effective over time and resistant against treatment (Gleick 2006). This requires more skills and resources, most terrorist actors are able and willing to provide. Even if some terrorists try to use such agents, a possible contamination of the protected water intake will be quickly detected by already implemented security systems. Hence, the cost-benefit-calculation of an attack on the water supply systems suggests that the costs do not match the benefits—failure is more likely than success.

It seems therefore rational that the European Union and Austria skipped the threat of a terrorist attack on water supply systems after reviewing its likelihood.

16.2.5.3 Austria and the Cyber Attack Threat

Although terrorist attacks are unlikely, cyber attacks receive more and more attention in the last decade by both industry and academia. Not only because "computer anarchists" (e.g. Anonymous) use the cyber world to increase their assertiveness, but also with the increasing use (see benchmark section above) and interconnectedness of control systems, critical infrastructures become more and more vulnerable to cyber-attacks. These concerns are also of high relevance for industrial control systems (ICS) (Igure et al. 2006). Here, a recent trend is the connection of ICSs to large-scale networks, such as the Internet, in order to enable remote control and maintenance. Since ICSs are applied to an increasing number of important critical infrastructure domains, including water supply, these systems become worthwhile targets to attackers. In addition, the use of cheap standard technologies, such as Ethernet and other connection oriented transport services, even in specialized domains, makes professional equipment widely available – unfortunately even to malicious attackers (Byres and Lowe 2004). It is therefore of paramount importance to develop concepts and methodologies that cope with arising challenges, including the detection, mitigation and remediation of cyber-attacks in order to ensure the safety and security of Austria's water supply infrastructures.

16.3 Approaches for Critical Water Infrastructure Protection

This section of the chapter describes several approaches developed by different institutions to identify critical control points and minimise the risks of interfering processes in the water supply sector. The approaches are presented with focus on water infrastructure, but can be adapted likewise for other sectors. Further, the benefits granted from these approaches and a link-up are discussed.

16.3.1 FEIS: Failure Experience Improvement System

The identification and assessment of risks requires the analysis of failure causes and their consequences. This task can be supported by exchange of experiences since failures in different water supply systems often have similar root causes and show similar failure propagation. The Failure Experience Improvement System (FEIS), which was funded by the Austrian programme for security research, is a method for failure documentation and analysis, and knowledge exchange within and between water utilities (Lukas et al. in press). The FEIS combines and tailors two concepts to the context of water supply: Failure reporting systems and Social Network Analysis (SNA). For the development of the FEIS, failure reporting questionnaires were used at six Austrian water utilities to systematically collect information on failures which have occurred in practice. In addition, a literature review and a survey of guidelines were carried out to collect information on potential failure events. The collected information was used to build the database of the FEIS, which basically consists of failure events and their interconnections in the water supply system. By linking failure causes and their effects, a failure network, which describes failure propagation in water supply systems, is created. In order to visualise the interconnected cause-effect chains and to analyse failure propagation, the concept of SNA was applied. SNA was originally developed to investigate social structures. In the FEIS, it is used to describe causal interrelationships and to localise vulnerable points in the water supply system.

The FEIS database is located on a server and its Flash based user interface can be accessed by the water utilities through an internet browser. After login, the user can select a specific failure event to visualise its cause-effect chains. For example, Fig. 16.2 shows the cause-effect chains upstream of the failure event “microbiological contamination at the well”.

Within the three-step neighbourhood, the FEIS database contains 47 causes for this particular failure event. In Fig. 16.2, the paths originating from two selected causative events are labelled. One of the causes is unregulated land use in the well head protection area, leading to microbiological contamination by animal husbandry. Another causative event is the intrusion of insects/vermin into the well. In order to identify the root causes of these failure events, the user can further navigate through the failure network.

The FEIS visualisations provide a concise overview of the failure systems in water supply infrastructures and help to understand the causes of failures and to identify possible corrective actions. Moreover, the size of a node in Fig. 16.2 represents the importance of the respective event in the failure network. The importance is determined by calculating indicators based on the network structure.

These indicators are relative values and evaluate events in comparison to each other. They can be used to rank failure events in order of importance and to identify high-consequence or high-dependence events. The FEIS can be used to plan corrective actions, to identify design improvements, to train staff and to develop awareness of the importance of failure management. Moreover, Lukas et al. (2011) demonstrate the use of the FEIS to support the WSP approach.

16.3.2 ACHILLES: Spatial Vulnerability Identification Tool

16.3.2.1 Introduction

I-VAM (Infrastructure Vulnerability Assessment Model), RAM-W (Risk Assessment Methodology for Water), RAMCAP (Risk Analysis and Management for Critical Asset Protection), VSAT (Vulnerability Self-Assessment Tool) are vulnerability assessment tools often used in the water security sector of North America. These tools support the functional description of critical infrastructure by containing instruments to identify system vulnerabilities and to determine the required level of protection. For instance, RAMCAP includes the designation of Subject Matter Experts for the assessment. IRA-WDS, another GIS-based risk analysis tool for water distribution systems, focuses on contamination. In this section the ACHILLES tool (Möderl and Rauch 2011a, b) is presented used for similar objectives as the tools mentioned before. The specific characteristic of ACHILLES is that it measures system performance based on hydraulic and water quality simulations and is implemented in GIS environment.

16.3.2.2 Features and Capabilities

Several types of hazardous events have an impact on urban water infrastructure. Anthropogenic hazards, but also natural hazards (see above: hazards typical for Austria) may cause system failures. Such failures can be e.g. interrupted power supply of pumps, burst of pipes and contaminated water intrusion, among others. The ACHILLES tool emulates such failures in combination with a sensitivity analysis. For that purpose, generic hazardous scenarios are implemented in the ACHILLES tool. Input parameters describe the magnitude of the impact. For instance, a contamination event is simulated by a source to each junction with a user-defined time and amount. The ACHILLES tool evaluates the sensitivity of each modified component of the infrastructure system by calculation of several hydraulic and water quality performance indicators of the entire system. The sensitivity is regarded as vulnerability as the parameter variation emulates a specific hazard impact. In a next step, a spatially referenced vulnerability map is created by joining the calculated indicator values of the entire system to the

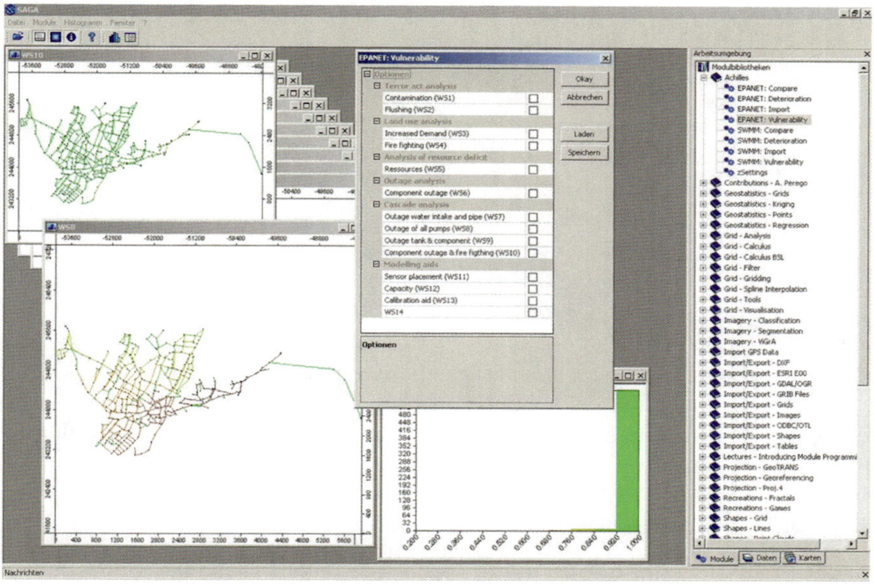

Fig. 16.3 Spatial vulnerability identification with ACHILLES

location of the corresponding modified component. Also cascade vulnerability is investigated using regional sensitivity analysis.

The ACHILLES tool is implemented in the open source environment SAGA GIS. This enables GIS post processing (e.g. export of vulnerability maps as shape files, etc.) without additional data management. It runs on any computing platform and is parallel coded to allow for a reduced calculation time on multi core machines. The user-friendly graphical user interface (Fig. 16.3) assists researchers, but also decision makers in protecting urban water infrastructure efficiently.

16.3.2.3 Output

Outputs of the ACHILLES tool are at present seven different vulnerability maps for water supply systems. For instance, the failure vulnerability map indicates locations where a total loss of functionality is most harmful. In addition, four cascade vulnerability maps are calculated automatically by evaluating the effect of two impacts that happen simultaneously (e.g. pipe burst and additional water demand for fire fighting).

16.3.3 ORTIS: Operational Risk Management Tool and Information System

To assist risk management teams in implementing an efficient and cost-effective risk management solution, the alps—Centre for Climate Change Adaptation Technologies developed a hands-on method for risk assessment in the public sector that follows the state of the art as well the practical needs of public authorities and infrastructure operators. The result of this development process is a blend of expert- and community-based approaches that considers multi-hazard risks as well as the principles of local participation and cost efficiency. Besides, a web application called ORTIS (Operational Risk Management Tool and Information System) has been developed based on the needs of the users. ORTIS is a strategic tool that assists municipalities and infrastructure companies in implementing a risk management system in their organization, by conducting the three steps risk analysis, risk control and risk monitoring.

The screenshot of Fig. 16.4 shows what the ORTIS tool looks like and how simple it is to go through the three steps:

The ORTIS risk assessment process is based on individual risk catalogues that provide the main structure for the individual workshops as well as the classification of risk-groups and master-risks. To follow the principle of flexibility, the catalogue can be adapted to the specific needs of the users.

With the aim of creating a sense of ownership, risk management experts visit the responsible teams in order to guarantee a high quality risk assessment. Local experts are brought together in one workshop and drive the process, guaranteeing

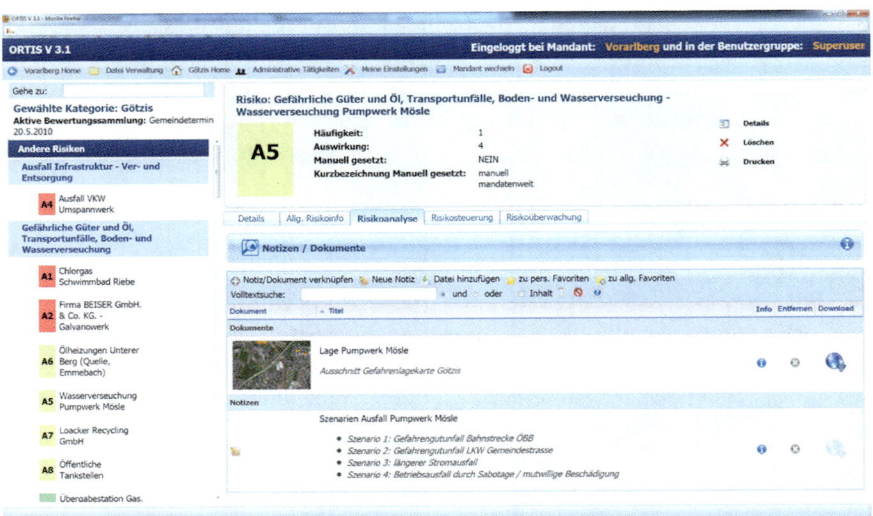

Fig. 16.4 Operational risk management tool and information system (ORTIS)

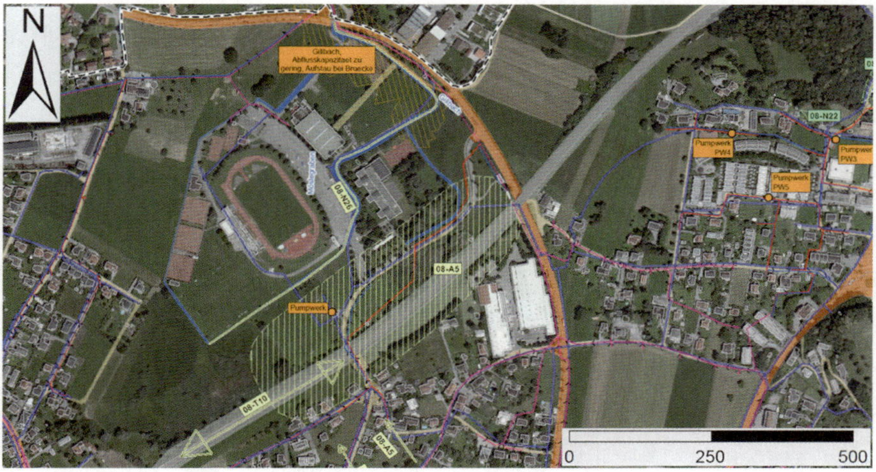

Fig. 16.5 Spatial risk map generated in the frame of ORTIS approach

optimal participatory action. During the workshop they are asked to contribute not only in their specific field of expertise, but also to think about potential parallel-events and cascading effects.

In the course of the ACHILLES project, risk landscapes for the municipalities of Göfis and Götzis as well as the City of Hall have been developed. By going through the process described above, the three authorities and their local emergency management teams successfully developed a risk list based on a specific risk catalogue. The risks of this list were subsequently evaluated according to frequency occurrence and effect. This made it possible to represent the risks in a so called risk matrix. The risks can also be shown in a spatial risk map (Fig. 16.5).

16.3.4 River Flood Forecasting and Infrastructure Protection

Flood forecasting is an essential component of preventive flood protection as well as an important part of measurement controls in a system of integrated floodwater management.

Already more than a hundred years ago, the necessity of reliable flood information was identified and an operational service was established on the occasion of the large Danube floods at the end of the nineteenth century. At that time the Austrian Hydrographical Service was made responsible for the operational distribution of flood information and it still is today.

During the past 100 years, the requirements concerning hydrological forecasts have changed to quite some extent, due to new settlements in river valleys,

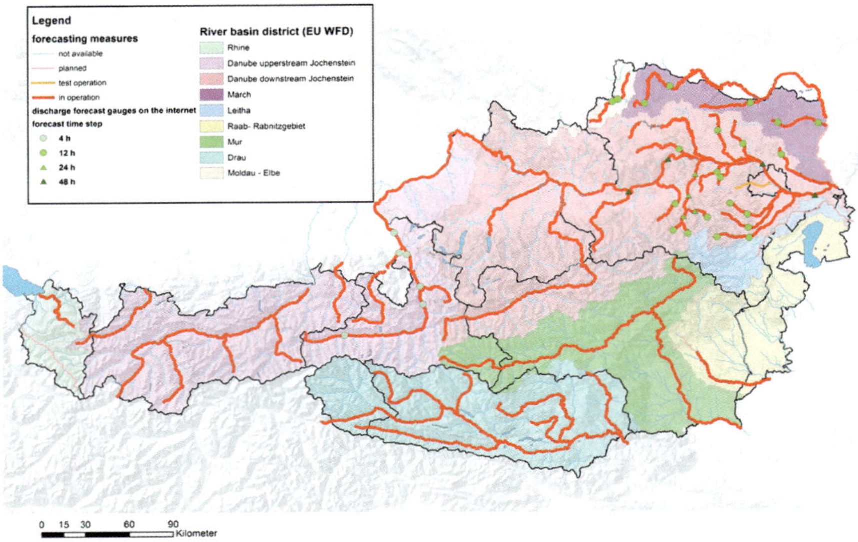

Fig. 16.6 River flood forecasting in Austria

reduction of flood plains and increased damage risks. In order to protect property and infrastructure, mobile floodwater protection measures have been installed more often. These mobile systems rely on hydrological forecasts, which enable the in-time installation of necessary elements and on well trained, friction free information paths, which are an essential part of floodwater management.

Hundred years ago the River Danube was regarded as the main task for flood information services. Nowadays complex water balance models, which permanently calculate the actual discharge as well as discharge up to 2 days ahead, are operated on almost every large Austrian river (see Fig. 16.6). The requirements, concerning necessary input data as well as data management itself, have increased enormously.

The huge damages of the 2002, 2005, 2006 and 2009 flood events also shown how important a well- functioning hydrographical monitoring network is for event analysis as well as for the development and improvement of forecasting models.

16.3.5 GEDES: Risk of Flood and Landfill Hazard Tool

To enable environmental risk management flood prone landfills must be identified to establish priorities for subsequent protection and mitigation measures. Therefore, the project GEDES (funded by the KIRAS initiative) developed two flood risk assessment approaches at different spatial scales. Both methodologies aim to determine the proportion of landfills endangered by flooding, and evaluate the impacts. The latter are expressed by means of risk categories (minor to serious) of impacts that flooded sites might have on humans and the environment.

The evaluation of 1,064 landfills in Austria yields roughly 30 % of landfills located within or close to flood risk zones. Material inventories of 147 sites exposed to flooding are established, and potential emissions during a flood event are estimated (Laner et al. 2008, 2009). Three representative case study areas were selected and investigated in detail based on 2D hydrodynamic models to calculate flow depths and shear stress and by developing emission scenarios (Nachtnebel et al. 2009). The applications of the developed methods outlined that hazardous emissions due to flooding can lead to significant impacts. However, uncertainty associated with related processes and data sources are considerably high. Nevertheless, the GEDES approach provides a decision support tool to identify landfills with imminent risk for humans and the environment.

16.3.6 AQUASEC-AUT: Austrian Crisis Management Laboratory

Immediate identification of unknown chemical, microbiological and to some extent radiological contaminants in the water sector is paramount to avoid breakout of a crisis or to get an emerging crisis under control. Deliberate contamination of water supply systems by terrorist attacks, accidents, natural disasters or technological failures are recognized scenarios which might lead to tragic public health consequences (Bundeskanzleramt 2008; Commission of the European Union 2006). AQUASEC-AUT was funded by the Austrian security research programme KIRAS and was led by the Environmental Agency Austria. The project was carried out between March 2010 and May 2012. Partners were the Austrian Agency for Health and Food Safety (AGES), the Austrian Ministry for Agriculture, Forestry, Environment and Water Management and the Ministry of Defence and Sports.

AQUASEC-AUT examined the degree of preparedness of a representative sample of Austrian water suppliers in terms of security plans, threat analysis or operating procedures and asked for experiences in crisis situations. Additionally, the laboratory resources for chemical, microbiological and radiological analysis were investigated. It was worked out which laboratories are able to offer special competences beyond routine analysis and which laboratories already have experience in crisis management. In order to establish a link between laboratory capacities and crisis management the structure of disaster management in Austria was compiled.

As a synthesis of elaborated deficiencies and available capacities AQUASEC-AUT proposes to establish a network among four existing laboratories. These accredited laboratories cover the major expertise in chemical, microbiological and radiological analysis in Austria and have experience in disaster situations. Moreover, their structure as semi-public institution may allow them to easily interact with crisis management structures in Austria. All four laboratories work in the field of environment and health assessment and therefore are equipped with state-of-the-art instrumentation. Bundling the expertise and resources to a network seems to be the most cost efficient way to establish a competence centre for this purpose. The main objective of the proposed laboratory network is to assist the governmental crisis management by identifying and quantifying (unknown) chemical contaminants (industrial chemicals, warfare agents), pathogens and radiological threats shortly in order to facilitate assessing the risk and to take appropriate measures. Horizontal scanning and observation of scientific developments are necessary to keep the skills of the laboratory network at state-of-the-art. Participation in proficiency tests is mandatory for accredited laboratories. Additionally, it seems to be helpful to also integrate the laboratory network in field exercises in order to train for emergencies and to evaluate their performance.

AQUASEC-AUT identified aspects to increase the security in the water sector and addressed recommendations to relevant authorities. Policy makers are asked to balance obligations for information (e.g. INSPIRE) and security aspects in terms of publishing pipe network data. Guidelines for the implementation of security plans in water works are issued by the Austrian Association for Gas and Water (ÖVGW 2008). Water suppliers are asked to further implement existing guidelines and standardized approaches to increase their preparedness for crisis situations.

A major outcome of AQUASEC-AUT is that in Austria there is a need for establishing a crisis laboratory in order to quickly respond to an emerging crisis. This can be formed by a network of four existing laboratories. An inter-ministerial agreement, however, is needed to initiate its implementation.

16.3.7 CAIS: Cyber Attack Information System

Information technology (IT) has been integrated massively in communication, production and decision-making processes in recent years, thus resulting in an enormous dependency on these technologies. The potential impact of these dependencies (for example if the IT systems are lacking appropriate security levels) are considerable—the loss of energy networks, the banking system, supply chains or public administration can cause enormous economic damage and massively affect entire nations (Lewis 2002). There are many different motivations and threats in "cyberspace", but the underlying technology is often similar. The common facts for cyber-attacks are that early warning is hard to do, tracking back to the originator is difficult and most often attacks have complex interactions which makes analysis challenging. To implement effective countermeasures it is necessary to get an overview of the current situation by analysing the extent and the impact of the specific threats.

Therefore, the CAIS project also funded by KIRAS aims to initiate the development of a "Cyber Attack Information System". The core, which is being developed in this project, consists of two methods and derived prototypical software implementations: a modelling and simulation tool for analysing the structure of large information and communication technology (ICT) systems, such as control networks for water supply infrastructures, in terms of their security and resilience against cyber-attacks, and an analysis and evaluation tool for the investigation of the current threat situation in networks. The project focuses on the modelling tool, which will be based on the experiences of other countries (e.g. Australian Government's Critical Infrastructure Protection Modelling and Analysis Program and others). The result is a scientific method and an accompanying software prototype, which allows operators of large ICT infrastructures to carry out detailed analysis and simulations (dependencies, cascade effects, etc.). Relevant data can also be made available to public authorities to get a national overview on strategic ICT infrastructures, of course placing a special emphasis on security and privacy issues.

To supplement this tool the project also develops an analysis tool for the investigation of data streams and attack patterns. Based on the various data collected in the current ICT networks, new algorithms for analysis and correlation of the data are being developed, for better and earlier detection of threats and anomalies. The algorithms developed are based on industry-proven tools for the investigation of large amounts of data. Different data sets are included in the analysis via interfaces to existing systems (netflow information, malware analysis, etc.). Interfaces for integrating additional data sources are included in the design from the beginning, and utmost importance is laid on data protection and privacy issues.

Major organizations with public security responsibilities participating in the project ensure the usefulness and integration for national initiatives on critical (information) infrastructure protection. The practical applicability of the project results is in the focus of key players from industry that operate large ICT networks or support those operators. Renowned research organizations in IT security and cyber strategy research provide the scientific expertise and are responsible for investigating appropriate methods for solving the problems, building upon the state of the art. Last but not least several experts in the field of privacy and data protection ensure compliance with relevant data protection issues and develop recommendations for the further development of the legal framework. Through cooperation with relevant agencies and programs (APCIP, EPCIP, etc.) the synchronization with national, European and international initiatives on cyber-security and critical infrastructure protection is guaranteed.

While this project focuses on the technical aspects of the overall security challenges of critical infrastructures, the information resulting from the tools has to be evaluated to derive appropriate responses and countermeasures. Organisations with specific responsibilities in the security domain, like the ones in this project, are able to derive their own assessment based on the information from CAIS, but since attacks in cyberspace are using the same techniques regardless of their motivation, a coordinated and joint action by all stakeholders (public and private) for preparing the data is required.

The project is coordinated by the Austrian Institute of Technology (AIT), Safety and Security Department. Further partners are Federal Chancellery,

Ministry of Defence, Ministry of the Interior, University of Applied Sciences of St. Pölten, Austrian Institute for International Affairs (OIIP), T-Mobile Austria, T-Systems Austria, and CERT.AT.

16.3.8 PiReM: Pipe Rehabilitation Management

PiReM Systems—Water Supply (2010) is a decision support tool for rehabilitation planning of water supply pipes and was developed with and applied to several utilities in Germany and Austria. The rehabilitation prioritisation process of PiReM Systems—Water Supply (see Fig. 16.7) follows a risk oriented approach. A proportional hazards model was applied to estimate future failure frequencies of pipe sections based on the current state of deterioration. Consequences of failure are monetarily estimated based on the damage and social costs caused by pipe breaks. The failure costs and the failure frequencies are aggregated over time for each pipe section, thus pipes are prioritized with the highest future failure risk. Further the software contains a module to calculate future performance according to yearly failure amounts. With a virtual rehabilitation of pipe sections the consequences of different rehabilitation scenarios are modelled. The consequences are expressed with performance indicators (e.g. failure rates) (Fuchs-Hanusch et al. 2008, 2011; Fuchs-Hanusch and Kasess 2012).

Figure 16.8 shows a visualisation of prioritized pipes according to the criteria failure risk, mean age of connection pipes and consumption density as key criteria for pipe replacement in Vienna.

16.3.9 Link-Up of Approaches

Quantitative risk analysis applied within the frame of e.g. ACHILLES, GEDES, ZuHaZu and PiReM approach aims at determining objective numerical values to represent the risk associated with a hazardous event. To carry out such an analysis,

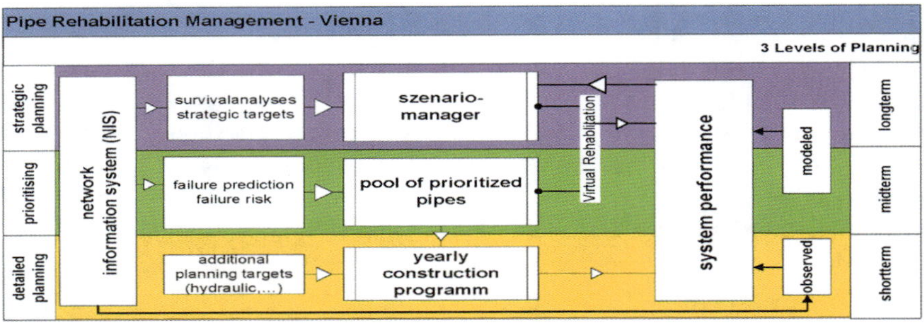

Fig. 16.7 Infrastructure rehabilitation management—PiReM approach

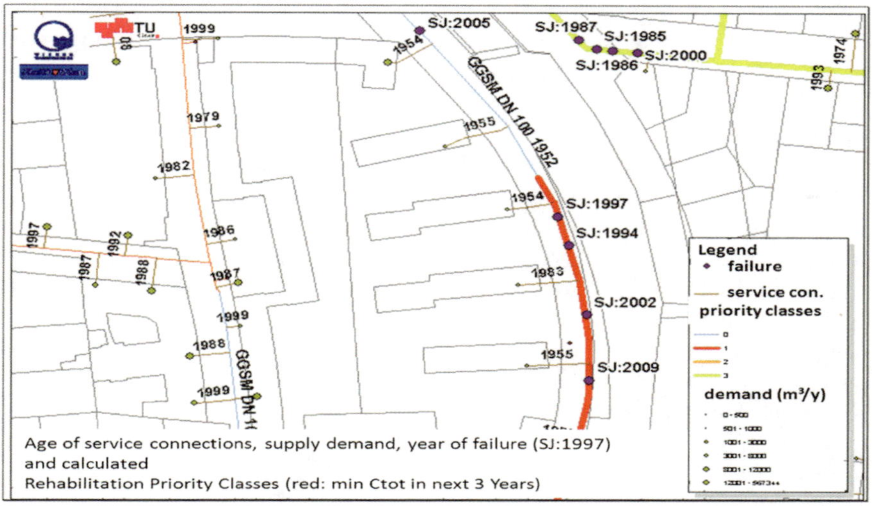

Fig. 16.8 Rehabilitation prioritisation for Vienna case study

the likelihood of occurrence and the consequences of the hazardous event have to be obtained. Moreover, mapping of the individual water supply system is needed for a quantitative risk analysis. However, since today many water utilities already have a Geographic Information System (GIS), the effort for mapping and modelling is in many cases manageable. In contrast to quantitative risk analysis, qualitative risk analysis applied within the frame of e.g. FEIS and ORTIS aim at prioritizing risks without determining absolute quantifications of the risks. The advantage of this approach is that complex or data intensive quantification of probabilities and consequences are not required.

Through linking the quantitative and qualitative as well as functional (e.g. FEIS) and spatial (e.g. ACHILLES and ORTIS) approaches, additional information can be derived to improve the risk assessment. FEIS aims at collecting causes and consequences of hazardous events from different water supply systems and other infrastructure in order to consolidate and qualitatively describe them in a generic system. ORTIS also collects causes and consequences of hazardous events. In contrast, the quantitative approach in ACHILLES, ZuHaZu and PiReM attempts to analyz risks associated with a specific failure scenario at an individual water supply system.

FEIS and ACHILLES can be linked through an information transfer. On the one hand, the FEIS provides failure events and cause-effect chains, which can be used as input scenarios for quantitative simulations in the ACHILLES. This way, information collected from experiences of many water utilities can be made available to other users and be used for individual quantitative risk analysis. On the other hand, the ACHILLES provides numerical values from modelling and quantitative risk analysis in individual systems, which can be used for the calibration of the functional failure network in the FEIS (Mayr et al. 2011).

One of the main goals of a cooperative project was to link-up ORTIS and ACHILLES. The results of this development process were illustrated with a dependency matrix for the components and the hazards. In this matrix the reasons behind a risk and possible cascading effects are shown. For example: an accident with dangerous goods on a railway track is a possible reason for contamination of a groundwater pump station. The matrix can also be used to build and interface between ACHILLES and GEDES. ACHILLES together with GEDES and ORTIS allows an in depth analysis of interfering process and identifies cascading effects. Besides, it highlights the effects of the planned measures on other risks.

ACHILLES can be used for hydraulic simulation of the impact of cyber attacks identified with CAIS to quantify the threat to the system. Thus, actions against such attacks can be designed on cyber and infrastructure level. Further, CAIS can use a risk index provided by ACHILLES to improve the analysis and evaluation of the current threat situation in networks.

Contamination vulnerability maps generated with ACHILLES, which display the diffusion of a pollutant in the relation to the injection place, can be used in emergency case to identify sufficient sample sites and thus supports AQUASEC-AUT. On the other hand, the proposed laboratory network should be considered in emergency plans for water supply systems. Also the river flood forecasting network should be considered in emergency plans. In addition, flood forecasting supports early warning systems for water infrastructure protection.

A link up between ACHILLES and ZuHaZu is planned which is expected to improve the risk assessment of ACHILLES according to the pipe failure likelihood. In addition to the age oriented failure likelihood the failure mode prediction developed in ZuHaZu is going to be linked to ACHILLES. A developing project for the link-up of ACHILLES, ZuHaZu and PiReM would lead to integrated rehabilitation planning including pro-active constructions to increase network redundancy when needed.

16.4 Summary and Conclusion

This chapter gives an overview of legal rules and standards concerning risk, safety and security management in the Austrian water sector, but includes as well interdependencies to other sectors. Further, best practices in management and planning established by Austrian utilities are discussed. Also results from utility benchmarks are shown. Research activities of relevant technical institutes in analysing and protecting critical infrastructure within the frame of EU and national activity plans are presented. A major part of the chapter deals with the description and linkup between several approaches and tools developed by different institutions for identifying critical control points and minimising the risks of interfering processes in the water sector. The approaches are presented with focus on urban water systems, but can likewise be used in other sectors.

ACHILLES identifies critical control points spatially and serves as a planning tool, using quantitative risk assessment based on hydraulic simulations. ORTIS serves for the qualitative assessment of risks on a municipal scale. FEIS provides a functional failure network by means of a qualitative risk assessment as database for the utility sector. CAIS consist of two tools for both, simulation of information technology infrastructure to analyse cyber-attacks, counter measures and interdependencies as well as for online detection of cyber-attacks on water control systems. ZuHaZu and PiReM, both pipe rehabilitation planning approaches, assess the structural condition influenced by deterioration and external hazards like stray current, ground motion or traffic load. With this information pipe breakage probability can be quantified. An important feature in the frame of critical infrastructure protection is the monetary ranking of rehabilitation and safety measures. The results of PiReM can be used for such tasks. Main objective of the AQUASEC-AUT is to assist the governmental crisis management by identifying and quantifying chemical contaminants, pathogens and radiological threats shortly in order to facilitate assessing the risk and to take appropriate measures. ACHILLES can support this objective by analysing contamination vulnerability maps for the identification of proper sample sites in the network.

The interfaces between ORTIS and ACHILLES are developed and tested. Two feasibility studies showed that a linkup of PiReM, FEIS and ACHILLES improves an integrated risk management approach. The results of this work showed that AQUASEC-AUT, CAIS, GEDES and the river flood forecasting network should be included in an integrated safety and security system.

Acknowledgments ACHILLES (nr. 824682), AQUASEC-AUT (nr. 824718), CAIS (nr. 832345), FEIS (nr. 813772), link-up (nr. 827659) are GEDES (nr. 813811) are outcomes of research projects funded by KIRAS (Federal Ministry for Transport, Innovation and Technology), a national program for security research. PiReM was developed within the Kompetenznetzwerk research project funded by the Austrian Ministry of Economy and Work.

References

Belevi H, Baccini P (1989) Long-term behavior of municipal solid waste landfills. Waste Manag Res 1(7):43–56

BMLFUW (1976). BGBl 436/1976—Verordnung des Bundesministers für Land-und Forstwirtschaft vom 30. Juli 1976 über die Gefahrenzonenpläne. Federal Ministry of Agriculture, Forestry, Environment and Water Management, Vienna, Austria (in German)

BMLFUW (2007) Wasser in Österreich—Zahlen und Fakten. Federal Ministry of Agriculture, Forestry, Environment and Water Management, Vienna, Austria (in German)

Brandt PT, Sandler T (2010) What do transnational terrorists target? Has it changed? Are we safer? J Conflict Resolut 54(2):214–236

Bundeskanzleramt (2008) Das österreichische Programm zum Schutz kritischer Infrastrukturen, Masterplan APCIP. Vienna, Austria (in German)

Byres E, Lowe J (2004) The myths and facts behind cyber security risks for industrial control systems. VDE Congress, Berlin, pp 213–218

Commission of the European Union (2006) European programme for critical infrastructure protection. Brussels, Belgium
Commission of the European Union (2004) Communication from the commission to the council and the European parliament: critical infrastructure protection in the fight against terrorism. Brussels, Belgium
Council of the European Union (2005) The European union counter-terrorism strategy. Brussels, Belgium
Council of the European Union (2008) Council directive 2008/114/EC of 8 December 2008 on the identification and designation of European critical infrastructures and the assessment of the need to improve their protection. Brussels, Belgium
Enders W, Sandler T (2004) What do we know about the substitution effect in transnational terrorism? research on terrorism: trends achievements and failures. Frank Cass, London
Friedl F, Möderl M, Rauch W, Liu Q, Schrotter S, Fuchs-Hanusch, D (2012) Failure propagation for "Large-Diameter Transmission Water Mains" using dynamic failure risk index. World environmental and water resources congress. American Society of Civil Engineers (ASCE), Albuquerque
Frisch H (2009) Strategic change in terrorist movements: lessons from Hamas. Stud Conflict Terrorism 32(12):1049–1065
Fuchs-Hanusch D, Gangl G, Kornberger B, Kölbl J, Hofrichter J, Kainz H (2008) PiReM—Pipe rehabilitation management, developing a decision support system for rehabilitation planning of water mains. Water Pract Technol 3(1)
Fuchs-Hanusch D, Kasess D (2012) Maintenance management—risk- and condition-oriented planning for Vienna's water pipe network. Water Loss 2012, Manila, Philipines
Fuchs-Hanusch D, Kornberger B, Friedl F, Scheucher R (2011) Whole of life cost calculations for water supply pipes. IWA LESAM 2011, Mülheim, Germany
Geller W, Ockenfeld K, Böhme M, Knöchel A (2004) Schadstoffbelastung nach dem Elbe-Hochwasser 2002. Final report of the ad-hoc-project 'Schadstoffuntersuchungen nach dem Hochwasser vom August 2002—Ermittlung der Gefährdungspotentiale an Elbe und Mulde. "UFZ—Umweltforschungszentrum, Leipzig-Halle GmbH, Magdeburg, Germany
Gleick PH (2006) Water and terrorism. Water Policy 8(6):481–503
Habersack H, Moser A (2003) Water and terrorism. Water POLICY Ereignisdokumentation—Hochwasser August 2002. Water and terrorism. Water Policy University of natural resources and applied life sciences, Vienna, Austria (in Germany)
Hernandez-Sancho F, Renard N, Theuretzbacher-Fritz H (2007) Water pricing: from theory to practice. In: Water utility management international, p 10
Höller P (2007) Avalanche hazards and mitigation in Austria: a review. Nat Hazards 43(1):81–101
Igure VM, Laughter SA, Williams RD (2006) Security issues in SCADA networks. Comput Secur 25(7):498–506
Klink RE, Ham RK (1982) Effects of moisture movement on methane production in solid waste landfill samples. Resour Conserv 8:29–41
Laner D, Fellner J, Brunner PH (2009) Flooding of municipal solid waste landfills—a long-term environmental hazard?. Sci Total Environ 407(12):3674–3680
Laner D, Fellner J, Brunner PH, Neuhold C, Kolesar C (2008) Environmental relevance of flooded MSW landfills in Austria. In: ISWA/WMRAS, ISWA/WMRAS world congress 2008—East meets Waste, Singapore, Nov 3–6
Lewis JA (2002) Assessing the risks of cyberterrorism, cyber war and other cyber threats. Center for Strategic and International Studies, Washington
Lukas A, Mayr E, Richard L, Perfler R (2011) Supporting the water safety plan (WSP) approach with the failure experience improvement system (FEIS). Water Sci Technol: Water Supply 11(3):288–296
Lukas A, Mayr E, Ruhri M, Katzmair H, Perfler R (in press) Failure experience improvement system (FEIS) for water supply systems. J Hydroinfor

Mayr E, Lukas A, Aichlseder A, Perfler R (2012) Experiences and lessons learned from practical implementation of a software-supported water safety plan approach. Water Sci Technol: Water Supply 12(1):101–108

Mayr E, Lukas A, Möderl M, Rauch W, Perfler R (2011) Integrales Risikomanagement für die Trinkwasserversorgung in Österreich. Österreichische Wasser- und Abfallwirtschaft 63(3–4): 82–86

Meinhardt PL (2005) Water and bioterrorism: preparing for the potential threat to U.S. water supplies and public health. Annu Rev Public Health 26:213–237

Möderl M, Rauch W (2011a) Spatial distributed risk assessment for urban water infrastructure. Handbook of Water and Wastewater Systems Protection. Springer, New York, ISBN 978-1-4614-0189-6

Möderl M, Rauch W (2011b) Spatial risk assessment for critical network infrastructure using sensitivity analysis. Frontiers of Earth Science 5(4):414–420

Nachtnebel HP, Holzmann H, Neuhold C, Haberl U, Kahl B, Bichler A (2009) GEDES: Gefährdung durch Deponien und Altablagerungen im Hochwasserfall Risikoanalyse und Minimierung—Teilbericht 2. University of Natural Resources and Life Sciences, Vienna, Austria

Neuhold C, Nachtnebel HP (2011) Assessing flood risk associated with waste disposals: methodology, application and uncertainties. Nat Hazards 56(1):359–370

Neunteufel R, Perfler R, Mayr E, Theurezbacher-Fritz H, Kölbl J (2010) Benefits from benchmarking—An Austrian case study. Water Pract Technol 5(2):

Neunteufel R (2011) Benchmarking: a performance comparison tool. Aqua Press Int 4:34–35

Neunteufel R, Theuretzbacher-Fritz H, Teix P, Kölbl J, Perfler R (2004) Benchmarking und Best Practices in der österreichischen Wasserversorgung—Stufe A.—Final report of ÖVGW pilot project 2003/04. Austrian Association for Gas and Water, Vienna, Austria (in German)

ÖVGW (2009). "Die Österreichische Trinkwasserwirtschaft – Branchendaten und Fakten. Österreichische Vereinigung für das Gas- und Wasserfach. Austrian Association for Gas and Water, Vienna, Austria (in German)

ÖVGW (2008) W 88—Anleitung zur Einführung eines einfachen Wassersicherheitsplanes. Austrian Association for Gas and Water, Vienna, Austria (in German)

ÖVGW (2007) W 100—Wasserverteilleitungen—Betrieb und Instandhaltung.Austrian Association for Gas and Water, Vienna, Austria (in German)

PiReM Systems—Water Supply (2010) Release 3.2. Decision support system for pipe rehabilitation management developed by KNet Waterpool. Graz, Austria, www.pirem.at

Rank G, Kardel K, Pälchen W, Greif A (2003) Schadstoffbelastungen im Mulde- und Elbe-Einzugsgebiet nach dem Augusthochwasser 2002. Statusseminar des BMBFAd-hoc-Verbundprojektes Schadstoffbelastung im Mulde- und Elbe-Einzugsgebiet.Freiberg, Germany

Rauch W, De Toffol S (2006) Climate change induced trends in high resolution rainfall. In: 7th International workshop on precipitation in Urban areas, St. Moritz, Switzerland

Reid RL (2009) Guiding critical infrastructure. Civil Eng 79(2):50–55

Soldati M, Corsini A, Pasuto A (2004) Landslides and climate change in the Italian dolomites since the Late glacial. CATENA 55(2):141–161

Stoffel M, Beniston M (2006) On the incidence of debris flows from the early little ice age to a future greenhouse climate: a case study from the Swiss Alps. Geophys Res Lett 33:187–202

Theuretzbacher-Fritz H, Neunteufel R, Kölbl J, Perfler R, Unterwainig M, Krendelsberger R (2006) Benchmarking und Best Practices in der österreichischen Wasserversorgung—Stufe B. Final Report of ÖVGW project 2005/06. Austrian Association for Gas and Water, Graz-Vienna-Wiener Neustadt, Austria (in German)

Wang Y, Au S-K (2009) Spatial distribution of water supply reliability and critical links of water supply to crucial water consumers under an earthquake. Reliab Eng Syst Saf 2:534–541

Young S, Balluz L, Malilay J (2004) Natural and technologic hazardous material releases during and after natural disasters: a review. Sci Total Environ 322(1–3):3–20

About the Author

Stefan Achleitner, Ph.D.
Stefan Achleitner studied Civil Engineering at the University of Innsbruck and majored in Hydraulic and Environmental Engineering at the University of New Orleans. He obtained his Ph.D. at the Institute of Environmental Engineering at the University of Innsbruck in 2006 and joined the alpS Center for Climate Change Adaptation Technologies in 2007 as postdoctoral fellow and lead of the hydraulic section. Since 2010, he has been at the Unit of Hydraulic Engineering at the University of Innsbruck as Assistant Professor. His research focuses include Hydrological and Hydraulic Modeling, Sediment Transport, Risk Analysis in the context of flooding, and flood prognosis.

Rakesh Bahadur, Ph.D
Dr. Rakesh Bahadur is Sr. Hydrologist and Project Manager at the Science Applications International Corporation (SAIC). He earned his M.S. (1988) and Ph.D. (1990) in Civil Engineering from the Colorado State University. He also holds an M.Sc. (1980) in Geology from the Punjab University. His experience is in developing and applying models to solve water sector problems related to water security and water quality. He has applied Geographic Information Systems to address various water issues related to risk assessment, environmental assessment, remediation, pathway analysis, and exposure assessment. Dr. Bahadur is working with SAIC since 1992. Prior to joining SAIC, Dr. Bahadur worked as a Research Scientist at the Atomic Energy Canada Ltd., Canada from 1990–1992; a Research Assistant at the Department of Civil Engineering, Colorado State University, Colorado, USA from 1985–1990; a Visiting Faculty at the University of Leads, UK in 1984; and with Haryana State Minor Irrigation Tubewells Corporation (HSMITC) in India from 1980–1985. Dr. Bahadur is well published and has regularly contributed to the scientific literature. He has authored and/or co-authored five chapters in books and papers published in refereed conference proceedings, and eight abstracts published in scientific and technical meetings.

Ralph Beuken
Ralph Beuken is a scientific researcher at KWR Watercycle Research Institute since 2001. His area of research includes asset management of water distribution

R. M. Clark and S. Hakim (eds.), *Securing Water and Wastewater Systems*,
Protecting Critical Infrastructure 2, DOI: 10.1007/978-3-319-01092-2,

systems, focusing on condition assessment of water mains, quantifying the performance of networks, and risk analyses. Mr. Beuken coordinates the asset management research theme executed by KWR within the Joint Research Programme for Dutch water companies. During his first years of employment Mr. Beuken was involved in research projects in water quality, leakage, and hydraulic modeling. Mr. Beuken holds an M.Sc. degree in Sanitary Engineering from the IHE in Delft, the Netherlands. He also holds a B.Sc. degree in Commerce and Technology from the Hogeschool Rotterdam, the Netherlands. He previously worked for 6 years as a civil engineer in development projects in Africa. Subsequently, he worked for 4 years as a sanitary engineer for a water company in the Netherlands.

Dave Birkett
Dave has worked in the public and private sectors in the water, power, mining, oil, gas, and transport industries. He has specialized in the security, risk, crisis, and emergency management planning areas of disaster management. Dave has applied his expertise with counter intelligence in Defence, Australia to advise organizations in the preparation, prevention, and control of emergencies and crises. Dave is dedicated and involved in the leading of project teams, implementing processes, and physical change. He has a broad understanding of government policies and procedures, complemented by private enterprise methods and practices. Dave has authored professional publications on water infrastructure protection in relation to natural disasters and human interventions in Australia and Europe. Dave has a BA in Politics and Psychology from the University of Adelaide, Australia, with post-graduate qualifications as a lead auditor with governance studies. He is currently studying toward his Ph.D. with the Global Terrorism Research Centre, Monash University, Australia, in the field of critical infrastructure protection.

Thomas Bleier, DI
DI Thomas Bleier, M.Sc. is the program manager for the “ICT Security” research program at the AIT Austrian Institute of Technology GmbH. The research program is focusing on applied research of ICT security aspects in distributed systems to improve the security of critical infrastructures. His current research interests include secure system design, national cyber defense, secure cloud computing, and security aspects of next generation energy networks (smart grids). Before joining AIT, Thomas Bleier was working in the industry for more than 10 years as a Systems Architect, Project Manager, Software Developer, and Technical Consultant. He holds a master’s degree in Information Security Management, and a master’s degree in Computer Science. Thomas Bleier is a Certified Information Systems Security Professional (CISSP), a Certified Project Manager (IPMA Level C), Certified SCRUM Master, Certified Ethical Hacker (CEH), and also holds several technical certifications.

Eyal Brill, Ph.D.
Eyal Brill is a senior lecturer at the Management of Technology faculty at Holon Institute of Technology. He earned his M.Sc. and Ph. D. degrees in Agricultural Economics from the Hebrew University in Jerusalem. His special areas of research and teaching include Abnormalities Detection in industrial processing using machine learning. Dr. Brill is leading the research group of water abnormalities detection of White Water technology Ltd, a worldwide leading company in the field of water infrastructure management. He is also the chairman of working group 9 of ISO technical committee 224 which is responsible for developing ISO standards for water abnormality detection.

Michael Brugger
Michael Brugger is currently working at the Austrian Federal Ministry for Transport, Innovation, and Technology (bmvit)—DG III/Staff Group for Technology Transfer and Security Research. Mr. Brugger is jointly responsible for the strategic governance of the Austrian security research program, KIRAS. He is an expert in the Program Committee for Security Research within EU's Seventh Framework Programme for research, technological development, and demonstration activities (FP7). Mr. Brugger holds a degree in Civil Engineering and Water Management (2003). Before joining bmvit he was working as a consultant to public and private entities.

Steven G. Buchberger, Ph.D.
Steven G. Buchberger is Professor of Civil and Environmental Engineering and Interim Director of the School for Advanced Structures in the College of Engineering and Applied Science at the University of Cincinnati (UC). Since joining the UC faculty in 1988, Dr. Buchberger has authored over 140 archived publications and directed nearly 10 million dollars in research projects. Dr. Buchberger specializes in water resources engineering and urban hydrology. He is chief architect of the Poisson Rectangular Pulse model for water use, a concept which led to the development of the first stochastic water demand simulator for urban water distribution systems. Dr. Buchberger has advised over 50 graduate students and served on numerous international Ph.D. juries. A number of his students have been recognized for their research, including three national best paper awards from the American Society of Civil Engineers. Dr. Buchberger has twice received the Neil Wandmacher teaching award from the UC College of Engineering and Applied Science. Dr. Buchberger served as Associate Editor of *ASCE Journal of Water Resources Planning and Management* for 10 years and was the Chief Editor of two special issues on Water Distribution Systems Analysis. Dr. Buchberger earned a B.S. degree in Civil and Environmental Engineering from the University of Wisconsin at Madison (1976), an M.S. degree in Civil Engineering with specialization in water resources and hydrology from Colorado State University (1979), and a Ph.D. in Civil Engineering from the University of Texas at Austin (1988). He worked for several years as a consulting engineer in Denver before returning to academia. Dr. Buchberger is a registered professional engineer in the State of Colorado.

He Chen. Ph.D.

He Chen is an associate professor in the School of Environment at Beijing Normal University. He earned M.A. and Ph.D. degrees in Environmental Sciences from Beijing Normal University. His special areas of research and teaching are environmental flow, ecological processes, and water resources management. Dr. Chen has published 30 scientific articles in leading ecology and water resources journals and edited 2 books. He is an inventor with 16 patent applications. He is constantly teaching classes on ecological modeling and water resources management at Beijing Normal University. In addition, he has written many reports and has given presentations at professional conferences. Dr. Chen has conducted several funded research and consulting projects for the National Nature Science Foundation of China, Ministry of Science and Technology of China, Ministry of Education of China, Ministry of Water Resources, and Ministry of Environmental Protection. In recognition of his work on these projects, Dr. Chen has been granted the National Science and Technology Progress Award in China.

Robert M. Clark, Ph.D.

Robert M. Clark is currently an independent consultant in environmental engineering and public health. He is an Adjunct Professor in Civil and Environmental Engineering at the University of Cincinnati and recently completed service as a member of the National Research Council's Committee on "Public Water Distribution Systems: Assessing and Reducing Risks." As a consultant Dr. Clark has worked on homeland security issues with Sandia National Laboratories, the US Environmental Protection Agency (USEPA) and Rutgers University (Newark Campus), among others. He served as an environmental engineer in the U.S. Public Health Service and the U.S. EPA from 1961 to August 2002 and was Director of the USEPA's Water Supply and Water Resources Division (WSWRD) for 14 years (1985–1999). In 1999 he was appointed to a Senior Expert Position in the USEPA with the title Senior Research Engineering Advisor and retired from the USEPA in August of 2002. Dr. Clark was a member of USEPA's Water Protection Task Force and was USEPA's liaison for homeland security research. Dr. Clark has published over 380 papers and 5 books and has been professionally active in several organizations where he served in numerous leadership positions. He is a lifetime member of both the American Water Works Association (AWWA) and the American Society of Civil Engineers (ASCE). Dr. Clark is recognized both nationally and internationally and has received numerous awards for his work. Dr. Clark holds B.S. degrees in Civil Engineering from Oregon State University (1960), and in Mathematics from Portland State University (1961), M.S. degrees in Mathematics from Xavier University (1964), and Civil Engineering from Cornell University (1968) and a Ph.D. in Environmental Engineering from the University of Cincinnati (1976). He is a registered engineer in the State of Ohio.

Henk De Kater

Henk de Kater (1957) earned his M.Sc. degree in Civil Engineering at the Technical University Delft in 1984. From 1985 until 1988 he worked as a sanitary engineer for a Dutch international cooperation project in Guiné Bissau, Africa. In 1989 he started working for the drinking water company of the city of Rotterdam, one of the predecessors of Evides Waterbedijf, his current employer. The focus of his work was hydraulic modeling, reliability studies, and asset management. Since 2004 he is involved in the joint research program of the Dutch water companies as the Evides counterpart for the water distribution program. In 2009 Henk was for one year a part time researcher with KWR. He is still involved in the challenge of applying research results, especially in the field of asset management performance indicators, in the practical world of a water company.

Edo Macan
Edo Macan. Senior Electrical/Energy Resources Engineer; Education: B.Sc. and M.Sc. in Electrical Engineering and Computer Science at the Massachusetts Institute of Technology, Cambridge, MA; Experience: Extensive 15 years' experience (Brattle Group Inc. Cambridge, MA; Duke Energy Corp. Houston, TX; Analysis Group Inc. San Francisco, CA) in the energy consulting and utilities industries, including analysis, economy, developing, and integrate planning. Also, successively experience (Coprogram d.o.o Zagreb, Croatia and Coprogram Inc. Oakland, CA) in the water and utilities industries, including balancing of energy use in water transport, distribution, planning, and management.

Franz Eder
Franz Eder is an Assistant Professor for International Relations at the Department of Political Science, University of Innsbruck. He studied Political Science and Contemporary History at the Universities of Innsbruck and Vienna. He was visiting research fellow at the German Orient Institute in Hamburg and the German Institute for International and Security Affairs in Berlin. His main research areas are foreign policy analysis as well as transnational terrorism and counterterrorism. Franz Eder is the co-editor of the books "Transatlantic Discord" and "European and Transnational Terrorism" and has published several journal articles in the field of terrorism studies, for example, "The European Union's counter-terrorism policy towards the Maghreb: trapped between democratisation, economic interests and the fear of destabilisation." in European Security.

Franz Friedl
Franz Friedl is currently a project assistant at Graz University of Technology, Austria, where he studied "Civil Engineering" and "Industrial Engineering," received his diploma (2007) and finished his Ph.D. thesis (2012) on the topic "comparison of statistical and physical models to calculate the probability of occurrence of specific failure modes on large-diameter transmission water mains." At the institute of Urban Water Management and Landscape Water Engineering he has been working since the year 2008 on the research topic "Sustainable Optimisation of Water Infrastructure." Franz has carried out several research projects and his expertise comprises deterioration modeling of water supply systems, multi-criteria decision support in rehabilitation planning, risk

management of water supply systems, optimizing inspection methods of water supply systems, and soil-mechanical finite element analysis.

Daniela Fuchs-Hanusch
Daniela Fuchs-Hanusch has 14 years of work experience as a researcher and consultant. She graduated from the University of Natural Resources and Life Science in Vienna. Afterwards she wrote her Ph.D. thesis on the topic "Decision Support Systems for Rehabilitation Planning of Water Supply Networks" at Graz University of Technology, where she worked as a researcher for several years and was project leader of projects mainly in the field of Pipe/Infrastructure Rehabilitation Management. From 2009 to 2010 she was working as an independent consultant for the Vienna Waterworks where she supported the network management team in implementing an integrated maintenance and rehabilitation system. At the moment she is vice head of the Institute of Urban Water Management at Graz University of Technology and is leading the research group "Sustainable Optimization of Water Infrastructure." Her main area of expertise is risk and cost-oriented rehabilitation and maintenance planning of water supply and sewerage systems, deterioration modeling of pipe networks, and uncertainty and sensitivity analysis.

Reinhold Godina
Reinhold Godina is working at the Austrian Federal Ministry of Agriculture, Forestry, Environment, and Water Management—Division VII/3 Water Balance and Hydrology. He graduated from the University of Natural Resources and Life Science (BOKU) in Vienna. He is head of the Hydrological Service in Austria since 2003. He is responsible for the national coordination of the hydrological network. His main working area is quality management and documentation of hydrological data as well as water balance analysis for Austria.

Simon Hakim, Ph.D.
Simon Hakim is a professor of economics and director of the Center for Competitive Government at Temple University. He is currently editing a book series on Protecting Critical Infrastructures with Springer Publisher. He earned M.A. and Ph.D. degrees in Regional Science from the University of Pennsylvania. He also holds an M.Sc. degree in City and Regional Planning from the Technion, Israel Institute of Technology, and a B.A. in Economics at Hebrew University in Jerusalem. His special areas of research and teaching are privatization, public policy, private/public police, and homeland security. Dr. Hakim has published 58 scientific articles in leading economic, criminal justice, security, and public policy journals. He has written over 40 professional articles and edited 14 books. He collaborated with Professor Blackstone on a major textbook dealing with the security industry. He is constantly invited to teach classes on privatization, and international economics in MBA programs worldwide. Dr. Hakim has conducted several funded research and consulting projects for the U.S. Departments of Justice and Labor, the Commonwealth Foundation, the Independent Institute, the Alarm Industry Research and Education Foundation, the City of Philadelphia, the

Philadelphia International Airport, ADT, Vector Security, and other leading security companies.

Philipp Hohenblum
Philipp Hohenblum is working at the Environment Agency Austria since 1998. He began his career in the laboratory dealing with method development for organic pollutants in water. He manages water-related projects and led the project Aquasec-Aut, which dealt with a crisis laboratory for the Austrian drinking water supply. Since 2012 he coordinates a working group at the European Commission's Joint Research Centre (Ispra, IT) aiming at enhancing the security of the critical infrastructure water in Europe. Before joining the Environment Agency Austria he worked as a civil engineer in the field of water control. In 1995 Philipp Hohenblum graduated from Vienna University of Technology (Chemical Engineering).

Robert Janke
Robert Janke has been with the U.S. EPA, Office of Research and Development, National Homeland Security Research Center, and Water Infrastructure and Protection Division in Cincinnati, Ohio since 2003. Mr. Janke is a founding member of the Threat Ensemble Vulnerability Assessment (TEVA) Research Program established within the National Homeland Security Research Center, Water Infrastructure and Protection Division. Mr. Janke is the contact person and one of the designers of the popular TEVA-SPOT, graphical user interface software program. Mr. Janke has published numerous articles on using water distribution system modeling to better understand and characterize the consequences to drinking water distribution systems from possible contaminant attacks and intrusions. Mr Janke's research is focused on designing, developing, and deploying software tools and methodologies to improve the security of water systems while also improving their management, operations, and water quality. Prior to joining EPA, Mr. Janke spent 12 years with the Department of Energy largely devoted to the design, construction, and deployment of real-time radiological detection instrumentation for monitoring and remediating large, outdoor surface areas contaminated with radionuclides. Mr. Janke earned his M.Sc. in Health Physics and B.Sc. in Chemistry, both from the University of Cincinnati.

Talis Juhna, Ph.D.
Talis Juhna is currently Vice-Rector for Research and a Full Professor in Water Engineering Riga Technical University (RTU). In 2002 he established the Water Research Laboratory which has become one of the leading laboratories in the Baltic States dealing with water security and safety. He has been involved in several pan-European research projects on water quality and technology issues in water distribution systems, decontamination of networks after deliberate attack, among others. As a consultant Dr. Juhna has worked with several large water companies on projects about surveillance of water biological and chemical contamination. Dr. Juhna has published over 60 papers and 2 books and has been professionally active in several organizations where he served in numerous

leadership positions. He is a member of both the American Water Works Association (AWWA) and the International Water Association (IWA). Dr. Juhna is recognized both nationally and internationally and has received numerous awards for his work. Dr. Juhna holds B.S. Degrees in Civil Engineering from Riga Technical University (1995), M.S. degrees in Environmental Engineering from Royal Institute of Technology in Stockholm (1997), and a Ph.D. in Environmental Engineering from the Lulea University (2002). He is an associated member of Latvian Academy of Science.

Johannes Lammel
Johannes Lammel is currently risk manager at TIWAG (Tyrolean Water Power Company), Innsbruck, Austria. As he contributed as author to the book chapter on Austrian activities he worked as researcher at the K1-Centre alpS—Centre for Climate Change Adaptation Technologies in Innsbruck. There, he was responsible for several projects focusing on the assessment of hazard and risk of municipalities and companies. He received his Master's in Civil Engineering from University Innsbruck in 2002 and became fully licensed as a professional engineer in 2009. He also wrote several research papers on risk communication and management.

Aditya Lukas, Ph.D.
Aditya Lukas is currently a scientist at the Institute of Sanitary Engineering and Water Pollution Control of the University of Natural Resources and Life Sciences (BOKU) Vienna, Austria. He received his M.Sc. and Ph.D. in Land and Water Management and Civil Engineering from BOKU University, Vienna. He also holds an M.Sc. degree in Natural Resources Management and Ecological Engineering from Lincoln University, New Zealand. His main research areas are risk management and efficiency in drinking water supply. Aditya has carried out several research projects on water safety plans in cooperation with Austrian water cooperatives and on risk assessment in water supply funded by the Austrian programme for security research of the Austrian Federal Ministry for Transport, Innovation, and Technology.

Helena Mala-Jetmarova
Helena has 13 years of experience in the water industry both in Australia and Europe. She has been involved in development and application of hydraulic models, management and design of projects of flood protections of towns, protective dykes, water structures, and waterways regulations. As a hydraulic modeler at Grampians Wimmera Mallee Water, she developed hydraulic models for the Wimmera Mallee Pipeline, a regional 8,800 km water distribution network in Victoria, Australia. Helena displays passion in connecting and networking water professionals from industry, academia, and research. In recent years, this is evident in her formation of the Victorian Modeling Group, the Australian Hydraulic Modeling Association, and supporting the initiation of the Modeling Groups in Queensland and New South Wales.
Helena has published scientific and professional articles on the subjects of water supply, optimization of water pipeline systems, and protection of water

infrastructure. Helena holds an Ing. from the Czech Technical University in Prague and is currently studying toward a Ph.D. from the University of Ballarat, Australia, in the area of optimization of water distribution systems operation.

Patrick Marchisella
Patrick is an applied mathematics major who is presently working as a research associate under Dr. Tularam from funds supplied by the Environmental Future Centre in the Environmental sciences. Patrick has an interest in partial differential equations and mathematical modeling related to population movements as well. He has written some papers with Dr. Tularam but is presently continuing his research work on mathematical logic in Information Technology. He is a highly trained base and double-base player as well as a musician who has research interests in mathematics of sound and vibrations in base bodies.

Ernest Mayr
Ernest Mayr is senior scientist at the Institute of Sanitary Engineering at BOKU-University of Natural Resources and Life Sciences, Vienna, Austria. His main research areas include water resources, risk management, and efficiency in drinking water supply. Ernest currently works on projects in the fields of water safety planning and risk management in Austria as well as bank filtration and soil aquifer treatment in Austria and India.

Edward A. Mcbean, Ph.D.
Ed McBean is currently a Professor of Water Resources Engineering, Canada Research Chair (Tier 1) in Water Supply Security, and A/Dean of the College of Physical and Engineering Science at the University of Guelph, in Ontario, Canada. Prior to 2003, Ed was a Vice-President of Conestoga-Rovers and Associates and President of CRA Engineering Inc. commencing in 1995. From 1974 to 1995, Ed was a faculty member at the University of Waterloo and University of California-Davis. Ed's research and engineering practice has primarily been focused on risk assessment and management, including emphasis on statistical interpretation of data. He is a Fellow of the Canadian Society of Civil Engineers and Engineers Canada and is the recipient of a number of awards including the Research and Development Medal from Professional Engineers of Ontario and Ontario Society of Professional Engineers. Ed has published 2 books, is an editor of 17 books, and has more than 275 papers in journals. Dr. McBean holds his B.A.Sc. (1968) in Civil Engineering from the University of British Columbia and his S.M. (1970), C.E. (1972), and Ph.D. (1973) from Massachusetts Institute of Technology. He is a registered engineer in eight states in the U.S. and the Province of Ontario.

Michael Möderl, D.Sc
Michael Möderl, D.Sc is a research fellow at the Faculty of Civil Engineering at the University of Innsbruck—Austria Institute of Infrastructure and consultant for water system modeling employed by hydro-IT GesmbH. Dr. Möderl is reviewer for the Journals Water Resource Management and Planning (ASCE), Water Science and Technology (IWA Publishing), and Water Research (Elsevier) among others. He obtained his master's degree in Civil Engineering from University of

Innsbruck (Austria) in 2006. Between 2006 and 2009, he worked as a researcher at the University of Innsbruck to commence his doctoral study under the supervision of Professor Wolfgang Rauch. The main goal of his Ph.D. was the development and application of methodical tools for water system analysis. In 2007 Dr. Möderl had an honorary appointment with the University of Exeter—UK Centre for Water Systems as researcher. Dr. Möderl's postdoctoral research activities are in the fields of water system security and automatic generation of virtual water systems.

Clemens Neuhold, Ph.D.
Clemens Neuhold is working at the Austrian Federal Ministry of Agriculture, Forestry, Environment, and Water Management—Division Flood Control Management. He earned his M.Sc. (2005) and Ph.D. (2010) degrees in Water Management and Environmental Engineering at the University of Natural Resources and Life Sciences (BOKU), Vienna and published roughly 40 contributions in the context of flood risk assessment and management. Dr. Neuhold is jointly responsible for the national coordination of the implementation of the EU-Floods Directive (2007/60/EG) and supports the development of national regulations, directives, and guidelines. He represents Austria in the flood protection expert group of the International Commission for the Protection of the Danube River and the working group F on floods established by the European Commission. Additionally, Dr. Neuhold is working as a consultant in the fields of flood risk management and water policy in the frame of research projects.

Roman Neunteufel, Ph.D.
Roman Neunteufel is currently a senior scientist, lecturer, and project manager with the University of Natural Resources and Life Sciences, Vienna. In his earlier professional career Dr. Neunteufel took responsibilities for planning, supervision, calculation, and project management in the sectors of landfill, contaminated land, and construction works. Nowadays, at his position with the Institute of Sanitary Engineering and Water Pollution Control, he is dealing with R&D projects on national and international Benchmarking activities of water supply services, risk analysis and risk control, statistics on water consumption, energy efficiency, infrastructure rehabilitation, and demand prognosis. Since 2003 Dr. Neunteufel is active within the expert committee of the Austrian Association for Gas and Water water (sector: economy; working group: benchmarking) and since 2010 member of the Academic Senate of the University. Dr. Neunteufel holds an M.Sc. degree in Land and Water Management and Engineering and Ph.D. related to benchmarking in the water supply sector from University of Natural Resources and Life Sciences, Vienna.

Reinhard Perfler, Ph.D.
Reinhard Perfler is a senior scientist and lecturer with the University of Natural Resources and Life Sciences, Vienna. He is coordinator of the "Freshwater/ Drinking Water Working Group" and vice-head at the Institute of Sanitary Engineering and Water Pollution Control. His actual working focus includes:

water supply, water treatment; sustainable water management; risk and quality management referring to water resources, infrastructure, rehabilitation, and operation; water safety planning. His educational background is Land and Water Management and Engineering at the University of Natural Resources and Life Sciences, Vienna (master's degree in 1982, Ph.D. in 1999). He finished his habilitation and got his venia docendi in Sanitary Engineering in 2011. His work in professional associations comprises the Austrian Association for Gas and Water (OVGW), Austrian Water and Waste Management Association (OWAV), Austrian Standards Organisation (ASI), International Association of Waterworks in the Danube River basin (IAWD), and the International Water Association (IWA).

Dr. Napoleón Puño

Dr. Napoleón Puño is a water resources and irrigation specialist with more than 30 years of experience. He is currently a Principal Professor at the Faculty of Agricultural Sciences, and the Professional College of Agronomy and Agricultural Engineering of the National University of Tumbes (State of Tumbes, Peru). Dr. Puño is also a professor of the graduate schools of the National University of Tumbes and National University of Piura, and the Private University of San Pedro in Chimbote, Peru. He holds chairs in irrigation, hydraulics, water resources optimization, environment and health, among others.

Dr. Puño earned Agricultural Engineering and M.S. degrees (both with concentration on water resources) from the National Agricultural University-La Molina (Lima, Peru), a Master's in Higher Education from the Catholic University of Los Angeles in Chimbote, and a Doctorate in Environmental Sciences at the National University of Piura. Dr. Puño has also received a Doctorate Honoris Causa from several Peruvian universities. He did post-graduate studies in irrigation at New Mexico State University, Las Cruces, New Mexico. He has recently concluded his term as Rector of the National University of Tumbes.

Dr. Puño has been a frequent national and international keynote speaker in water resources and irrigation and has published numerous papers, journals, and books in water resources and soils. Dr. Puño has held several positions with the Government of Peru such as senior agricultural consultant for the Regional Directorate of Agriculture in Tumbes, co-director of the National Project of Irrigation of Tumbes (sponsored by the European Union), executive director of the Bi-national Puyango-Tumbes Project, and technical administrator of the Tumbes and Chira (Piura) Water Districts.

Dr. Puño lives with his wife and children in Tumbes, the northernmost State in Peru. In his free time, he enjoys working on his small rice farm (where he experiments with new and sustainable crop varieties and irrigation techniques) and hiking the beautiful mangroves of Northern Peru.

Wolfgang Rauch, Ph.D.

Wolfgang Rauch studied Civil Engineering at the Technical University of Graz, Austria. and the Swiss Federal Institute of Technology, where he also graduated in 1985. In 1991 he could finish his doctoral work on the topic of temperature anomalies in groundwater. After joining the Institute of Environmental

Engineering in Innsbruck as Assistant Professor in 1992 he felt it necessary to further his knowledge on modeling and analysis of environmental systems. In 1994 he was awarded a post-doctoral fellowship which he spent at the Department of Environment and Resources, Technical University of Denmark. In 1997 he was again awarded a fellowship, this time for research at BIOMATH, Department of applied mathematics, biometrics, and process control at the University Gent, Belgium. In 1999 he joined the environmental engineering group at the Swiss Federal Institute for Environmental Science and Technology. In 2002 he returned to the University of Innsbruck as professor for sanitary engineering and head of the Institute of Environmental Engineering. Wolfgang Rauch has published more than 80 papers in peer reviewed journals, conference, and book contributions. Moreover, he serves as editor of one of the most important journals in the field of Water Research.

Leonīds Ribickis, Ph.D.
Academician, Prof., Dr. habil.sc. ing. Leonids Ribickis is a rector of Riga Technical University (RTU), head of Division of Industrial Electronic equipment, faculty of Power and Electrical Engineering, director of Institute of Industrial Electronics and Electrical Drives of Faculty of Power and Electrical Engineering, and a scientific head of Electromechatronics Scientific Laboratory. L. Ribickis received an engineer degree from Riga Technical University (Faculty of Electrical Engineering) in 1970 and Ph.D. in 1980. He is a Dr. habil. sc. ing., in Riga Technical University since 1994.
During his tenure as Vice Rector for Science Professor Ribickis achieved an increase of 10 times in the number of RTU scientific personnel, he managed the transition of doctoral study programs to the length of 4 years. RTU is the only higher education institution in Latvia implementing 4-year doctoral study programs. He has created the Doctoral School in Riga Technical University. L. Ribickis is a Full Member of Latvian Academy of Science, member of Latvian Council of Science, member of the Board of National Economy of Latvian Ministry of Economy, member of European Power Electronics and Drives Association (EPE), IEEE Latvia Section Chair, Board member of WEC LNC (World Energy Council Latvian National Committee), Delegate and Expert of Republic of Latvia in the Energy commission of European Union 7th Framework Research workgroup, and Chairman of Latvian University Association. The scientific work of L. Ribickis is connected with such directions of power and electrical engineering as power electronics, electric drives, electrical machines, process control systems and transport as well. As a visiting lector he lectured at Polytechnico di Torino, Italy, in 2001, at Tokio Denki Technical University in 2002, and at Tallinn Technological University in 2006–2010, Estonia. He presented the results of researches at 69 international conferences in 25 countries.

L. Ribickis has more than 320 scientific publications including 15 scientific monographs and 18 textbooks. He is the owner of 64 patents, including 2 US patents.

Janis Rubulis

Janis Rubulis (born 2 June 1978) is a senior researcher in Division of Water Engineering and Technology (www.wrl.bf.rtu.lv) at Riga Technical University. As a researcher Dr. Rubulis has worked on projects of European Commission Research and Technological Development 5th–7th Framework Programmes (since 2002). Dr. Rubulis has conducted several national research projects for Ministry of Education and Science Republic of Latvia, as well as European Regional Development Fund and with industrial partners. Dr. Rubulis was a member of International Water Association and is member of Baltic Water Research Association. He earned an M.Sc.ing and Dr.Sc.ing. degrees in Heat, Gas, and Water Technology from the Riga Technical University for drinking water accounting of domestic sector; and the influence of phosphorus on the formation of biofilm in drinking water distribution network, respectively. His special areas of research are accumulation/resuspension processes of particles and microbiological quality of drinking water in distribution network. Dr. Rubulis has published more than 10 scientific papers. He was born in the Latvian town of Varaklani, and is married to Sanita Rubule. They have two children, Gustavs and Liliana.

Slobodan Macan

Slobodan Macan. Senior Civil/Water Resources Engineer:Education: B.Sc. in Civil Engineering at the University of Zagreb, Croatia; Experience: Extensive 30 years experience (Hidroprojekt d.o.o, Zagreb and Coprogram d.o.o, Zagreb, Croatia) in the water and utilities industries, including planning and mathematical modeling of large and complex inter-regional water distribution system, operational water supply management, and safety of water supply system. In the last 15 years Mr. S. Macan has also worked intensively on the informatization of the water distribution systems, including GPRS telemetry, AMR/AMI, and water leak detection system.

William B. Samuels, Ph.D.

Dr. Samuels is the Director of the Center for Water Science and Engineering at Science Applications International Corporation (SAIC). He is an Assistant Vice President for Technology and a SAIC Technical Fellow. He supervises a modeling and analysis team, including civil engineers, computer scientists, hydrologists, and GIS analysts. He has over 34 years of experience in water quality analysis, waterborne trans-port modeling, hydrology, oceanography, and geographic information systems. Dr. Samuels is the principal investigator for SAIC's Integrated Water and Wastewater Security System. He has worked with more than 25 water utilities in implementing the Pipeline Net water distribution and consequence assessment tool. He has provided water quality security analyses for the Winter Olympics in Salt Lake City and the Republican National Convention in New York City. He is a member of EPA's Expert Review Panel for the Water

Contaminant Information Tool (WCIT) and consults with EPA's National Homeland Security Research Center. He also led the development effort for the Incident Command Tool for Drinking Water Protection (ICWater), which is currently operational for all public water supplies on U.S. streams and rivers. His international experience includes projects in Jordan and the Philippines; and he has presented papers at international conferences in Peru, South Africa, China, Australia, Singapore, Croatia, UK, India, Australia, and Thailand. Dr. Samuels is a member of professional organizations, including the American Society of Civil Engineers, Sigma Xi Scientific Research Society, and the New York Academy of Science. He has more than 60 published papers in peer-reviewed scientific journals, books, and proceedings and has made numerous presentations at national and international conferences such as the Water Distribution System Analysis Symposium and the NATO Advanced Research Workshop on Water Security. Prior to joining SAIC, Dr. Samuels worked for the US Geological Survey performing oil spill trajectory modeling. He holds a B.S degree in Biology and Geology from the University of Rochester, an M.S. degree in Marine Science from Long Island University, and a Ph.D. in Biology from Fordham University

Hailiang Shen, Ph.D.
Hailiang Shen is working in the Integrated Water resources Management Group in Conestoga-Rovers & Associates since 2011. His study areas include various water-related modeling aspects: water distribution system hydraulic and water quality modeling for network expansion, rehabilitation with InfoWater and EPANET2; waterbodies hydrodynamic and water quality (thermal and ice cover) modeling with RAM2, EFDC, CE-QUAL-W2, and Delft3d; and hydrological modeling with HEC-HMS and PCSWMM. He developed numerous tools in VBA/ArcObjects, Python, and C# to aid various modeling purposes. Dr. Shen has published over 15 papers in water distribution systems. Dr. Shen holds B.Eng. in Environmental Engineering from Hebei University of Engineering (2005), M.Sc. in Environmental Engineering from Tianjin University (2007), and Ph.D. in Water Resources Engineering from University of Guelph (2011).

Manish Shrestha
Manish Shrestha is an Environmental Engineer. He earned his M.S. degree in Environmental Engineering from the University of Cincinnati in 2012. He worked as a Graduate Research Assistant in the Environmental Engineering Department for 2 years from 2010 to 2012. His special areas of research include hydraulic and hydrologic modeling. He also holds an MBA degree from Clemson University and a B.S. degree in Civil Engineering from Tribhuvan University, Nepal. He is a member of the American Society of Civil Engineers (ASCE).

Florian Skopik, Ph.D.
Florian Skopik is currently working as a Scientist at the AIT Austrian Institute of Technology, focusing on applied research of security aspects of distributed systems and service-oriented architectures. Current research interests include secure smart grids and the security of critical infrastructures, especially in course

of national cyber defence. Before joining AIT, he received a college degree in Electrical Engineering in 2001, as well as a bachelor's degree in Technical Computer Science 2006, and master's degrees in Computer Science Management and Software Engineering and Internet Computing from the Vienna University of Technology in 2007. He was with the Distributed Systems Group at the Vienna University of Technology as a research assistant and postdoctoral research scientist from 2007 to 2011, where he was involved in a number of international research projects. In the context of these projects, he also finished his Ph.D. studies. Florian Skopik further spent a sabbatical at IBM Research India in Bangalore for several months. He published around 40 scientific conference papers and journal articles, and is a member of various conference program committees and editorial boards. Parallel to his academic activities, he was working in the industry as a firmware developer for microcontroller systems for more than 10 years.

Kristina Tihomirova, Ph.D.
Kristina Tihomirova is a docent in Riga Technical University, Division of Heat, Gas and Water, Department of Water Engineering and Technology. She received her MChem at the University of Latvia in 2000 and Ph.D. at Riga Technical University in 2011. At the moment she is a researcher at the Water Research laboratory. Her main research interests are water treatment and quality, optimization of treatment processes, and rapid detection of contaminants in network. K. Tihomirova has written 6 scientific publications and has presented her research findings at 15 international conferences in the last 5 years. Her doctoral thesis was published as a monograph.

Michael Tryby, Ph.D.
Michael Tryby is an environmental engineer with the U.S. Environmental Protection Agency's (US EPA) Water Supply and Water Resources Division (WSWRD). He is currently working on the development of simulation tools for urban water infrastructure. Dr. Tryby's research interests include modeling water supply, stormwater, and wastewater infrastructure systems; source characterization in infrastructure networks and watersheds; and the optimal design and operation of water supply systems. He holds a Ph.D. in Civil Engineering from North Carolina State University, an M.S. in Environmental Engineering, and a B.S. in Civil Engineering from the University of Cincinnati.

Gurudeo Anand Tularam
Gurudeo Anand Tularam is a senior lecturer in mathematics and statistics presently based at Griffith University. He has taught at other universities including Queensland University of Technology and University of Wollongong lecturing mathematics and time series to engineering, science, education, health, finance, and economics students. Presently, his work is mainly to do with applied mathematics and mathematical modeling. Recently, he has written many papers and book chapters and is presently working on the modeling of population movements caused by critical water and food crises. He earned his bachelor's

degree in pure Mathematics from Massey University (NZ) and his master's and doctoral degrees in Mathematics from the Queensland University of Technology (Australia). His special areas of research and teaching are diffusion models including groundwater flow and pollution and modeling large datasets using time series methods. However, he has also worked on diffusion models and stochastics in financial stock price and superannuation investments. Dr. Tularam has presented more than 40 seminars internationally and published more than 80 articles in conference proceedings, journals, book chapters, and books.

Jan Vreeburg, Ph.D.
Jan Vreeburg (1960) is principal researcher in the field of distribution of drinking water at KWR Watercycle research institute since 1989. He holds a Ph.D. earned at the Technical University Delft at which university he also held a part-time position as assistant professor in the period 2001–2012. Recently, he switched to Wageningen University, where he is involved in research and teaching the field of Managing Urban Infrastructures as associate professor.

Yi Wang
Mr. Wang holds Bachelor's and Master's degrees in Environmental Engineering from the Tianjin University, P.R. China (2006 and 2008). Yi Wang is currently a Ph.D. candidate in School of Engineering, University of Guelph, and is working with Dr. Ed McBean on identification of extreme rainfall changes in the Province of Ontario. Mr. Wang was enrolled in a Ph.D. program in Tianjin University during 2009–2010, and he applied for a patent with his advisor for the methodologies they developed to optimize flow patterns in small lakes. From 2008 to 2009, he served at Tanggu Sino-French Water Supply Co. Ltd., P.R. China, being responsible for developing and calibrating hydraulic models of their water distribution system. He conducted analyses on the risk of pipe failures in water distribution systems during his master's studies, and engaged in the development of several GIS systems for water supply companies.

Karl Weber, Ph.D.
Prof. Weber studied law at the University of Innsbruck. He obtained his Ph.D. also at the University of Innsbruck in 1975 and then joined the Department of Public Law, State and Administrative Theory. Between 1976 and 1977 he studied in Zurich. From 1990 until 1999 Prof. Weber was head of the Institute of Environmental Law. From 1994 to 2004 he was Dean of the Faculty of Law. His research interests are among others, water law and civil law liability.

Joesph Weiss
Joseph Weiss is an industry expert on control systems and electronic security of control systems, with more than 35 years of experience in the energy industry. Mr. Weiss spent more than 14 years at the Electric Power Research Institute (EPRI) where he led a variety of programs including the Nuclear Plant Instrumentation and Diagnostics Program, the Fossil Plant Instrumentation & Controls Program, the Y2K Embedded Systems Program, and the cyber security for digital control systems. As Technical Manager, Enterprise Infrastructure Security (EIS) Program,

he provided technical and outreach leadership for the energy industry's critical infrastructure protection (CIP) program. He was responsible for developing many utility industry security primers and implementation guidelines. He was also the EPRI Exploratory Research lead on instrumentation, controls, and communications. Mr. Weiss serves as a member of numerous organizations related to control system security. These include the North American Electric Reliability Corporation (NERC), Control Systems Security Working Group (CSSWG), the International Electrotechnical Commission (IEC), Technical Committee (TC) 57 Working Group 15 - Data and Communication Security, the Process Controls Security Requirements Forum, CIGRÉ WG D2.22 - Treatment of Information Security for Electric Power Utilities (EPUs), and other industry working groups. He served as the Task Force Lead for review of information security impacts on IEEE standards. He is also a Director on ISA's Standards and Practices Board. He has provided oral and written testimony to three House subcommittees, one Senate Committee, and a formal statement for the record to another House Committee. He has also responded to numerous Government Accountability Office (GAO) information requests on cyber security and Smart Grid issues. He is also an invited speaker at many industry and vendor user group security conferences, has chaired numerous panel sessions on control system security, and is often quoted throughout the industry. He has published over 60 papers on instrumentation, controls, and diagnostics including a chapter on cyber security for Electric Power Substations Engineering, Protecting Industrial Control Systems from Electronic Threats (ISBN 978-1-60650-197-9), and Cyber security Policy Guidebook (ISBN: 978-1-118-02780-6). He supported MITRE and NIST in extending NIST SP800-53 to include control systems and the development of NIST SP800-82. He was tasked to write the White Paper on Industrial Control Systems Security for the Center for Strategic and International Studies Blue Ribbon Panel preparing cyber security recommendations for the Obama administration. Mr. Weiss has conducted SCADA, substation, plant control system, and water systems vulnerability and risk assessments and conducted short courses on control system security. He is a member of Transportation Safety Board Committee on Cyber Security for Mass Transit. He also established and chairs the annual Industrial Control System (ICS), Cyber Security Conference. Mr. Weiss has received numerous industry awards, including EPRI Presidents Award (2002) and is an ISA Fellow, Managing Director of ISA Fossil Plant Standards, ISA Nuclear Plant Standards, ISA Industrial Automation and Control System Security, a Ponemon Institute Fellow, and an IEEE Senior Member. He is a Voting Member of the TC65 TAG and a U.S. Expert to TC65 WG10, Security for industrial process measurement and control—network and system security and IEC TC45A Nuclear Plant Cyber Security. He has two patents on instrumentation and control systems, is a registered professional engineer in the State of California, a Certified Information Security Manager (CISM), and Certified in Risk and Information Systems Control (CRISC).

Robert B. Wenger, Ph.D.

Robert B. Wenger is Professor Emeritus of Natural and Applied Sciences at the University of Wisconsin-Green Bay (UWGB). He was on the faculty at that institution from 1969 to 1999; he started his academic career there as an assistant professor and at the end of his tenure was the Barbara Hauxhorst Cofrin Professor of Natural Science. He served as Chair of the Natural and Applied sciences Department for a number of years and, in this capacity, was a leader of its Environmental Science Program. Dr. Wenger is a mathematician whose research has focused on applications of mathematics to environmental problems. This research work has taken place in an interdisciplinary context at UWGB where he has worked with physicists, engineers, and ecologists on solid waste management problems, air quality problems, water quality management problems, and, most recently, climate change issues. Dr. Wenger has extensive international experience, having held visiting professorships at Aalborg University in Denmark in 1981–1982 and at Beijing Normal University in China in 1987–1988. Since his initial experience as a visiting professor at the Institute of Environmental Science (now called the School of the Environment) at Beijing Normal University, Dr. Wenger has returned on numerous occasions to present lectures and work with faculty and graduate students on environmental science problems and issues. In 2011 he organized and led an international seminar on Environmental and Water Resources Management that was hosted by the School of the Environment. He has also worked with professionals in Peru and Chile on environmental and water quality management issues. Dr. Wenger has approximately 30 publications in environmental science and mathematics journals. Nearly all of these publications are jointly authored with collaborators from disciplines other than his own. In addition, he has written many reports and given presentations at professional meetings and to various interest groups. In recent years, he has devoted much of professional work to editing and reviewing manuscripts and serving on proposal review panels (including work for the U.S. Environmental Protection Agency). Dr. Wenger holds a B.S. degree in Mathematics from Eastern Mennonite College (now Eastern Mennonite University), an M.A. degree in Mathematics from Pennsylvania State University, and a Ph.D. degree in Mathematics from the University of Pittsburgh.

Hans Wiesenegger, DI

Dipl. Ing. Hans Wiesenegger, head of the hydrology department at the regional government office of Land Salzburg, studied Civil Engineering at the University of Natural Resources and Life Sciences, Vienna. He started his career in the hydrology department in 1984 as an expert in hydrology / hydrometry, became head of the Salzburgian flood forecasting service in 1996: In 2006 Hans Wiesenegger was elected as head of the hydrology department. His major interest and expertise focuses on operational flood forecasting, integrated water resources management, climate change effects in hydrology as well as operational glacier monitoring in high alpine regions. Occasionally he lectures at the Department of Geography and Geology at the University of Salzburg and at ZGIS International

Summer School events and cooperates in several scientific projects. Hans Wiesenegger holds the position of vice president of the Austrian hydrological society and represents Austria in international flood forecasting affairs. As an expert member of Austrian Standards he also represents Austria in international hydrometry standard committees, e.g., ISO TC 113 and CEN TC318.

Anatolijs Zabašta
During 2004–2009 Anatolijs Zabašta managed the Electronic Government department at the Secretariat of eGovernment Affairs of Latvia, which was responsible for coordination of ICT implementation in public governance institutions of Latvia. He was responsible for elaboration of the Latvia eGoverment Development Plan 2005–2009; in 2005–2009 he implemented several projects that resulted in creation of the first national public e-services portal and state information systems integrator for cross-institution e-services for business and consumers; research studies aiming to identify threats and provide recommendation to secure information systems of public governance institutions were made. In 2004–2009 he was a member of the e-Government working group leading by DG INFSO of European Commission resolving issues of security of information systems, cross-border e-services, and future technologies for e-Government. In 2009 he joined Micro Dators, which provides research and piloting in the field of monitoring and operational management of water distribution systems using ICT tools. In 2010–2012 as a researcher of Ventspils University and Riga Technical University, he participated in the piloting of the model for monitoring and control of water distribution system in Kurzeme Region. Since 2012 he manages TEMPUS program project Development of training network for improving education in energy efficiency. A. Zabašta holds M.Sc. degrees in Management Information Systems (in 2000), Business Administration (in 2004), and a primary degree in Telecommunication Engineering. Since 2010 he studies Computerized control of electrical Technologies at Riga Technical University, Faculty of Power and Electrical Engineering, with the purpose to obtain a Ph.D. degree. He has published more than 12 scientific papers.

Christopher J. Ziemniak
Mr. Christopher Ziemniak is a software engineer at Science Applications International Corporation (SAIC). His responsibilities include software development for various projects; testing and debugging software; writing/editing requirements documents, design documents, user guides, help files, and software testing documentation; installing and configuring Windows, UNIX, and web services on servers for project use. His software development projects include ICWater (Incident Command Tool for Drinking Water Protection), SewerNet, which is a GIS-based software tool that interacts with SWMM (Storm Water Management Model) and the application of the Geo Spatial Stream Flow Model. Prior to joining SAIC, Mr. Ziemniak worked at the Naval Research Laboratory in the Acoustics Division where he conducted applied research in underwater acoustic propagation and the effects on the surrounding environment.

Mr. Ziemniak holds B.S. in Applied Mathematics from the Rochester Institute of Technology and an M.S. in Systems Engineering from George Washington University.

Index

R. M. Clark and S. Hakim (eds.), *Securing Water and Wastewater Systems*, Protecting Critical Infrastructure 2, DOI: 10.1007/978-3-319-01092-2,

Printed by Printforce, the Netherlands